Springer Complexity

Springer Complexity is a publication program, cutting across all traditional disciplines of sciences as well as engineering, economics, medicine, psychology and computer sciences, which is aimed at researchers, students and practitioners working in the field of complex systems. Complex Systems are systems that comprise many interacting parts with the ability to generate a new quality of macroscopic collective behavior through self-organization, e.g., the spontaneous formation of temporal, spatial or functional structures. This recognition, that the collective behavior of the whole system cannot be simply inferred from the understanding of the behavior of the individual components, has led to various new concepts and sophisticated tools of complexity. The main concepts and tools – with sometimes overlapping contents and methodologies – are the theories of self-organization, complex systems, synergetics, dynamical systems, turbulence, catastrophes, instabilities, nonlinearity, stochastic processes, chaos, neural networks, cellular automata, adaptive systems, and genetic algorithms.

The topics treated within Springer Complexity are as diverse as lasers or fluids in physics, machine cutting phenomena of workpieces or electric circuits with feedback in engineering, growth of crystals or pattern formation in chemistry, morphogenesis in biology, brain function in neurology, behavior of stock exchange rates in economics, or the formation of public opinion in sociology. All these seemingly quite different kinds of structure formation have a number of important features and underlying structures in common. These deep structural similarities can be exploited to transfer analytical methods and understanding from one field to another. The Springer Complexity program therefore seeks to foster cross-fertilization between the disciplines and a dialogue between theoreticians and experimentalists for a deeper understanding of the general structure and behavior of complex systems.

The program consists of individual books, books series such as "Springer Series in Synergetics", "Institute of Nonlinear Science", "Physics of Neural Networks", and "Understanding Complex Systems", as well as various journals.

Understanding Complex Systems

Series Editor

J.A. Scott Kelso
Florida Atlantic University
Center for Complex Systems
Glades Road 777
Boca Raton, FL 33431-0991, USA

Understanding Complex Systems

Future scientific and technological developments in many fields will necessarily depend upon coming to grips with complex systems. Such systems are complex in both their composition (typically many different kinds of components interacting with each other and their environments on multiple levels) and in the rich diversity of behavior of which they are capable. The Springer Series in Understanding Complex Systems series (UCS) promotes new strategies and paradigms for understanding and realizing applications of complex systems research in a wide variety of fields and endeavors. UCS is explicitly transdisciplinary. It has three main goals: First, to elaborate the concepts, methods and tools of self-organizing dynamical systems at all levels of description and in all scientific fields, especially newly emerging areas within the Life, Social, Behavioral, Economic, Neuro- and Cognitive Sciences (and derivatives thereof); second, to encourage novel applications of these ideas in various fields of Engineering and Computation such as robotics, nano-technology and informatics; third, to provide a single forum within which commonalities and differences in the workings of complex systems may be discerned, hence leading to deeper insight and understanding. UCS will publish monographs and selected edited contributions from specialized conferences and workshops aimed at communicating new findings to a large multidisciplinary audience.

New England Complex Systems Institute Book Series

Series Editor

Dan Braha
New England Complex Systems Institute
24 Mt. Auburn St.
Cambridge, MA 02138, USA

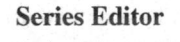

New England Complex Systems Institute Book Series

The world around is full of the wonderful interplay of relationships and emergent behaviors. The beautiful and mysterious way that atoms form biological and social systems inspires us to new efforts in science. As our society becomes more concerned with how people are connected to each other than how they work independently, so science has become interested in the nature of relationships and relatedness. Through relationships elements act together to become systems, and systems achieve function and purpose. The study of complex systems is remarkable in the closeness of basic ideas and practical implications. Advances in our understanding of complex systems give new opportunities for insight in science and improvement of society. This is manifest in the relevance to engineering, medicine, management and education. We devote this book series to the communication of recent advances and reviews of revolutionary ideas and their application to practical concerns.

D. Braha   A.A. Minai   Y. Bar-Yam (Eds.)

Complex Engineered Systems

Science Meets Technology

With 112 Figures

 Springer

NECSI

Dan Braha
University of Massachusetts
Charlton School of Business
North Dartmouth, MA 02747-2300
USA
Email: dbraha@umassd.edu

Yaneer Bar-Yam
New England Complex Systems Institute
Mt. Auburn St. 24
Cambridge, MA 02138-3068
USA
Email: yaneer@necsi.org

Ali A. Minai
University of Cincinnati
Department of Electrical and
Computer Engineering, and Computer Science
P.O.Box 210030, Rhodes Hall 814
Cincinnati, OH 45221-0030
USA
Email: Ali.Minai@uc.edu

This volume is part of the NECSI Studies on Complexity collection.

ISSN 1860-0840

ISBN 978-3-642-06937-6   e-ISBN 978-3-540-32834-6

Springer is a part of Springer Science+Business Media
springer.com
© NECSI Cambridge/Massachusetts 2006
Softcover reprint of the hardcover 1st edition 2006

Cover design: *design & production* GmbH, Heidelberg

New England Complex Systems Institute

President
Yaneer Bar-Yam
New England Complex Systems Institute
24 Mt. Auburn St.
Cambridge, MA 02138, USA

NECSI

For over 10 years, The New England Complex Systems Institute (NECSI) has been instrumental in the development of complex systems science and its applications. NECSI conducts research, education, knowledge dissemination, and community development around the world for the promotion of the study of complex systems and its application for the betterment of society.

NECSI was founded by faculty of New England area academic institutions in 1996 to further international research and understanding of complex systems. Complex systems is a growing field of science that aims to understand how parts of a system give rise to the system's collective behaviors, and how it interacts with its environment. These questions can be studied in general, and they are also relevant to all traditional fields of science.

Social systems formed (in part) out of people, the brain formed out of neurons, molecules formed out of atoms, and the weather formed from air flows are all examples of complex systems. The field of complex systems intersects all traditional disciplines of physical, biological and social sciences, as well as engineering, management, and medicine. Advanced education in complex systems attracts professionals, as complex systems science provides practical approaches to health care, social networks, ethnic violence, marketing, military conflict, education, systems engineering, international development and terrorism.

The study of complex systems is about understanding indirect effects. Problems we find difficult to solve have causes and effects that are not obviously related. Pushing on a complex system "here" often has effects "over there" because the parts are interdependent. This has become more and more apparent in our efforts to solve societal problems or avoid ecological disasters caused by our own actions. The field of complex systems provides a number of sophisticated tools, some of them conceptual helping us think about these systems, some of them analytical for studying these systems in greater depth, and some of them computer based for describing, modeling or simulating them.

NECSI research develops basic concepts and formal approaches as well as their applications to real world problems. Contributions of NECSI researchers include studies of networks, agent-based modeling, multiscale analysis and complexity, chaos and predictability, evolution, ecology, biodiversity, altruism, systems biology, cellular response, health care, systems engineering, negotiation, military conflict, ethnic violence, and international development.

NECSI uses many modes of education to further the investigation of complex systems. Throughout the year, classes, seminars, conferences and other programs assist students and professionals alike in their understanding of complex systems. Courses have been taught all over the world: Australia, Canada, China, Colombia, France, Italy, Japan, Korea, Portugal, Russia and many states of the U.S. NECSI also sponsors post-doctoral fellows, provides research resources, and hosts the International Conference on Complex Systems, discussion groups and web resources.

The New England Complex Systems Institute is comprised of a general staff, a faculty of associated professors, students, postdoctoral fellows, a planning board, affiliates and sponsors. Formed to coordinate research programs that transcend departmental and institutional boundaries, NECSI works closely with faculty of MIT, Harvard and Brandeis Universities. Affiliated external faculty teach and work at many other national and international locations. NECSI promotes the international community of researchers and welcomes broad participation in its activities and programs.

Preface

Recent advances in science and technology have led to a rapid increase in the complexity of most engineered systems. In many notable cases, this change has been a qualitative one rather than merely one of magnitude. A new class of Complex Engineered Systems (CES) has emerged as a result of technologies such as the Internet, GPS, wireless networking, micro-robotics, MEMS, fiber-optics and nanotechnology. These engineered systems are composed of many heterogeneous subsystems and are characterized by observable complex behaviors that emerge as a result of nonlinear spatio-temporal interactions among the subsystems at several levels of organization and abstraction. Examples of such systems include the World-Wide Web, air and ground traffic networks, distributed manufacturing environments, and globally distributed supply networks, as well as new paradigms such as self-organizing sensor networks, self-configuring robots, swarms of autonomous aircraft, smart materials and structures, and self-organizing computers. Understanding, designing, building and controlling such complex systems is going to be a central challenge for engineers in the coming decades.

A fertile source of ideas and methods for CES are natural complex systems such as brains, insect colonies, immune systems, and ecosystems, as well as human systems such as societies, economies and markets. The issues that these systems have evolved to address are precisely the same as those confronted by complex engineered systems today: Scalability, adaptability, self-organization, resilience, robustness, durability, reliability, self-monitoring, and self-repair. The existing paradigms of goal-oriented design, centralized control and reductionistic analysis fail completely when faced with systems that have millions of components and billions of interactions distributed over an extended area. It is instructive to note that the most successful complex engineered systems — the Internet and the World Wide Web — are self-organizing and have almost no centralized control or planning. The issue is whether this self-organized paradigm can be extended to other systems, and with what consequences.

A primary obstacle to the systematic study of complex engineered systems is the lack of an appropriate technical framework — an essential terminology, a set of central concepts, and a consensus on important issues. The emerging discipline of complex systems research offers the possibility of such a framework using well-developed concepts such as chaos, fractals, power laws, self-similarity, emergence, self-organization, networks, adaptation, evolution, etc. One advantage of such an approach is to put natural and complex engineered systems within the same discipline, thus allowing a "closing of the loop" whereby the study of natural complex systems leads to better methods for complex engineered systems, while experience with building and manipulating complex engineered systems enhances understanding of how natural complex systems function.

The objective of this book is to demonstrate, for the first time, the potential of complex systems perspectives to understanding and improving the design, implementation, and dynamics of complex engineered systems. The book will consist of an opening chapter (Chapter 1) that lays out the case for CES and discusses the relevant issues, followed by 15 chapters covering specific issues or applications.

Cambridge, MA, USA   *Dan Braha*
March 2006   *Ali Minai*
                                                                        Yaneer Bar-Yan

Contents

Chapter 1

Complex Engineered Systems: A New Paradigm

Ali A. Minai
Department of Electrical & Computer Engineering and Computer Science
University of Cincinnati
Cincinnati, OH 45221
Ali.Minai@uc.edu

Dan Braha
University of Massachusetts
North Dartmouth, MA 02747
New England Complex Systems Institute
Cambridge, MA 02138
dbraha@umassd.edu
braha@necsi.org

Yaneer Bar-Yam
New England Complex Systems Institute
Cambridge MA 02138
yaneer@necsi.org

1. Introduction

Human history is often seen as an inexorable march towards greater complexity — in ideas, artifacts, social, political and economic systems, technology, and in the structure of life itself. While we do not have detailed knowledge of ancient times, it is reasonable to conclude that the average resident of New York City today faces a world of much greater complexity than the average denizen of Carthage or Tikal. A careful consideration of this change, however, suggests that most of it has occurred recently, and has been driven primarily by the emergence of technology as a force in human life. In the 4000 years separating the Indus Valley Civilization from 18th century Europe, human transportation evolved from the bullock cart to the hansom, and the methods of communication used by George Washington did not differ significantly from those used by Alexander or Rameses. The world has moved radically towards greater complexity in the last two centuries. We have moved from buggies and letter couriers to airplanes and the Internet — an increase in capacity, and through its diversity also in complexity, orders of magnitude greater than that accumulated through the rest of human history. In addition to creating iconic artifacts — the airplane, the car, the computer, the television, etc. — this change has had a profound effect on the *scope* of experience by creating massive, connected and multi-level systems — traffic networks, power grids, markets, multinational corporations — that defy analytical understanding and seem to have a life of their own. This is where complexity truly enters our lives.

Everyone would agree that a microprocessor, with its millions of electronic components, is an extremely complicated system. The same can be said of the U.S. economy. Both the microprocessor and the economy are human constructions, but there is clearly a significant difference between them. The complicated microprocessor was carefully designed and tested by a team of engineers who placed every electronic component in its place with the utmost precision, and that is why it works. But no one designed the U.S. economy, and no one can claim to entirely understand or control it — and yet it works! And while the microprocessor can be augmented only through a careful redesign by competent engineers, the economy grows (and shrinks) on its own, without explicit control by anyone, and yet shows little sign of catastrophic strain. Also, the successful operation of a microprocessor is highly dependent on the successful operation of every one of its core sub-components, while the efficiency of the economy is much more robust to perturbations and failures at the level of its constituent elements. Looking around, one can see many other systems with the same characteristics: Communication networks, transportation networks, cities, societies, markets, organisms, insect colonies, ecosystems. What is it that unites these systems, and makes them different from airplanes and computers? And can something be learned from them that would help us build not only better airplanes and computers, but also smarter robots, safer buildings, more effective disaster response systems, and better planetary probes? What, one might ask, can engineers learn from the birds and the bees? A complementary goal is to utilize the knowledge of engineering in gaining insight into natural phenomena. For example, the ultra-robustness of inter- and intra- cellular activities may be attributed, in part, to a highly sophisticated hierarchy of feedback loops—an elementary concept in engineering control theory—operating at multiple layers and multiple scales. The increased demand for reliable and disturbance-free power systems, in turn, could lead to the development of sophisticated self-healing and recovery technologies that embody biologically-inspired procedures. More generally, we can ask how we can understand the relationship of structure to function in nature through engineering concepts, for the benefit of science. Before addressing this question explicitly, it is useful to look in greater detail at the systems being considered. These systems— markets, insect colonies, etc.—have come to be called *complex systems* [3], not to be confused with merely very *complicated* systems such as microprocessors and aircraft carriers. Such a designation is useful because these systems arguably share fundamental characteristics [4, 16], and this is where we begin.

2. Fundamental Characteristics

Perhaps the single most important characteristic shared by all complex systems is *self-organization*: Self-organization can be described as the spontaneous appearance of large-scale organization through limited interactions among simple components. Nature is replete with examples of self-organization. From galaxies to tornadoes, from canyons to crystals, from ecosystems to cells, self-organization is responsible for most, if not all, of the order we see around us. Upon reflection, this is not surprising, since most forces in Nature act over short distances and can, therefore, only support limited interactions among components such as subatomic particles, molecules, stars, and organisms. Yet, large-scale structure is ubiquitous. In a broad sense, some form of self-organization must underlie almost everything, though some examples of it seem more profound than others (today we are much more surprised and impressed by termite nests than by crystals!) But if self-organization is so common, why is it a useful concept at all? The answer to this is that, while many natural systems show some form of self-organization, almost none of the systems explicitly engineered by humans do so. Thus, the effort to understand and build self-organizing systems is, in a sense, an attempt to create systems analogous to "natural" ones—with all the profound strengths and subtle weaknesses of these systems. Not only does this promise a leap in the scope of engineering, it would also allow a better understanding of those human systems—such as economies and cities—that do demonstrate self-organization. The latter are, therefore, both a source of ideas about self-organization and a target for the application of these ideas.

While many man-made systems fail to exhibit spontaneous self-organization, we suggest that self-organization does occur in the human processes associated with design. For example, between the time specifications (requirements and constraints) are assigned to design teams and the time the artifact achieves its final form. During this period, the design process is a highly social process consisting of hundreds of designers, customers, and other participants. These actors—each a complex biological system in itself! — are involved in creating and refining a shared meaning of requirements and potential solutions through continual negotiations, deliberations, explanations, evaluations, and revision [16, 23]. Another form of self-organization refers to the act of successive changes or improvements made to previously implemented man-made systems. This ecology of evolving man-made systems is often driven by a multitude of locally-operating, noisy, socio-technical processes, and frequently involves adaptation processes that lead to more-fit new systems. Moreover, when engineering is considered as embedded in the socio-economic marketplace, the role of self-organization is apparent. Characterizing the real-world structure, and eventually the dynamics of these complex design/redesign processes, may lead to the development of guidelines for coping with complexity. It would also suggest ways for improving the multi-agent decision making process, and the search for innovative engineered systems. By contrast, the more conventional approaches to systems engineering often strive to eliminate self-organization processes in favor of reductive piece by piece design characteristic of the way complicated rather than complex systems arise.

The first fundamental insight provided by self-organizing complex systems is that *non-trivial, large-scale order can be produced by simple processes involving interactions operating locally on simple agents or components.* This insight, modifies the simplistic but common assumption that cause and effect must operate at the same scale. Over time, it is possible for the effect of local interactions to aggregate together in creating large scale order. The principle of self-organization explains the origin of collective patterns in complex systems as well as many aspects of their functioning through collective acts and collective response. Indeed, it demonstrates that the emergence of large-scale order is a process just as general (and opposite) to the process of increasing entropy, and that pattern formation is an integral part of most complex systems' functionality. One might say that, for a complex system, "becoming" is "being." This contrasts sharply with the classical paradigm in engineering with its clear distinction between the design and production phase on the one hand and the functional phase on the other. Even systems considered to be adaptive (such as adaptive controllers or most neural networks) follow this two-phase paradigm, allowing adaptation only in the superficial sense of parameter adjustment whereas complex systems change not only their parameters but also their fundamental structures and processes. Thus, complex systems have been described as "operating far from equilibrium" — in our context this implies that they undergo major changes in the context of operational activity, a notion that is anathema in classical engineering.

The second fundamental insight provided by complex systems is that *highly complex functional systems (more complex than their creators) can only arise through evolutionary processes of selection in the context of actual tasks.* This insight is based both upon the fundamental fallacy of the concepts of spontaneous generation in biological and other complex systems, and theorems that prove the inadequacy of testing to fully characterize complex systems [3, 6, 7, 16, 17, 27]. This statement contrasts fundamentally with the ongoing efforts to design large real-time response systems by specification followed by implementation. The increasing tendency to spiral and recursive implementation is only a partial adoption of the fundamental need to implement parallel in-situ evolutionary processes that are capable of creating much more complex systems than those that can be planned by conventional specification driven processes.

Complex systems emerge and function in complex, dynamic environments, and their characteristics reflect this reality. As technology seeks to produce systems that can operate in similar situations, it seems appropriate to turn to the principles underlying existing complex systems. However, this will require a drastic re-evaluation of many fundamental assumptions and methods of the classical engineering paradigm. This is indeed one of the primary focuses of this book.

3. Engineering Complex Systems

The structure of a complex system is not the result of a historic design process, but a contingent process of evolution. Thus, it does not reflect the principle of static optimality and rational decision-making often used as the basis of engineering design, but of an evolving fitness constrained by a dynamic (perhaps co-evolving) space of possibilities. This is precisely what makes complex systems suitable for operation in complex, dynamic environments, but it also means that the criteria used to determine the quality and correctness of engineered systems do not apply. It is interesting to consider this issue in some detail, since it underlies virtually all the material included in this book.

3.1 The Classical Engineering Process

The goal of the classical engineering process is to produce efficient and reliable systems that meet pre-specified constraints and pre-specified standards of performance in pre-specified situations. It is fundamentally a goal-oriented process, seeking to achieve *known* specific ends using *well-understood* means. The principle underlying this process is what one might call the *tool paradigm*: *Every engineered system is a tool made to serve the ends of its user.* Not surprisingly, the dominant themes are predictability, reliability, stability, controllability and precision. After all, a tool that cannot be controlled completely by its user or varies greatly in performance over time is not very useful.

Broadly, the classical engineering process may be seen in terms of the following steps:

- **Functional Specification**: The first step is usually a specification of what the system is expected to do. It is worth noting that this usually includes constraints and tolerances that, implicitly, represent a prediction of the circumstances in which the system will need to operate.

- **Design:** This is the main component of the engineering process, where the system is designed carefully in terms of its components – often by several teams of engineers. The design process may occur at many levels sequentially or simultaneously, with different teams working at each level. Today there is significant feedback and interaction between these teams, resulting in a process with multiple loops and a complex network of influences [14, 15, 23, 34]. In a broad sense, however, the process is fundamentally "top-down" since it moves logically from a desired functionality towards a design that implements that functionality. Levels of design are defined in terms of level of detail. The prime motivation at every level is always, "How can subtask X be done using the components and methods available?" Each team might ask this question at its own level, but in this too, the functions desired at lower (more detailed) levels typically flow from the needs already articulated at higher levels, and ultimately from the pre-specified functionality desired from the system overall, and the component decompositions that preceded it.

- **Testing and Validation:** Once designed, the system is tested under a set of conditions designed to mimic reality to ensure that it performs as needed, to discover flaws and to correct them. Both simulation and fabricated prototypes may be used. Today, this process often operates in a loop with the design process, and may even involve changes in specifications under extreme circumstances. Still, in the traditional engineering process testing intrinsically follows design and precedes implementation. This process assumes that both the task specified and the environment in which performance is to occur are sufficiently well known to be embodied in a reasonable (as measured by time and effort) number of tests. Once the system is tested and validated, it is deemed to meet the specifications to which it was built. We note that the inadequacy of conventional sequential specification, design and testing for highly complex systems is manifest today in that many systems — e.g., software — undergo additional field-testing (e.g. market-testing), where prototypes of the system are made available to customers for testing in actual applications. The results are then fed back into the design process. However, as common bugs in even highly tested systems show, the testing process for complex systems is never complete [6, 14, 34].

- **Manufacturing:** Once the system has been designed and tested, *exact copies* of it are manufactured in appropriate numbers, ranging from a few to millions or even billions. The users of these copies purchase them in the expectation that each copy functions *precisely* like the original design, and satisfies the desired functionality. Thus, the skill and diligence expended by the engineers on designing and testing the system becomes the guarantor of the system's reliability to the end-user. This is true even in the very exceptional cases (such as the space shuttle) where only one system is built, as that system still relies on the quality of many mass-produced components.

The process described above, with minor variations, underlies the production of almost all modern engineered artifacts from automobiles to paint, from computers to houses, and from widgets to satellites. The basic mode can be described in a sentence: *Given a problem to solve, figure out how to do it once, and then do it the same way each time.* Like the scientific method, this "engineering method" has developed over thousands of years with contributions from ancient cultures and modern ones. The shipwrights of ancient China, the builders of ancient Egypt and the swordsmiths of ancient Rome used essentially the same methodology, though with less precision and, therefore, greater variation. With each advance in mathematics, physics, chemistry and mechanization, the process became better understood and more precise, leading to today's robotic assembly lines and precision fabs. With the possibility of engineering complex systems, we are facing a whole new paradigm in engineering, and it is instructive to reflect on what it offers in comparison with the existing paradigm.

3.2 The Logic of the Classical Paradigm

The classical engineering process described above has several notable characteristics that define its scope, determine its logic, and circumscribe its possibilities. Before turning to complex systems, we look explicitly at some of these characteristics. The most important assumption is that the problem to be solved is uniquely and clearly specified from the outset. In a sense, this specification is extrinsic to the engineering process, but this assumption is necessary as a basis for traditional engineering. The engineering steps engage in the solution of this prespecified problem.

3.2.1 The search for a single solution

The goal of traditional engineering is to seek one solution, which often revolves around a unique design concept, for the specified problem. Though engineers understand fully that every problem admits of multiple design concepts, it is always assumed that, in the end, the engineering process will produce a single acceptable—perhaps optimal—design. Multiple design concepts may be considered during the process, but the result is often a single final design, though a certain amount of customization might be left in this design for reasons that address varying market/customer demands. The need for converging onto a single design concept is motivated by several factors. Most importantly, it is the basis of well-characterized systems that can be patented, branded and marketed as distinct, well-defined products that are the best possible solution to a well-defined problem. And, finally, it produces economies of scale through mass production that make products more affordable.

3.2.2 Seeking well-behaved systems

The classical engineering process seeks systems whose behavior can be predicted and encapsulated by precise description. This is reflected in the characteristics that are seen as the *sine qua non* of all engineered systems: stability, predictability, reliability, transparency, controllability, and—ideally—optimality. Under the current paradigm, these systems lack the ability to adapt, evolve, innovate or grow after release. Even such characteristics as robustness and resilience are seen in terms of the ability of the system's performance to be insensitive to pre-specified sources of uncertainty rather than to the possibility that the system might adapt itself to faults or changing circumstances. Adaptation, even when included in the system, is carefully circumscribed within predictable limits. The purpose of good design is seen as the elimination of the unforeseen, the unexpected and the unintended, not as the consideration of the unforeseeable, the unthinkable and the unknown. Indeed, the choice of optimality as the ultimate goal reflects the essential optimistic reductionism of the classical engineering paradigm. Since complexity often complicates the search for optimality, there is a strong tendency to control or limit complexity instead of embracing it. This has worked remarkably well, but is becoming untenable as engineering expands its scope to systems that are inherently complex.

3.2.3 Engineering as top-down problem-solving

The classical top-down design process depends fundamentally on the reductionistic assumption that any system can be described wholly by describing the behavior of its parts and their interactions. This assumption enables designers to work at different levels of abstraction with the confidence that subsystems at each level can be analyzed and synthesized completely in terms of subsystems at the next lower level. Thus, a VLSI chip designed at the level of functional modules can be translated into a register-level description, and thence to gate-level, device-level and wafer-level descriptions. At each step, the desired higher-level functionality is specified *before* its lower-level implementation is designed, so that the design process is reduced to a series of problem-solving activities. As mentioned earlier, this corresponds to an inherently problem solving view of engineering where the goal is to produce tools for specific predefined purposes whose utility is taken as given. This can be contrasted with what one might call a "meta-utilitarian" view where utility itself is subject to reassessment as the environment changes and as what can be done changes due to unanticipated innovations or insights, including bottom-up self-organization that generates unexpected emergent phenomena [36].

3.2.4 What the Classical Approach Offers

The classical engineering approach has been remarkably successful, and is responsible for virtually all the technological innovations we see around us. In particular, it confers several crucial attributes on the systems it produces—attributes that have come to embody the very notion of an engineered system. These are:

Stability: The system's performance is insensitive to pre-specified variations in the system's parameters and external environment.

Predictability: The system works in predictable ways.

Reliability: The ability of the system to perform a required function under stated conditions for a stated period of time.

Transparency: All the structures and processes in the system can be described explicitly.

Controllability: The design process and the system can be controlled directly.

4. A New Paradigm for Engineering Complex Systems

Classical engineering requires prediction of the environment in which the system will operate, the conditions it will face, and the tasks it will be required to perform. Very clever designers must then determine how these tasks can be performed as desired, and by what components put together in what fashion. Thus, the designers determine not only the *behavior* or *functionality* of the system but also the *process* or *procedure* by which that is achieved. The ultimate performance of the system depends wholly on the knowledge, competence, skill and imagination of the designers. Nothing, as far as possible, is left unspecified. All loops are closed, all contingencies considered. The result is a well-designed, reliable system that operates exactly as advertised within the limits of its tolerances. It is not expected to change (beyond wear-and-tear or accidents that are detrimental) or to "grow," and its quality is measured by the *stability* of its performance: How long can it continue to function at the same level as when it was new? This is the question we ask when judging the quality of a computer, a dishwasher, an automobile or an airplane. Significantly, this is not the way we judge the performance of an employee, a pet, an economy, or a society. What is the difference?

The primary difference is that systems designed through the classical engineering process are expected to perform foreseeable tasks in a bounded environment, whereas complex systems such as humans or other organisms are expected to function in complex, open environments with unforeseeable contingencies. One can argue that the classical engineering process is an ideal one for the former case, and is unlikely to be superseded by a process based on self-organization or adaptation. Barring radical changes in concept, aircraft and automobiles will continue to be designed the old-fashioned way because it offers predictability, reliability, controllability, etc. However, the classical engineering process suffers from serious drawbacks when applied to complex systems.

Many engineering applications, such as real-time decision support, communications and control, are reaching the point where classical methods are no longer feasible for reasons of system interdependencies and complexity. At the same time, it is increasingly clear that existing complex systems, both natural and artificial, handle these problems with ease and efficiency. Complex systems, once understood, promise a much wider repertoire of techniques and algorithms needed to engineer large systems that can work in complex, dynamic environments.

Ultimately the need to go beyond conventional engineering practice arises from the recognition that only complex systems can perform complex tasks. Equivalently, in a highly uncertain (complex) environment, planning the response of a system is guaranteed to lead to failure, precisely because we cannot anticipate all of the possibilities that may be encountered.

4.1 The Logic of Complex Systems Engineering

The key difference in the logic underlying the classical and complex engineering paradigms is in the definition of the objective. Complex systems engineering [5, 7, 8] does not primarily seek to produce predictable, stable behavior within carefully constrained situations, but rather to obtain systems capable of adaptation, change and novelty—even surprise. Some of the key concepts underlying this approach are:

Local action, global consequences. The scalability of a wide variety of complex systems arises primarily from the fact that most of the relevant processes—processes with high information requirements—are performed locally and, therefore, are low cost. Global consequences arise through self-organization or adaptation rather than explicit design, aided in many cases by non-specific global processes such as modulatory signals or global threshold-setting. The complex systems engineer, therefore, does not seek to design the system in all its details, but focuses instead on configuring the context and the local interactions that may lead to effective global behavior. Generic methods for the design of such local interactions are, indeed, one of the biggest challenges facing complex systems research.

Expectation of the unexpected. Complex systems are required in order to function in environments that are themselves too complex to be completely predicted or constrained. In such operating environments, the exact analytic relationships between the system parameters and the system behaviors are unknown or cannot be practically determined. As such, complex systems have to incorporate capabilities necessary to handle novel circumstances with incomplete information (low observability) and limited control (poor controllability). Also, these systems must be able to dynamically modify the way information about the uncertain external environment (including changes in the system's initial requirements) is represented within the system, and to reconfigure itself based on these modifications. And, since the problems faced by the system are not wholly predictable, all the solutions for all possible contingencies cannot be entirely determined in advance. Unlike standard engineered systems, complex systems must explicitly leave room for unforeseen changes in their behavior. All loops *cannot* be closed before production. Indeed, they cannot be closed even during operation. In addition to the functional dynamics necessary for any utilitarian system, a complex system also has a "meta-dynamics" that keeps the space of its behavioral *possibilities* in flux as well. One important consequence of this for engineers building such systems is the necessity of their partial ignorance about their own system—a kind of "residual irreducibility." Unlike the designers and manufacturers of standard systems, complex systems engineers must appreciate the inherent limitations of their knowledge and capabilities. For the builder of a chip, it may be embarrassing to admit ignorance of its precise behavior. However, for an engineer trying to build an autonomously intelligent robot, or a real time system involving many hardware and software components as well as people (e.g., the air traffic control system) such ignorance can establish the conditions for suggesting viable solutions.

The inherent uniqueness of individual systems. As pointed out in the discussion of the classical paradigm, the main idea in that context is to find *one* solution to the problem at hand, and often to mass-produce identical copies of that design. For engineered complex systems, some degree of replication may be effective, but in general the existence of a variety of types enables multiple approaches to be tried and for progressive improvement to arise from information obtained during operation. Changing environments may yield variable benefits for different designs, but the presence of variety allows for rapid adaptation to changing demands. Moreover, each individual adaptive system operating in its own unique complex environment will, over time, develop unique structural and behavioral characteristics. This is very different from every car of the same model developing its own quirks. Such quirks are seen as undesirable from an engineering viewpoint and good design seeks to minimize them. In contrast, individual complex systems are *required* to develop individual techniques to cope with their complex environments. Conformity, for a complex system, is not often a virtue, and novelty is not at all a vice.

Redundancy and degeneracy at all levels. The concept of redundancy is a well-established one for fault-tolerant design. In complex systems, the need for redundancy increases, as safety or performance constraints must be reliably satisfied in changing environments. Recently, the notion of *degeneracy*—multiple processes with identical consequences—has also been suggested as an important one in complex systems (for a recent discussion see [29, 37]). Redundancy relies on internal duplication of process modules, so that when a few fail, others can take their place. However, redundancy provides no protection against disruptions that attack the inherent functioning of the modules, disabling them all simultaneously. Degeneracy, in contrast, provides multiple processes for achieving the same end, and a single type of disruption is extremely unlikely to disable all these different processes. It has been suggested that degeneracy allows the genome to withstand enormous change (necessary for evolution) without disrupting basic viability for the phenotype. Engineered complex systems will also need such degeneracy to perform in complex, dynamic environments.

Off-label utilization of modules. Complementary to the attribute of degeneracy, which involves doing the same thing in multiple ways, complex systems also engage in prodigious and promiscuous re-use of the same modules and processes for multiple, often novel, purposes — perhaps with minor modifications. This allows complex systems to build on what has been achieved rather than re-inventing a new wheel each time one is needed. The resulting cumulative compounding of adaptation is key to these systems' success in handling otherwise daunting complexity.

Opportunistic leveraging of the combinatorial explosion. The "curse of dimensionality" or the "combinatorial explosion" of the solution space is dreaded by engineers as a harbinger of failure in their quest for optimality [27]. Complex systems, not seeking to be optimal in the first place, actually *benefit* from the combinatorial explosion. In combination with a mechanism for selective reinforcement, the diversity provided by exponential possibilities represents an opportunity rather than a problem. The extreme diversity of configurations makes it *likelier* that solutions to difficult sub-problems are present within this space, and complex systems—notably exemplified by biological evolution—have discovered ways to "mine" it.

Robustness-by-structure. The classical engineering approach defines system robustness as the ability of a product or process to function close to ideal specifications under actual environmental and use conditions. Designers then seek to find the right combination of parameter values that minimize the design's sensitivity to noise factors. These methods are based on statistically designed experiments that reveal sensitivities of the output response to the input variable values. The robustness of complex systems goes far beyond optimal settings of system parameters. One remarkable feature of complex systems is that their underlying structural properties have a major effect on their functionality, dynamics, robustness, and fragility. Robustness-by-structure can be achieved by appropriately designing the interactions among the system's elementary components [14]. Not only does this strategy inhibit system-wide catastrophe, it also enables the development of highly robust systems by effectively utilizing imperfect or faulty components. The latter property is also shared by many biological and ecological systems (e.g., food-webs and neural networks), which seem to posses a spectacular ability of fault-tolerance despite numerous faulty components or the deletion of some of their constituent members altogether.

4.2. Enlightened Evolutionary Engineering

When the overall process of complex systems engineering is considered, we come to a broader view of system creation that can be understood best by analogy with biological evolution or technological development in a market economy [5, 16]. Traditionally, in engineering, evolutionary methods have been considered as just another optimization technique where human designers create the meta-process of problem specification and interpretation. This is not what is intended here. In complex systems engineering, evolution serves as the meta-process and human engineers/designers create the *components* on which this meta-process operates. These components are still produced through traditional problem-solving methods that are quite effective if the individual components are not overly complex. The evolution of large complex systems takes place primarily in their functional environment, enabling the system to adapt to real world tasks through changes in components and their interactions over time. Large engineering systems should be considered as hybrids of people and equipment. Thus, people too serve as components in the system, both during the operation and in the design of the system. The existence of variety in the components at multiple levels of organization enables evolutionary selection to occur. Selection changes the population of components so that the introduction of more effective components leads to their wider adoption over time. When viewed with a wide lens, this is the process that has been used historically for engineering within a free market economy, as different products compete with each other for market share. With the need to develop larger and more complex systems, engineering must explicitly recognize the role of evolutionary change, not only in the human-centered process of innovation and design, but also *within* the systems being engineered and deployed. Evolution, in its broadest sense, permeates all levels of a complex system.

4.3. A New Set of Challenges

Given its radical redefinition of the classical engineering paradigm, it is not surprising that complex systems engineering poses several significant challenges of its own. Here, we identify only a few of the more fundamental issues that researchers in complex systems engineering must address if this discipline is to establish itself as a viable paradigm.

4.3.1 Configuring Viable Configuration Spaces

In a series of papers, John Doyle and colleagues have recently contrasted the purely self-organized view of complex systems with the optimization-based paradigm rooted in engineering [18, 19, 25]. One essential insight to emerge from this work is that, while most work in complex systems has focused on *generic* or *typical* systems within ensembles, well-designed and optimized systems are likely to be *rare* and *atypical* within their configuration spaces. Attributing the focus on ensembles to standard practice in mathematical physics, the authors conclude that most work on the broad principles of complex systems [2, 10, 11, 30] has not contributed much to the understanding of real systems such as the Internet [25] and metabolic networks [31], which are best understood in terms of optimal design. Leaving aside the specifics of the critique, this observation clarifies a fundamental issue for engineering complex systems: *The need for solution-rich configuration spaces*. For the inherently stochastic process of self-organization to produce high-performance instantiations of a system, such instantiations must not be too rare in the configuration space. This is not true of the configuration spaces in which most design currently occurs, and the classical engineering paradigm can be seen essentially as an algorithm for finding the rare, atypical configurations that provide peak performance, i.e., optimal designs. The challenge for complex systems engineers is to devise the components of their systems and the interactions between them in such a way that stochastic processes such as relaxation, annealing, swarming, evolution, etc. can find near-optimal configurations relatively quickly, which is only possible if such configurations are not too rare or completely atypical. Just as traditional engineering seeks optimal solutions, complex systems engineering must seek "optimal" configuration spaces where near-optimal configurations for an infinite number of as-yet unforeseen circumstances are numerously implicit.

Currently, almost all complex systems engineering research has focused on specific domains such as multi-agent systems [32, 33], collective robotics [24, 26], swarms [12], and networks [1, 10, 11, 13, 14, 15, 22, 35], and the emergence of good algorithms have relied heavily on the ingenuity of human researchers. However, a clue towards a general strategy comes from biological systems, where evolution's profound success is supported by the meta-attribute of *evolvability*: The ability of the configuration space (in this case, the space of genotypes or phenotypes) to produce an endless supply of viable configurations with remarkably few obvious dead-ends. Redundancy — often cited as a source of this evolvability [20] — is only a partial explanation, and factors such as degeneracy [29] may play a key role. Indeed, it has been suggested that evolvability itself is an evolved quality [20]. Uncovering the characteristics that make evolution so efficient may well enable complex systems engineers to devise systems that have the attribute of *self-optimizability*.

4.3.2 Obtaining Specific Global Functionality from Local Processes

While the methods and processes of complex systems engineering may differ from those of classical engineering, they still share the ultimate goal of *utility*: The need for the engineered system to demonstrate specific functionality. In the classical paradigm, this is assured by the explicit design and testing of the processes that produce the desired functionality, and the pathway from component behavior to system behavior is clear. This is not the case in engineered complex systems where, by definition, system functionality is emergent and too complex to be described explicitly in terms of component behavior. The engineering process defines components and their interactions, but ensuring that the design produces the desired global functionality is the primary challenge for complex systems engineering.

Several approaches have been tried to address this issue, but they rely primarily on three ideas. The first is the idea of *coordination*, where global behavior is related to component behavior using a set of coordination variables. In this approach, which is primarily control-theoretic, components or agents coordinate their choices through a shared set of quantities that can be related more directly to global functionality. Thus, the components seek to achieve certain desirable coordination states (e.g., synchronization, partitioning) that correspond to the desired functional state of the system. The coordination-based approach has been used very successfully in multi-agent systems [32, 33], collective robotics [24, 26], and swarms [12, 21]. Another approach has been developed where a high-level language is used to describe system functionality. Functional programs can then be "compiled" into a specification for the behavior of individual components [28].

The second important idea is that of *analogy*, where the functionality of existing complex systems such as insect colonies, ecosystems, economies, etc., is used to infer desirable behaviors for components/agents. For example, since insect swarms perform brood-sorting through local and stigmergic interactions [21], the behavior of individual insects during the process presumably has the effect of sorting, and can be used to design a system of sorting agents. This analogy-based approach has been especially fruitful for complex systems, and underlies paradigms such as neural networks, swarms, artificial worlds and artificial life. In particular, analogies with processes at all levels in biological systems promise a comprehensive framework for engineering a variety of useful complex systems. It is important to note that designing by analogy is not necessarily distinct from the coordination-based approach (above), and analogies often yield a set of suitable coordination mechanisms.

The third important idea in controlling the functionality of complex systems is *selective plasticity*. This can take the form of learning, fitness-based selection, adaptation, or a combination of these. In many cases, while it may not be clear a priori how a specific global functionality may be obtained from component behavior, it can be arrived at through an adaptive process. This relates to the earlier discussion of defining good configuration spaces: If the configuration space implied by the components and their interactions includes the desired global functionality, then an adaptive process that makes this functionality its attractor will lead to a suitable design. Of course, this is easier said than done, but the process can be aided by choosing a configuration space likely to be rich in "good" designs. A systematic method for specifying such configuration spaces is a fundamental challenge for complex systems engineering.

4.3.3 Defining Functional and Meta-Functional Performance Metrics

The definition of performance metrics is a necessity for any engineering process, since the goal of design is to produce a high-performance system. In engineering complex systems, however, care is needed to focus on the appropriate performance domain. The fundamental question to be answered is: What makes a good complex system? Is it a system that performs a specific task very well in a particular situation, or is it one which can adapt to perform well in a variety of situations? Implicit in most work on complex systems is the notion that complex systems should be judged on their *meta-attributes* such as robustness, evolvability, adaptivity, scalability, etc, rather than on narrowly-defined tasks. However, defining and measuring these properties is still far from being an exact science. Interestingly, this is a problem prevalent even in the evaluation of "real" complex system such as organizations or individuals. For example, students' capabilities are evaluated both by using rigidly specified examinations with well-defined correct responses and through open-ended challenges such as research projects. Current methods for evaluating engineered systems correspond mainly to the first of these modes, and metrics to assess the meta-attributes that make a complex system worth its complexity are still in development.

4.4 What the Complex Engineering Paradigm Offers

In order to engineer useful complex systems, the complex engineering paradigm seeks to provide behavior-rich systems that, when confronted with a problem-rich environment, discover a variety of potential solutions in their repertoire. These potential solutions can then be selected through an evolutionary adaptation process to produce progressively better (and continuously improving) solutions. The promise of complex systems engineering is, therefore, one of open-ended discovery rather than predetermined performance. The contrast between the limitations of traditional systems and the power of complex systems can be seen in terms of several key attributes including scalability, flexibility, evolvability, adaptability, resilience, robustness, durability, reliability, self-monitoring, and self-repair. Overarching these, the essential property of complex systems is their complexity, which enables them to perform highly complex tasks without running into insuperable capacity constraints. As the complexity of tasks facing engineered systems grows, the complex systems approach to engineering will increasingly become the default option rather than just another interesting alternative.

5. About this Book

The first of its kind, the objective of this book is to demonstrate the potential of complex systems perspectives to understanding and improving the design, implementation, and dynamics of complex engineered systems. The book provides essential terminology, a set of central concepts, and an appropriate technical framework for the systematic study of complex engineered systems. In particular, the book inspires discussion about fundamental questions such as: Is there any place in the complex systems paradigm for explicitly sought design characteristics, or must everything be emergent? If the default paradigm must be trial-and-selection rather than specification-and-design, how can complex systems engineering attain the systematic aspect of classical engineering?

The book is conceptually divided into two parts: The first part (Chapters 2-12) is devoted to understanding the general characteristics of Complex Engineered Systems, whereas the second part (Chapters 13-16) addresses specific systems and technologies that embody key features, characteristic of complex adaptive systems, as part of their behavior. In particular:

Bar-Yam (Chapter 2) demonstrates the fundamental limitations of decomposition-based engineering for the development of highly complex systems. Recognizing these limitations, it is argued that a new strategy for constructing many highly complex systems should be modeled after biological evolution, or market economies, where multiple design efforts compete in parallel for adoption through testing in actual use.

In the next two chapters, Braha and Bar-Yam (Chapter 3) and Valverde and Solé (Chapter 4) examine the statistical properties of software architectures and product development organizational networks, respectively. They show that the structure of these man-made information flow networks have properties that are similar to those displayed by other social, biological and technological networks. These statistical structural properties are shown to have a major effect on the functionality, dynamics, robustness, and fragility of Complex Engineered Systems. Braha and Bar-Yam (Chapter 3) further present a model and analysis of product design dynamics on complex networks, and show how the underlying network topologies provide direct information about the characteristics of this dynamics.

In Chapter 5, Anderson considers a few aspects associated with choosing an initial strategy towards designing a particular desired self-organized system. In particular, he discusses at a broad level some of the general pros and cons of approaches such as bottom-up simulation, top-down engineering, analogy and mimicry, and interactive evolution. Some of the key criteria, decisions, and constraints that *might* help pinpoint an initial useful approach to tackling the design of specific self-organized systems are extracted.

Maier and Fadel suggest in Chapter 6 that in design, the semantic, non-rational, non-algorithmic, impredicative, subjective, and unpredictable nature of humanity is inescapable. This is so because artifacts are always designed for human use, usually designed by humans themselves (using computers and other tools), and situated within a larger context of a complex world economy. Consequently, they argue that design in general is a member of the class of systems that are formally described as open and complex, and not a member of the class of systems that are formally described as closed and algorithmic.

Mihm and Loch (Chapter 7) and Klein et al (Chapter 8) present dynamic models of complex product development projects that are characterized by decomposition into an interrelated set of localized development tasks. They show how a 'rugged performance landscape' arises from simple interdependent components (local design teams) that have 'simple' performance functions. Consequently, they discuss the circumstances under which projects exhibit persistent problems or reach satisfactory performance levels (convergence of the development process).

Baldwin and Clark (Chapter 9) present a model of design and industrial evolution by emphasizing the role of modularity — building complex products from smaller subsystems that can be designed independently yet function together as a whole – as a financial force that can change the structure of an industry. They explore the value and costs that are associated with constructing and exploiting a modular design, and examine the ways in which modularity is exploited through the use of design rules and modular operators that correspond to search paths in the "value landscape" of a complex engineering system.

Norman and Kuras (Chapter 10) argue, based on their experience with developing the Air and Space Operations Center for the US Air force, that the methods for the engineering of complex systems should be based on a view of complex systems as having the characteristics of an *ecosystem*. This includes the use of processes which take advantage of emergence and which deliberately mimic *evolution* to accomplish and manage the engineering outcomes desired.

Klein et al show in Chapter 11 that collaborative design negotiation, involving many interdependent issues, has properties that are substantially different from the independent issue case that has been studied to date in the negotiation literature, and requires as a result different protocols to achieve near-optimal outcomes in a reasonable amount of time. Consequently, They describe a family of negotiation protocols that make substantial progress towards achieving near-optimal outcomes for complex negotiations.

Complex Engineered Systems comprise agents (animate or inanimate) that are intrinsically idiosyncratic and bounded-rational. This general characteristic introduces a long-running difficulty of applying conventional game theory. In Chapter 12, Wolpert shows how to modify conventional game theory to accommodate the bounded rationality of all real-world players. To this end, he presents a statistical physics approach, known as Product Distribution (PD) theory, as a principled formulation of bounded rationality.

As discussed earlier, one of the central challenges in engineering complex systems is to determine local rules of interaction that lead, via self-organization, to a desired global behavior. Chapters 13 through 16 address this issue in various contexts, presenting well-developed, general approaches to solving the problem.

In Chapter 13, Nagpal addresses the issue of specifying local behaviors to achieve pre-specified global results using the idea of *global-to-local compilation*. The global behavior is specified in terms of primitive behaviors at the agent level and this "program" is then "compiled" into a common behavioral specification for all agents, ensuring the emergence of the desired global effect. The idea of global-to-local compilation is inspired in part by the processes seen in living cells, and is applicable in principle to a large class of distributed systems.

Dahl, Mataric and Sukhatme (Chapter 14) address the issue of global organization emerging from local behaviors in the specific context of multi-robot systems. They present an approach based on behavior-based decision-making, reinforcement learning and vacancy chains, demonstrate its efficacy, and extend it to heterogeneous groups of robots. The approach represents a systematic and powerful method for the organization of complex global behavior in multi-robot systems.

In Chapter 15, de Croon, Nolfi and Postma describe how pro-active embodied agents can be produced by using neural controllers optimized through evolutionary learning. This extends the traditional embodied cognitive science framework by allowing agents to learn complex behaviors using internal states rather than simply exhibiting primitive stimulus-response behaviors. This extension opens up the possibility of achieving very complex emergent global behaviors in multi-robot systems.

An extremely useful attribute of natural complex systems is the ability to reconfigure themselves in response to their situation. However, achieving global self-reconfigurability requires robust mechanisms that correctly lead to desired configurations without getting trapped in sub-optimal ones. In Chapter 16, Salemi, Will and Shen describe a practically implementable, general approach to this problem using the CONRO self-reconfigurable robot to demonstrate its utility.

It is hoped that this book will contribute towards putting natural and engineering complex systems within the same discipline, thus allowing a new kind of "closing of the loop" whereby the study of natural complex systems leads to better methods for complex engineered systems, while experience with building and manipulating complex engineered systems enhances understanding of how natural complex systems function.

References

[1]   Albert, R. and A-L. Barabasi (2002) "Statistical mechanics of complex networks", Review of Modern Physics **74**, pp. 47-97.

[2]   Bak, P. (1996) How Nature Works: The Science of Self-Organized Criticality, Springer-Verlag Telos.

[3]   Bar-Yam, Y. (1997) Dynamics of Complex Systems, Perseus Press.

[4]   Bar-Yam, Y. (2002a) General features of complex Systems, Encyclopedia of Life Support Systems (EOLSS), EOLSS Publishers.

[5]   Bar-Yam, Y. (2002b) Enlightened Evolutionary Engineering / Implementation of Innovation in FORCEnet, Report to Chief of Naval Operations Strategic Studies Group (May 1, 2002) http://www.necsi.org/projects/yaneer/SSG_NECSI_2_E3_2.pdf

[6]   Bar-Yam, Y. (2003a) Unifying principles in complex systems, in Converging Technology (NBIC) for Improving Human Performance, M. C. Roco and W. S. Bainbridge eds, Kluwer.

[7] Bar-Yam, Y. (2003b) When systems engineering fails—toward complex systems engineering, Proceedings of the International Conference on Systems, Man & Cybernetics, 2003, Vol. 2, 2021- 2028, Piscataway, NJ: IEEE Press.

[8] Bar-Yam Y. and Kuras M, (2004) Complex systems and evolutionary engineering, AOC Concept Paper, http://www.necsi.org/projects/yaneer/AOCs_as_complex_systems.pdf

[9] Bar-Yam, Y. (2005) Making Things Work, NECSI Knowledge Press.

[10] Barabasi, A.-L. and E. Bonabeau (2003) "Scale-free networks" Scientific American **288**, pp. 60-69.

[11] Barabasi, A.-L. (2002) Linked: the New Science of Networks, Perseus Books.

[12] Bonabeau, E., M. Dorigo and G. Theraulaz (1999) Swarm intelligence: from natural to artificial systems, Oxford, UK: Oxford University Press.

[13] Bornholdt, S. and H. G. Schuster (Eds.) (2002) Handbook of graphs and networks: from the genome to the Internet, Berlin: Wiley-VCH.

[14] Braha, D. and Y. Bar-Yam (2006) "The Statistical Mechanics of Complex Product Development: Empirical and Analytical Results," Management Science (forthcoming).

[15] Braha, D. and Y. Bar-Yam (2004) "The Topology of Large-Scale Engineering Problem-Solving Networks," Physical Review E. **69**, 016113.

[16] Braha, D. and O. Maimon (1998a) A Mathematical Theory of Design: Foundations, Algorithms and Applications, Kluwer, Boston.

[17] Braha, D. and O. Maimon (1998b) "The Measurement of a Design Structural and Functional Complexity" IEEE Transactions on Systems, Man and Cybernetic. Part A. **28** (4), pp.527-535.

[18] Carlson, J.M and J.C. Doyle (2000) "Highly optimized tolerance: robustness and design in complex systems", Physical Review Letters **84**, pp. 2529-2532.

[19] Carlson, J. M. and J.C. Doyle (2002) "Complexity and robustness", Proceedings of the National Academy of Science USA **44** (suppl. 1), pp. 2539-2545.

[20] Dawkins, R. (1986) The blind watchmaker: why the evidence of evolution reveals a universe without design, London, UK: W.W. Norton and Co.

[21]   [21] Dorigo, M., G. Di Caro and L.M. Gambardella (1999) "Ant algorithms for discrete optimization", Artificial Life 5, pp. 137-172.

[22]   Dorogovtsev, S.N. and J.F.F. Mendes (2003) Evolution of networks: from biological nets to the Internet and WWW, Oxford, UK: Oxford University Press.

[23]   Klein, M., H. Sayama, P. Faratin, and Y. Bar-Yam. (2002) A Complex Systems Perspective on Computer-Supported Collaborative Design Technology. Communications of the ACM, 45, No. 11, pp. 27-31.

[24]   Kube, C.R. and H. Zhang (1992) "Collective robotic intelligence", Proceedings of the Second International Conference on Simulation of Adaptive Behavior, pp. 460-468.

[25]   Li, L., D. Alderson, R. Tanaka, J.C. Doyle and W. Willinger (2004) "Towards a theory of scale-free fraphs: definition, properties and implications (extended version)", arXiv:cond-mat/0501169v1.

[26]   Mataric, M.J. (2001) "Learning in behavior-based multi-robot systems: policies, models, and other agents", Cognitive Systems Research (special issue on multi-disciplinary studies of multi-agent learning) 2, pp. 81-93.

[27]   Maimon, O. and D. Braha (1996) "On the Complexity of the Design Synthesis Problem," IEEE Transactions on Systems, Man and Cybernetic. Part A. 26 (1), pp.142-150.

[28]   Nagpal, R. (2001) Programmable Self-Assembly: Constructing Global Shape using Biologically-inspired Local Interactions and Origami Mathematics, PhD thesis, Department of Electrical Engineering and Computer Science, MIT.

[29]   Sole, R.V., R. Ferrer, J. M. Montoya, and S. Valverde. (2002) "Selection, tinkering, and emergence in complex networks," Complexity 8, pp. 20-33.

[30]   Stauffer, D. and A. Aharony (1994) Introduction to Percolation Theory, London: Taylor & Francis.

[31]   Tanaka, R. and Doyle, J.C. (2004) "Scale-rich metabolic networks: background and introduction", arXiv:q-bio.MN/0410009 v1.

[32]   Weiss, G. (1999) Multiagent Systems: A Modern Approach to Distributed Artificial Intelligence, Cambridge, MA: MIT Press.

[33]   Wooldridge, M. (2002) Introduction to MultiAgent Systems, John Wiley & Sons.

[34] Yassine, A. N. Joglekar, D. Braha, S. Eppinger, and D. Whitney (2003) "Information Hiding in Product Development: The Design Churn Effect," Research in Engineering Design **14** (3), pp. 131-144.

[35] Zhao, F. and L. Guibas (2004) Wireless sensor networks, San Francisco, CA: Morgan Kaufmann.

[36] Braha, D. and Y. Reich (2003) "Topological Structures for Modeling Complex Engineering Design Processes," Research in Engineering Design **14** (4), pp. 185-199.

[37] Edelman, G.M. and J.A. Gally, (2001) "Degeneracy and complexity in biology systems," Proc. Natl. Acad. Sci. USA **98**, pp. 13763–13768.

Chapter 2

Engineering Complex Systems: Multiscale Analysis and Evolutionary Engineering

Yaneer Bar-Yam
New England Complex Systems Institute
Cambridge MA 02138
yaneer@necsi.org

1. Overview

We describe an analytic approach, multiscale analysis, that can demonstrate the fundamental limitations of decomposition based engineering for the development of highly complex systems. The interdependence of components and communication between design teams limits any planning based process. Recognizing this limitation, we found that a new strategy for constructing many highly complex systems should be modeled after biological evolution, or market economies, where multiple design efforts compete in parallel for adoption through testing in actual use. Evolution is the only process that is known to create highly complex systems.

2. Introduction

The idea that highly complex system design and engineering requires new insights and tools has become a topic of increasing interest and importance as the number of active elements in systems and the real time demands on systems increase.[1-5]

One of the central realizations about highly complex systems is that analysis and synthesis do not follow the same processes. This is dramatically different from the case of conventional engineering analysis and design. When a system is sufficiently simple, analysis and synthesis occur by decomposition. The system is broken down into parts and each part is described. Then, the function of the entire system can be realized through recomposition of the parts. When a system is highly complex, this approach is not feasible.[1-3] We will demonstrate this both through a formal analytic treatment and through historical experience in this chapter.

We have developed an analytic approach to the study of complex systems called Multiscale Analysis [1,6-13], which directly addresses the complexity of the system and its relationship to structure and function. This approach provides basic insights into design trade-offs. However, it also enables us to demonstrate quantitatively that decomposition design strategies are unable to create systems beyond a certain level of complexity. The level of complexity is limited by a single agent's ability (i.e. a human being's ability) to understand the interdependencies between the components. When higher levels of complexity are needed to design systems, it is necessary to transition to an alternative synthesis strategy: the strategy of evolutionary engineering.

Evolutionary engineering abandons many of the highly valued conventional systems engineering strategies for arriving at well planned and fully understood systems. It replaces this with the creation of a planned environment that fosters learning by doing and enables unanticipated advances. The evolutionary approach is the natural strategy for developing highly complex systems because their behavior is ultimately untestable[6,9]. Discovery is a key part of ongoing improvement and the necessary time scale for use and improvement is far shorter than what can be achieved by traditional cycles of planning and implementation. The false sense of security in planning is inferior to the recognition that creating the right environment is a better guarantor of rapid change improvement and innovation.

Aspects of the evolutionary approach we describe [1-3] can be found in various more traditional and recent approaches. Incremental engineering [14] and experience-based learning [15,16] are very traditional approaches in certain contexts. Recent extensions include spiral development and evolutionary acquisition [17] and adaptive programming; [18] various modifications of conventional engineering are relevant for different engineering contexts.[19] There are key differences between the evolutionary approach we describe and other strategies. These include an emphasis on parallel competitive development teams and the importance of creating an ongoing fielded implementation strategy, where coexistence of multiple types of components and interactions is possible. This evolutionary process is most commonly associated with the formation of complex biological organisms. A free market system is also an example of an evolutionary system with particular features that are not present in all evolutionary contexts.

In this chapter we will describe both the analysis of the limitations of conventional systems engineering and the new ideas of evolutionary engineering. In Section 3, we will describe the framework of multiscale analysis, mention its implications for design decisions and use it to prove the limitations of decomposition based design. In Section 4 we will describe historical experience with complex engineering projects, and some of the steps we have taken toward defining an enlightened evolutionary engineering strategy. Readers who are less interested in the formal mathematical proof of the limitations of conventional systems engineering, and more interested in the new concepts of evolutionary engineering can proceed directly to Section 4.

3. Analysis

Multiscale analysis [1,10] builds on the twin recognitions that scale and variety/complexity both play a role in effective system performance:

• Scale: A task requires a system to have sufficient "scale" of action, where scale refers to the number of elementary components that are coordinated to perform a task.

• Variety: A task requires that a system have the ability to perform many distinct actions, this known as variety and is measured as the logarithm of the number of distinct actions that can be taken in a specified interval of time.

Thus, for example, a system can be effective at some tasks via brute force, but for others it must carefully choose the right action to take. When designing a system for its tasks, recognizing the degree to which scale and complexity are essential in the effective functioning of the system is also directly relevant to the design process.

To understand the design implications of this analysis conceptually, we note that when components are acting in a coordinated way, they cannot act independently. When large scale action is required, the components must act coherently. Conversely, when high variety is required, the components must be able to act independently. More generally, there are various degrees of coordination that may be necessary to achieve the particular amount of variety needed at each scale of action to perform ongoing tasks.

The key to multiscale analysis of the variety of any system is that each of the components has a limit on its variety—the logarithm of the number of its distinguishable states. Components can be individuals that act in performing tasks, or they can be individuals that serve to manage or coordinate tasks, or they can be physical or computer based communication channels. We do not assume anything a-priori about the specific tasks or actions of the components, so that they could be widely different kinds of entities such as biological cells, human beings or artificial devices. The components could also be the same as each other or different. The key is that each of them has a bound on its variety. If we have a system that is formed of many components, and some of these components are responsible for coordinating other components, then we can establish limits on what particular organizational structures can do. It may be that the variety associated with the coordination exceeds the variety of the components. This is true even if the components that must be coordinated are relatively simple. It is also true if the components have a high variety. The key is that quite generally the coordination may require of order N times the variety of the individual components, even in a fixed configuration of coordination. This means we may need N coordinating entities.

To understand the organizational limitations that are established by such an analysis consider a hierarchical system, such as a human organization with hierarchical chains of communication or a hierarchically decomposed engineering system with hierarchical specification. Indeed, these two representations are synergistic, in that hierarchical organizations are generally the mechanism by which hierarchically decomposed systems are generated. The difficulty with this architecture is that there is a bandwidth limitation in the communication channels. This limitation is manifest most clearly in that the component at the apex of the hierarchy must perform all coordinations between the large groups that branch from the apex. More generally, the channels of communication pass through individual components, and if we assume that each component has a limit on its variety, then we can see that the communication channels are limited by the variety of their components.

This is a severe limitation on the variety of the system behavior because in a more networked structure it is possible for the components at the bottom of the hierarchy to coordinate with each other directly in a way that would increase dramatically the variety of possible pairwise actions well above what can be coordinated through the hierarchy. This illustrates the well known phenomenon in engineering of the explosion of interface specification, and the dramatic efforts that are devoted to coordination of components. Thus, while the conventional decomposition strategy presumes that the components are the entities that require engineering, when systems become highly complex it is the coordination that requires the effort of engineering. Therefore, the conventional decomposition strategy breaks down, requiring other engineering strategies.

The formal proof of this statement requires one subtlety, which is quantifying the coordination above the level of behavior of an individual. The variety that is the most limiting variety for a hierarchical organization is the variety that has a scale that requires more than one individual to perform a task, but less than that of the system as a whole. It is the existence of large varieties at intermediate scales that are not possible for hierarchical organizations. Either a completely independent or a completely dependent organizational behavior can be readily achieved. We describe the formalism that captures variety at all scales in several steps.

Quantitatively, the understanding of the requirements of variety was articulated in Ashby's Law of Requisite Variety.[20] Recently,[10] this law has been generalized to consider the issue of scale as well as variety. In the generalization, it is assumed that the system is composed of a number of components and that these components can be combined to perform specific tasks that might require more than a single component to perform.

More specifically, we assume that a responding system is composed of a number of subsystems, N, that are variously coordinated to respond to external contexts. The number of possible actions that the system can take, M, is not more than, m^N, the product of the possible actions of each part, m. We could directly apply the Law of Requisite Variety for that case, but we further constrain the problem of effective function by assuming that effective actions require a sufficient variety at each scale of action corresponding to the requirements for action at that scale. At every scale, the variety necessary to meet the tasks must be larger for the system than the task requirements. It is conventional to measure variety, like information, in logarithmic units so that the total variety of a set of independent components $V = \log(M)$ is the sum of the variety of the components, $V = Nv$, where $v = \log(m)$. If we assume a simple coordination mechanism such that the system is partitioned into groups that are fully coordinated, and different groups are independent of each other, then the variety of actions of each group is the same as the variety of actions of any individual of that group, and the scale of action equals the number of individuals in that group. For the entire system, the variety at scale k is $D(k) = vn(k)$ where $n(k)$ is the number of different k-member fully coordinated groups needed to perform the entire task, which therefore at a minimum requires $N = \sum kn(k)$ components to perform. The total variety of the task is proportional to the total number of subsets of any scale $V = \sum D(k)$.

With these assumptions, given a predetermined number, N, of components, the system can, in the extreme, perform a task of scale N, with variety equal to that of one component, or a task of scale one with variety N times as great. More generally the equation (obtained from $N = \sum kn(k)$)

$$Nv = \sum kD(k) \tag{1}$$

can be considered a constraint on the possible behavior patterns (sum rule) of a system due to different mechanisms of organization. It is often convenient to think about the variety of a system, $V(k)$, that has a scale k or larger, as this is the set of possible actions that can have at least that scale,

$$V(k) = \sum_{k'=k}^{N} D(k'). \tag{2}$$

Then the total variety of the system is $V(1)$, and the sum rule can be written as:

$$\sum_{k=1}^{N} V(k) = Nv. \tag{3}$$

The sum rule given by equation (1) or (3) describes the existence of a tradeoff between variety at different scales. Increasing the variety at one scale, by changing the organizational form, must come at the expense of variety at other scales. Our generalization of the Law of Requisite Variety is directly relevant to an analysis of whether or not coordination mechanisms of an organization are well or ill suited to the tasks being performed. Given the constraint imposed by the number of components, a successful organization has a coordination mechanism that ensures that the groups are coordinated at the relevant scale of tasks to be performed. The multiscale version of the Law of Requisite Variety captures this simple and intuitive statement.

In considering the requirements of multiscale variety, we can state that in order for a system to be effective, it must be able to coordinate the right number of components to serve each task, while allowing the independence of other sets of components so that they can perform their respective tasks without binding the actions of one such set to another. This now serves as a key characterization of system organization. Specifically, the Multiscale Law of Requisite Variety implies that for a system to be successful its coordination mechanisms must allow for independence and dependence between components so as to allow the right number of sets of components acting at each scale.

In order to formalize the analysis we must define how a manager functions in an organization in terms of the coordination of subordinates. How do we describe a coordinator/manager? A manager specifies the state of the subordinates and a coordination mechanism. We assume that at any particular time the manager can only coordinate a particular subset, indexed by w, of the subordinates, and at that time these subordinates are fully coordinated, while the others act independently (one cannot be in two places at the same time). $q(w)$ is the number of subordinates that are being coordinated, which, for values of zero or one corresponds to no coordination. A specification of the manager at a particular time thus can be written (s_m, w), where the state of s_m specifies the states of all the coordinated subordinates, while w specifies which subordinates are coordinated. For simplicity we do not count the redundancy provided by the manager (who we assume does not do the action only specifies it) and therefore s_m is not needed in the description of the system since it is redundant to the actions of the subordinates. We also neglect the information in specifying w by treating the information as conditional on the coordination mechanism. These assumptions can be relaxed without changing the conclusions. Then we have the multiscale variety for a particular coordination state given by:

$$D(k \mid w) = v(N - q(w))\delta_{k,1} + v\delta_{k,q(w)} .$$

(4)

Combining coordination states, each with a probability $P(w)$ we have:

$$D(k) = \sum_w P(w)\delta_{q(w),k}v + \delta_{k,1}\sum_w P(w)(N - q(w))v .$$

(5)

This gives the expected bound on the total coordination:

$$V(2) = \sum_{k=2}^{N} D(k) = \sum_{k=2}^{N}\sum_w P(w)\delta_{q(w),k}v \le v .$$

(6)

Where the inequality provides the quite reasonable statement that the variety of the system for scales larger than one individual cannot be greater than the variety of the manager.

This coordination limitation is recursively applied to each level of managers for the set of individuals under their supervision so that the mutual information between individuals (workers or managers) at one level of organization is limited by the manager that supervises them. This implies, for example, that the combined mutual information between all workers is no more than the variety of the first level supervisors. Assuming that the variety of a manager is typically no more than the variety of a worker, we would expect that the limit of mutual information to be N/B where B is the branching ratio, i.e. the number of workers supervised by a single manager. Higher level managers are similarly restricted in their ability to coordinate the managers at the lower level. When an upper level manager in a conventional hierarchy coordinates parts of the organization, information must be communicated through the lower level managers. This also reduces the degree to which their own inter-worker coordination can be performed (i.e. to the extent that the higher level manager performs coordination, this reduces the capacity of the lower level managers to coordinate).

We can make a more direct connection to multiscale variety if we consider a somewhat generalized version of hierarchical control. In the generalized version of the hierarchy, managers exist at a certain level of authority, i.e. supervising a certain fraction of the organization, but do not have a particular set of subordinates that they supervise (the "matrix organization" [21] is an intermediate case). By not including the constraint of a strict hierarchy, i.e. that a manager has a particular subset of the individuals and cannot coordinate others outside of this subset, we obtain an upper bound on the coordination of a more conventional hierarchy. For the conventional hierarchy, the coordination of the system is further limited since even only two individuals that are in different divisions of the organization require coordination by the CEO. For the generalized hierarchical model, we can generalize the equations above and reach a conclusion that

$$V(2) = \sum_{k=2}^{N} D(k) \leq Cv. \tag{7}$$

Where C is the number of managers. This states quite reasonably that the total variety of actions greater than the scale of one individual is not greater than the total variety of the managers. For managers having a certain limit on how many subordinates they can control, so that managers at level l can coordinate up to B^l subordinates, we further limit the number of those coordinated at larger scales by

$$V(B^{l-1}+1) = \sum_{k=B^{l-1}+1}^{N} D(k) \leq \sum_{l' \geq l} C_{l'} v \tag{8}$$

which reasonably states that the variety of behaviors associated with a number of individuals is only as great as the variety of the managers that can coordinate that number of individuals.

For example, we consider the role of the CEO and assign him/her the obligation of determining those issues that are of relevance to the actions of a large proportion of individuals that are part of the organization. If we consider 10% to be the threshold fraction, then all decisions involving 10% of the individuals of the organization are coordinated by the CEO. The maximal possible variety of such portions (at this scale of action) is ten times the variety of a single individual. However, this cannot be done when coordinated by a single individual, the maximum is the CEO's variety. More generally, we can categorically state: to the extent that a single individual is coordinating the behavior of an organization, the coordination defined by mutual information cannot have a higher variety than that of a single individual.

We see that for a hierarchically coordinated system, the combined conditional mutual information of subunits of a manager cannot be greater than the variety of that manager. This is not a problem for either of two cases (dictated by environmental conditions associated with the task to be performed): 1) if the system has a simple coherent behavior, or 2) if the manager exercises very little control over subordinates so that the workers are almost totally independent of each other. It is a problem, however, when the behaviors of subunits themselves have a high variety (greater than that of an individual) and these must be coordinated. Thus, a hierarchical control system is well designed for relatively simple large scale behaviors, or for systems with very distributed control, but not for highly coordinated behaviors, i.e. when the coordination of these behaviors is more complex than a human being can communicate.

In summary, a generalization of the Law of Requisite Variety suggests that the effectiveness of a system organization can be evaluated by its variety at each scale of tasks to be performed. In its simplest form, a system with a high degree of coordination is large scale. When it is not coordinated, allowing for independent component action, it has high variety. The tradeoff of large scale action, as compared to the variety possible when actions of components are independent, provides a direct analysis of system organization. While this analysis does not specify that a particular system is capable of performing a task, it can provide a necessary condition for effectiveness. In considering biological and social systems, such analysis provides a way of classifying their behavior and considering the functional role they play in survival and societal function. [1,6-11] In engineering this analysis can provide an approach to guiding system design, by comparing the scale and analysis of tasks to the system capabilities. However, the analysis of hierarchical control also points to the inability of hierarchical decomposition to achieve desired coordination of the components, and thus the need for evolutionary approaches when the complexity of systems above the scale of the individual components is too high.

4. Enlightened Evolutionary Engineering

The failure rate of major engineering projects in recent years has been remarkably high costing many billions of dollars and much wasted time and effort. [1,3,4] In the conventional systems engineering approach, the project is recursively broken into subparts. The parts are then put together, with the task of defining and coordinating the subprojects being in the domain of the systems engineer developing the design at each level of decomposition. In order to perform the decomposition at a particular level, the systems engineer must perform the coordination that is necessary of the components at that level of decomposition. According to the multiscale analysis given above this may exceed the capacity of an individual. To understand the existing approach it is helpful to review its conceptual paradigm.

The traditional approach to large engineering projects follows the paradigm established by the Manhattan project and the Space program. There are several assumptions inherent to this paradigm: 1) substantially new technology will be used; 2) the new technology to be used is based upon a clear understanding of the basic principles or equations that govern the system (i.e. the relationship between energy and mass, $E=mc^2$, for the Manhattan project, or Newton's laws of mechanics and gravitation $F=-GMm/r^2$ for the space program); 3) the goal of the project and its more specific objectives and specifications are clearly understood; 4) based upon these specifications, a design will be created essentially from scratch and this design will be implemented and, consequently the mission will be accomplished.

Large engineering projects today generally continue to follow this paradigm. Projects are driven by a need to replace old "obsolete" systems with new systems, and particularly with the desire to use new technology. The time line of the project involves a sequence of stages: a planning stage at the beginning giving way to a specification stage, a design stage, and an implementation stage. The various stages of the process all assume that managers know what needs to be done and that this information can be included in a specification. Managers are deemed successful or unsuccessful depending upon whether or not this specification is achieved. On the technical side, modern large engineering projects generally involve the integration of systems to create larger systems, their goals include adding multiple functions that have not been possible before, and they are expected to satisfy additional constraints, especially constraints of reliability, safety and security.

The images of success of the Manhattan and Space Projects remain with us. However, the reality of most large engineering projects is much less satisfactory. Many projects end up in failure and are abandoned. This is true despite the tremendous investments that are made. The largest documented financial cost for a single project was the government's effort to improve air traffic control in the United States, the Federal Aviation Administration (FAA) Advanced Automation System. Many of the major difficulties with air traffic delays and other limitations are blamed on the antiquated / obsolete air traffic control system. This system, originally built in the 1950s, used remarkably obsolete technology, including 1960s mainframe computers and equipment based upon vacuum tubes [22], with functional limitations that would compel any modern engineer into laughter. Still, an effort that cost 3-6 billion dollars between 1982 and 1994 was abandoned without improving the system. While the failure of government projects is frequently blamed on specific issues related to government acquisition, a general survey of large software engineering projects in 1994 [32] showed that such failures were widespread in both private and public sector projects. This study classified projects according to whether they met stated goals of the project, the time table, and cost estimates. They found that under 20% of the projects were on-time, on-budget and on-function (projects at large companies had a lower rate of under 10% success). Over 50% of the projects were "challenged" which means that they were over budget typically by a factor of two, were over schedule by a factor of two, and did not meet about two thirds of the original functional specifications. The remaining 30% of the projects were called "impaired" which meant that they were abandoned. When considering the major investments these projects represent in time and money, the numbers are staggering, easily reaching $100 Billion each year in direct costs. The high percentage of failures and the remarkable percentage of challenged projects suggest that there is a systematic reason for the difficulty involved in large engineering projects beyond the specific reasons for failure that one might identify in any given case.

Indeed despite various efforts to improve acquisition of large systems, successors of the Advanced Automation System that are being worked on today are finding the process to be slow and progress limited [33]. From 1995 until today, major achievements in air traffic control modernization include replacing mainframe computers, replacing communications switching systems, and the en-route controller radar stations. The replacement of the Automated Radar Terminal System at Terminal Radar Facilities responsible for air traffic control near airports (the Standard Terminal Automation Replacement System (STARS) program), faced many of the problems that affected the Advanced Automation System: cost overruns, delays, and safety vetoes of implementation, and was implemented beginning in 2002 in particular airports by FAA emergency decree, and the full implementation at all airports is expected to take at least a decade if not longer. Even with the limited modernization taking place, the new equipment continues to be used in a manner that follows original protocols used for the old equipment.

A fundamental reason for the difficulties associated with modern large engineering projects is their inherent complexity. Complexity implies that different parts of the system are interdependent so that changes in one part may have effects on other parts of the system. Complexity may cause unanticipated effects that lead to failures of the system. These "indirect" effects can be discussed in terms of feedback loops in the system, and in terms of emergent collective behaviors of the system as a whole [6]. Such behaviors are generally difficult to anticipate and understand. Despite the superficial complexity of the Manhattan and Space Projects, the tasks that they were striving to achieve were relatively simple compared to the problem of air traffic control. To understand complexity of Air Traffic Control (ATC), it is necessary to consider the problem of 3-dimensional trajectory separation --- ensuring that paths of any two planes do not intersect at the same time, considering the many airplanes taking off and landing in a short period of time, and taking into account the remarkably low probability of failure that safety constraints impose. Failure in any one case may appear to have a specific cause, but the common inability to implement high cost systems can be attributed to their intrinsic complexity.

While the complexity of engineering projects has been increasing, it is important to recognize that complexity is not new. Indeed, engineers and managers are generally aware of the complexity of these projects and have developed systematic techniques to address them. There are several strategies that are commonly used including modularity, abstraction, hierarchy and layering. These methods are useful, but at some degree of interdependence they become ineffective. Modularity is a well recognized way to separate a large system into parts that can be individually designed and modified. However, modularity incorrectly assumes that complex system behavior can be reduced to the sum of its parts. As systems become more complex, the design of interfaces between parts occupies increasing attention and eventually the process breaks down. Abstraction simplifies the description or specification of the system. However, abstraction assumes that the details to be provided to one part of the system (module) can be designed independently of details in other parts. Modularity and abstraction are generalized by various forms of hierarchical and layered specification, whether through the structure of the system, or through the attributes of parts of a system (e.g. in object oriented programming). Again, these two approaches either incorrectly portray performance or behavioral relationships between the system parts or assume details can be provided at a later stage. Similarly, management has developed ways to coordinate teams of people working on the same project through various carefully specified coordination mechanisms.

4.1. Evolve highly complex systems

One way to address the difficulty of complex projects is to simplify what is attempted. However, simplifying the function of an engineered system is not always possible because the necessary or desired core function is itself highly complex. When the inherent nature of a complex task is too large to deal with using conventional large engineering processes, a better solution is to use an evolutionary process [7], that is to create an environment in which continuous innovation can occur.

Evolutionary processes, commonly understood to be analogous to free market competition, are based on incremental iterative change. However, there are basic differences between evolution and the notion of incremental engineering. Among these is that evolution assumes that many different systems exist at the same time, and that changes occur to these systems in parallel. The parallel testing of many different changes, which can be combined later is distinctly different from conventional incremental engineering. The use of parallel initial exploration has been advocated in engineering [31]. However, this approach is unlike evolution as it leads to the selection of a single option rather than multiple parallel implementation. Multiple parallel implementation is more similar to the parallel and largely independent exploration of product improvements by different companies in a market economy, especially when there are many small companies. Another basic idea of evolution is that much testing is done "in the field"; the process of learning about effective solutions occurs through direct feedback from the environment. There are many more aspects of evolution that should be understood to make effective use of this process in complex large engineering projects, some of which are discussed below. Even the conventional concepts of evolution as they are currently taught in basic biology courses are not sufficient to capture the richness of modern ideas about evolution [6 (ch. 6),1,26-8].

Many of the more recent programming strategies, e.g. spiral development, extreme programming, and the open source movement, embody features of evolutionary processes. Still, a better understanding is necessary to realize the promise of evolutionary methods. The objective revolves around mimicry of the processes that promote rapid innovation through competition. The creation of an effective ``artificial ecology" or ``artificial economy" requires design. In and of itself, a competitive system is not self-sustaining as it tends to become stuck through monopolization, or self-destructive behavior.

To introduce the concepts of evolution, it is helpful to start from the conventional perspective and then augment it with some of the modern modifications. Evolution is about the change in a population of organisms over time. This population changes not because the members of the population change directly, but because of a process of generational replacement by offspring that differ from their parents. The qualities of offspring are different from their parents, in part, because some parents have more offspring than others. The process by which the number of offspring are determined, termed selection, is considered a measure of organism effectiveness / fitness. Offspring tend to inherit traits of parents. Traits are modified by sexual reproduction and mutation that introduce novelty / variation. This novelty allows progressive changes over many generations. Thus, in the conventional perspective, evolution is a process of replication with variation followed by selection based upon competition. In contrast to a traditional engineering view, where the process of innovation occurs through concept, design, specification, implementation and large scale manufacture, the evolutionary perspective suggests that we consider the population of functioning products that are in use at a particular time as the changing population that will be replaced by new products over time. The change in this population occurs through the selection of those products that increase their proportion in the population. This process of evolution involves the decisions of people as well as the changes that occur in the equipment itself.

It may be helpful to point out that this approach (the treatment of the population of engineered products as evolving) is quite different than the approach previously used to introduce evolution in an engineering context through genetic algorithms or evolutionary programming (GA/EA) [29,30]. The GA/EA approach has considered automating the process of design by transferring the entire problem into a computer. According to this strategy, we develop a representation of possible systems, specify the utility function, implement selection and replication and subsequently create the system design in the computer. While the GA/EA approach can help in specific cases, it is well known that evolution from scratch is slow. Thus it is helpful to take advantage of the capability of human beings to contribute to the design of systems. The objective of the use of evolutionary process described here is to avoid relying upon an individual human being to design systems that can perform highly complex tasks. A computer by itself cannot solve such problems either. Our objective here is to embed the process of design into that of many human beings (using computers) coordinated through an evolutionary process.

4.2. Environment for evolution

The basic concept of designing an evolutionary process is to create an environment in which a process of innovation and creative change takes place. To do this we develop the perspective that tasks to be performed are analogous to resources in biology. Individual parts of the system, whether they are hardware, software or people involved in executing the tasks, are analogous to various organisms that are involved in an evolutionary process. Changes in the individual parts take place through introducing alternate components (equipment, software, training or by moving people to different tasks). All of these changes are part of the dynamics of the system. Within this environment it is possible for conventional engineering of equipment or software components to occur. The focus of such engineering efforts is on change to small parts of the system rather than on change to the system as a whole. This concept of incremental replacement of components (equipment, software, training, tasks) involves changes in one part of the system, not in every part of the system. Even when the same component exists in many parts of the system, changes are not imposed on all of these parts at the same time. Multiple small teams are involved in design and implementation of these changes. It is important to note that this is the opposite of standardization—the explicit imposition of variety. The development environment should be constructed so that exploration of possibilities can be accomplished in a rapid (efficient) manner. Wider adoption of a particular change, corresponding to reproduction in biology, occurs when experience with a component indicates improved performance. Wider adoption occurs through informed selection by the individuals involved. This process of "selection" explicitly entails feedback about aggregate system performance in the context of real world tasks.

Thus the process of innovation involves multiple variants of equipment, software, training or human roles that perform similar tasks in parallel. The appearance of redundancy and parallelism is counter to the conventional engineering approach, which assumes specific functional assignments rather than parallel ones. This is the primary difference between evolutionary processes and incremental approaches to engineering. The process of overall change consisting of an innovation that, for example, replaces one version of a particular type of equipment with another, occurs in several stages. In the first stage, a new variant of the equipment (or other component) is introduced. Locally, this variant may perform better or worse than others. However, overall, the first introduction of the equipment does not significantly affect the performance of the entire system because other equipment is operating in parallel. The second stage occurs if the new variant is more effective: others may adopt it in other parts of the system. As adoption occurs there is a load transfer from older versions to the new version in the context of competition, both in the local context and in the larger context of the entire system. The third stage involves keeping older systems around for longer than they are needed, using them for a smaller and smaller part of the load until eventually they are discarded 'naturally'. Following a single process of innovation, is, however, not really the point of the evolutionary engineering process. Instead, the key is recognizing the variety of possibilities and subsystems that exist at any one time and how they act together in the process of innovation.

The conventional development process currently used in large engineering projects is not entirely abandoned in the evolutionary context. Instead, it is placed within a larger context of an evolutionary process. This means that individuals or teams that are developing parts of the system can still use well known and tested strategies for planning, specification, design, implementation and testing. The important caveat to be made here is that these tools are limited to parts of the system whose complexity is appropriate to the tool in use. Also, the time scale of the conventional development process is matched to the time scale of the larger evolutionary process so that field testing can provide direct feedback on effectiveness. This is similar to various proposals suggested for incremental iterative engineering. What is different, is the importance of parallel execution of components in a context designed for redundancy and robustness so that the implementation of alternatives can be done in parallel and effective improvements can be combined. At the same time, the ongoing variety provides robustness to changes in the function of the system. Specifically, if the function of the system is changed because of external changes, the system can adapt rapidly because there are various possible variants of subsystems that can be employed.

Understanding a complex system approach to design and implementation involves recognizing the many differences between the natural evolutionary process and traditional engineering practices. Evolutionary engineering employs, among others, the following key concepts,[31] that may be contrasted to traditional engineering practices:

- ° Focus on creating an environment and process rather than a product.
- ° Continually build on what already exists.
- ° Operational components are modifiable in situ.
- ° Operational systems include multiple versions of functional components.

○ Utilize multiple parallel development processes.
○ Evaluate experimentally in-situ.
○ Increase utilization of more effective components, gradually.
○ Effective solutions to specific problems cannot be anticipated.
○ Conventional systems engineering should be used for not-to-complex components.
Brief notes on each of these points follow.

4.2.1 Focus on creating an environment and process rather than a product:

Ongoing change in a system is the underlying mechanism of creation, not the formulation and execution of a particular plan. The environment must include rules for a structured competition between design teams and the components they develop. Encouraging and safeguarding the ongoing change and monitoring its outcomes are the absolute essentials of an evolutionary-based process.

4.2.2 Continually build on what already exists:

Off-line engineering of complex systems is impractical because the complexities of their environment and true functional requirements do not permit practical specification or testing prior to implementation. In complex systems, correct expectations and testing both depend on the immediate consequences of current operations.

4.2.3 Individual components must be modifiable in situ

The interdependencies between system components must be such that individual components can be modified in situ. In practice this requires point 4.2.4.

4.2.4 Operational systems include multiple versions of functional components

Complex systems should be understood as populations rather than as rigid assemblies of unique components. Individual components can overlap substantially in terms of both functionality and interaction. Evolutionary processes impact both populations and individuals. Redundancies are not always unwanted inefficiencies.

4.2.5 Utilize multiple parallel development processes

The existence of populations of components allows multiple parallel efforts to explore modifications that might (but that are not guaranteed) to improve system components and/or total system capability.

4.2.6 Evaluate experimentally in-situ

Testing and experimentation increasingly overlap. Off-line qualification testing becomes a prelude to active field testing for components in a large variety of operational environments. Results (including unexpected results) are ratified or rejected as they occur based on then-current overall system capability.

4.2.7 Increase utilization of more effective components, gradually

The replacement of components cannot be abrupt as testing is never complete and operation is continuous. Augmentation and parallel operation is the preferred approach.

4.2.8 Effective solutions to specific problems cannot be anticipated

Specification efforts cannot assume that the most efficient or effective solutions can be anticipated in advance of an exploration and discovery process involving multiple parallel development efforts.

4.2.9 Conventional systems engineering should be used for not-to-complex components.

Conventional systems engineering is a highly effective discipline when it is used for systems that are not too complex, it provides an important acceleration for the evolutionary engineering process over that of naive evolution where randomness is the only player.

4.3 The Integration of Systems, and the Systems of Systems Concept

Frequently, engineering today is concerned with the integration of multiple components into larger aggregate systems. Such aggregations are often called system of systems, and are characterized by conceptual properties such as interoperability. The ideas presented here suggest that conventional engineering is unlikely to be effective in achieving such aggregate systems that involve many interactions and interdependencies. Ultimately, the challenge that has to be met in order to create such systems is to avoid or abandon conventional control and planning and institute evolutionary engineering processes.

In order to operate a evolutionary engineering process, the prevailing concept of system integration must be radically rethought. Effective application of the ideas in this chapter involves a paradigm shift from complete system specification to the creation of environments that are conducive to ongoing change in components of systems, while supporting the more or less constant evaluation of their overall effectiveness through virtual as well as real world testing.

5. Conclusions

It is important to appreciate that there are fundamental reasons that highly trained systems engineers have been unable to successfully complete the very highly complex engineering projects undertaken in recent years. Efforts to improve the existing decomposition based approach will not solve these problems. The application of multiscale analysis reveals that the coordination between components that is required to develop such systems is incompatible with decomposition. This can be most easily understood as an underlying bandwidth limitation in the hierarchical structure in which the decomposition of the design is performed.

The solution to this problem is to develop an environment for parallel design teams to develop components that can be field tested and compete for wider adoption. This approach underlies both the creation of complex biological systems and some complex social system, for example, those that utilize market competition process. Its implementation in large engineering projects may be a challenge, however, according to our understanding of the science of complex systems, it is the only way to successfully engineer highly complex systems.

References:

[1]   Y. Bar-Yam, Making Things Work; Solving complex problems in a complex world, (NECSI Knowledge Press, 2004).

[2]   D. Braha, A. Minai, and Y. Bar-Yam, eds. Engineered Complex Systems, NECSI Knowledge Press, 2004.

[3]   Y. Bar-Yam, Enlightened Evolutionary Engineering / Implementation of Innovation in FORCEnet, Report to Chief of Naval Operations Strategic Studies Group, 2002 (Brief 2000).

[4]   Y. Bar-Yam, When Systems Engineering Fails --- Toward Complex Systems Engineering, International Conference on Systems, Man & Cybernetics, 2003, Vol. 2, 2021- 2028, IEEE Press, Piscataway, NJ, 2003.

[5]   D. Braha and O. Maimon, A Mathematical Theory of Design: Foundations, Algorithms and Applications, Kluwer, Boston, 1998.

[6]   Y. Bar-Yam, Dynamics of Complex Systems, (Perseus, Reading, MA, 1997).

[7]   Y. Bar-Yam: General Features of Complex Systems, in Encyclopedia of Life Support Systems (EOLSS), UNESCO, EOLSS Publishers, Oxford, UK, 2002

[8]   Y. Bar-Yam: Complexity rising: From human beings to human civilization, a complexity profile, in Encyclopedia of Life Support Systems (EOLSS), UNESCO, EOLSS Publishers, Oxford, UK, 2002

[9]   Y. Bar-Yam, "Unifying Principles in Complex Systems" in Converging Technology (NBIC) for Improving Human Performance, M. C. Roco and W. S. Bainbridge, Dds., Kluwer, 2003.

[10]   Y. Bar-Yam, Multiscale Variety in Complex Systems, Complexity 9:4, pp. 37-45, 2004.

[11]   Y. Bar-Yam, A Mathematical Theory of Strong Emergence using Multiscale Variety, Complexity 9:6, pp. 15-24, 2004.

[12] Y. Bar-Yam: Multiscale Complexity / Entropy, Advances in Complex Systems 7, pp. 47-63, 2004.

[13] S. Gheorghiu-Svirschevski and Y. Bar-Yam, Multiscale analysis of information correlations in an infinite-range, ferromagnetic Ising system, Phys.Rev. E 70, 066115 (2004)

[14] Standish Group International, The CHAOS Report, 1994.

[15] G. S. Lynn, J. G. Morone and A. S. Paulson, "Marketing and discontinuous innovation: The probe-and-learn process," California Management Rev., Vol. 38, No. 3, pp. 8–36, 1996.

[16] [R. W. Veryzer, "Discontinuous innovation and the new product development process," J. Product Innovation Management, Vol. 15, pp. 304–321, 1998

[17] DoD Directive 5000.1, "The Defense Acquisition System," May 12, 2003.

[18] See e.g. Agile Software Development, CrossTalk, Vol. 15, No. 10, Oct. 2002.

[19] M. T. Pich, C. H. Loch and A. De Meyer, "On Uncertainty, Ambiguity and Complexity in Project Management" Management Science 48, 1008-1023, 2002.

[20] W. R. Ashby, An Introduction to Cybernetics, Chapman and Hall, London, 1957.

[21] J. Galbraith, Designing Complex Organizations, (Addison-Wesley, 1973)

[22] Committee on Transportation and Infrastructure Computer Outages at the Federal Aviation Administration's Air Traffic Control Center in Aurora, Illinois [Field Hearing in Aurora, Illinois] hpw104-32.000 HEARING DATE: 09/26/1995.

[23] Standish Group International, The CHAOS Report, 1994.

[24] U.S. House Committee on Transportation and Infra-structure, FAA Criticized For Continued Delays In Modernization Of Air Traffic Control System, Mar. 14, 2001.

[25] D. K. Sobek, A. C. Ward and J. K. Liker, "Toyota's principles of set-based concurrent engineering," Sloan Management Rev., Vol. 40, pp. 67–83, 1999.

[26] E. Rauch, H. Sayama and Y. Bar-Yam, "The role of time scale in fitness,"Phys. Rev. Lett. 88, 228101-4 2002.

[27] J. K. Werfel and Y. Bar-Yam, The evolution of reproductive restraint through social communication, PNAS 101, 11019-11024 (2004).

[28] Y. Bar-Yam: Formalizing the gene centered view of evolution, Advances in Complex Systems 2, 277-281 (1999).

[29] L. J. Fogel, A. J. Owens and M. J. Walsh, *Artificial Intelligence through Simulated Evolution,* Wiley, New York, 1966.

[30] J. H. Holland, *Adaptation in Natural and Artificial Systems,* 2d ed. MIT Press, Cambridge, 1992.

[31] Y. Bar-Yam, *About Engineering Complex Systems: Multiscale Analysis and Evolutionary Engineering*, in Engineering Self Organising Systems: Methodologies and Applications, S. Brueckner, G. Di Marzo Serugendo, A. Karageorgos, R. Nagpal (Eds.), ESOA 2004, LNCS 3464, pp. 16–31, 2005.

Chapter 3

The Structure and Dynamics of Complex Product Design

Dan Braha
University of Massachusetts, Dartmouth MA
New England Complex Systems Institute, Cambridge MA
braha@necsi.org

Yaneer Bar-Yam
New England Complex Systems Institute, Cambridge
MA
yaneer@necsi.org

1. Introduction

The usefulness of understanding organizational network structure as a tool for assessing the effects of decisions on organizational performance has been illustrated in the social science and management literatures [39-42]. There it has been shown that informal networks of relationships (e.g., communication, information, and problem-solving networks) – rather than formal organizational charts – determine to a large extent the patterns of coordination and work processes embedded in the organization. In recent years, networks have also become the foundation for understanding numerous and disparate complex systems outside the field of social sciences (e.g., biology, ecology, engineering, and internet technology, see [9, 10, 49]).

The goal of this chapter is to examine, for the first time, the statistical properties of an important class of large-scale information-carrying networks – new product development. We discuss the significance of these statistical properties in providing insight into ways of improving the strategic and operational decision-making of the organization. In general, information-carrying networks constitute the infrastructure for exchanging knowledge that is important to the achievement of work by individual agents. We believe that our results will also be relevant to other information-carrying networks.

Distributed product development (abbreviated as 'PD'), which often involves an intricate set of interconnected tasks carried out by hundreds of designers, is fundamental to the creation of complex manmade systems [1]. The interdependence between the various tasks makes system development fundamentally iterative [2]. Iterations are driven by the repetition (rework) of tasks due to the availability of new information generated by other tasks such as changes in input, updates of shared assumptions, components, boundaries, or the discovery of errors. In such a network of interactions, iterations occur when some development tasks must be attempted even though the complete predecessor information is not available or known with certainty [3]. As this missing or uncertain information becomes available, the tasks are repeated to come closer to the design specifications or goals. This iterative process proceeds until convergence occurs [3-5].

Design iterations, which are the result of the PD network structure, might slow down the PD convergence or have a destabilizing effect on the system's behavior. This will delay the time required for product development, and thus compromise the effectiveness and efficiency of the PD process. For example, it is estimated that iteration costs about one-third of the whole PD time [6] while lost profits result when new products are delayed in development and shipped late [7]. Characterizing the *real-world* structure, and eventually the dynamics of complex PD networks, may lead to the development of guidelines for coping with complexity. It would also suggest ways for improving the decision making process, and the search for innovative design solutions.

The last few years have witnessed substantial and dramatic new advances in understanding the large-scale structural properties of many real-world complex networks [8-10]. The availability of large-scale empirical data on one hand and the advances in computing power and theoretical understanding have led to a series of discoveries that have uncovered statistical properties that are common to a variety of diverse real-world social, biological and technological networks including the world-wide web [11], the internet [12], power grids [13], metabolic and protein networks [14, 15], food webs [16], scientific collaboration networks [17-20], citation networks [21], electronic circuits [22], and software architecture [23]. These studies have shown that many complex networks are sparse; that is, they have only a small fraction of the possible number of links. Despite being primarily locally connected, such networks exhibit the "small-world" property of short average path lengths between any two nodes. Studies also have found that complex networks are characterized by an inhomogeneous distribution of nodal degrees (the number of nodes a particular node is connected to) with this distribution often following a power law (termed "scale free" networks in [29]). Scale-free networks have been shown to be robust to random failures of nodes, but vulnerable to failure of the highly connected nodes [24]. A variety of network growth processes that might occur on real networks, and that lead to scale-free and small-world networks have been proposed [9, 10]. The dynamics of networks can be understood to be due to processes propagating through the network of connections; the range of dynamical processes include disease spreading and diffusion, search and random walks, synchronization, games, Boolean networks and cellular automata, and rumor propagation. Indeed, the raison d'être of complex network studies might be said to be the finding that topology provides direct information about the characteristics of network dynamics [48]. In this chapter, we study network topologies in the context of large-scale product development; and discuss their relationship to the functional utility of the system as well as to the dynamics of the underlying distributed design problem-solving.

Planning techniques and analytical models that view the PD process as a network of interacting components have been proposed before [2-5, 25, 26, 45]. However, others have not yet addressed the large-scale statistical properties of real-world PD task networks. In the research we report here, we study such networks. We show that task networks have properties (sparseness, small world, scaling regimes) that are like those of other biological, social and technological networks. We discover a distinctive asymmetry between the distributions of incoming and outgoing information flows (links) of PD networks, which has implications for their functionality, sensitivity, and robustness (error tolerance) properties.

We further present a model of product development dynamics embodying interactions through the network. Using analysis and simulation we study its behavior to determine the conditions under which all the PD development tasks are completed, and the rate of convergence to the completed state. We show that network topology provides key information about the characteristics of convergence, both whether and how rapidly convergence occurs. We find, quite reasonably, that the PD network dynamics will converge unless the total rate at which a task is affected by its neighboring tasks exceeds the 'internal completion rate' of the task. Convergence is impeded by the existence of nodes that have both high numbers of incoming and outgoing information flows, i.e. convergence is controlled by the joint distribution of incoming and outgoing links. A more general result [See 50] shows that the characteristics of convergence depend upon the incoming and outgoing information flows among *multiple* tasks.

The chapter is organized as follows: In Section 2, we review the basic structural properties of real-world complex networks. In Section 3, we describe the PD data used in this chapter. In Section 4, we present an analysis of the PD task networks, their small-world property, and node connectivity distributions. We demonstrate the distinct roles of incoming and outgoing information flows in distributed PD processes by analyzing the corresponding in-degree and out-degree link distributions. In Section 5, we present a dynamical model of PD processes on complex networks, and show analytically and numerically how the empirical structural properties bear upon the PD dynamics. In Section 6, we present simulation results. In Section 7 we present our conclusions.

2. Structural Properties of Complex Networks

Complex networks can be defined formally in terms of a graph $G = (V, E)$, which is a pair of nodes $V = \{1, 2, ..., N\}$, and a set of lines $E = \{e_1, e_2, ..., e_L\}$ between pairs of nodes. If the line between two nodes is non-directional, then the network is called *undirected*; otherwise, the network is called *directed*. A network is usually represented by a diagram, where nodes are drawn as points, undirected lines are drawn as edges and directed lines as arcs connecting the corresponding two nodes. Three properties have been used to characterize 'real-world' complex networks [9, 10]. The first characteristic is the average distance (geodesic) between two nodes, where the distance $d(i, j)$ between nodes i and j is defined as the number of edges along the shortest path connecting them. The characteristic path length ℓ is the average distance between any two vertices:

$$\ell = \frac{1}{N(N-1)} \sum_{i \neq j} d_{ij} \tag{1}$$

The second characteristic measures the tendency of vertices to be locally interconnected or to cluster in dense modules. The clustering coefficient C_i of a vertex i is defined as follows. Let vertex i be connected to k_i neighbors. The total number of edges between these neighbors is at most $k_i(k_i - 1)/2$. If the actual number of edges between these k_i neighbors is n_i, then the clustering coefficient C_i of the vertex i is the ratio

$$C_i = \frac{2n_i}{k_i(k_i - 1)} \tag{2}$$

The clustering coefficient of the graph, which is a measure of the network's potential modularity, is the average over all vertices,

$$C = \frac{1}{N} \sum_{i=1}^{N} C_i \tag{3}$$

The third characteristic is the distribution of degrees of vertices. The degree of a vertex, denoted by k_i, is the number of nodes adjacent to it. The mean nodal degree is the average degree of the nodes in the network,

$$\langle k \rangle = \frac{\sum_{i=1}^{N} k_i}{N} \tag{4}$$

If the network is directed, a distinction is made between the *in-degree* of a node and its *out-degree*. The in-degree of a node, $k_{in}(i)$, is the number of nodes that are *adjacent to* i. The out-degree of a node, $k_{out}(i)$, is the number of nodes *adjacent from* i.

Regular networks, where all the degrees of all the nodes are equal (such as circles, grids, and fully connected graphs) have been traditionally employed in modeling physical systems of atoms [8]. On the other hand, many 'real-world' social, biological and technological networks appear more random than regular [8, 9-10]. With the scarcity of large-scale empirical data on one hand and the lack of computing power on the other hand scientists have been led to model real-world networks as completely random graphs using the probabilistic graph models of Erdös and Rényi [46].

In their seminal paper on random graphs, Erdös and Rényi considered a model where N nodes are randomly connected with probability p. In this model, the average degree of the nodes in the network is $\langle k \rangle \cong pN$, and a Poisson distribution approximates the distribution of the nodal degree. In a Poisson random network, the probability of nodes with at least k edges decays rapidly for large values of k. Consequently, a typical Poisson random network is rather homogenous, where most of the nodal degrees are concentrated around the mean. In particular, the average distance between any pair of nodes ℓ_{random} scales with the number of nodes as $\ell_{random} \sim \ln(N)/\ln(\langle k \rangle)$. This feature of having a relatively short path between any two nodes, despite the often large graph size, is known as the *small-world* effect. In a Poisson random graph, the clustering coefficient is $C_{random} = p \cong \langle k \rangle / N$. Thus, while the average distance between any pair of nodes grows only logarithmically with N the Poisson random graph is poorly clustered.

Regular networks and random graphs serve as useful models for complex systems; yet, many real networks are neither completely ordered nor completely random. Watts and Strogatz [13] found that social, technological, and biological networks are much more highly clustered than a random graph with the same number of nodes and edges (i.e., $C_{real} \gg C_{random}$), while the characteristic path length ℓ_{real} is close to the theoretically minimum distance obtained for a random graph with the same average connectivity. Small-World Networks are a class of graphs that are highly clustered like regular graphs ($C_{real} \gg C_{random}$), but with a small characteristic path length like a random graph ($\ell_{real} \approx \ell_{random}$). Many real-world complex systems have been shown to be small-world networks, including power-line grids [13], neuronal networks [13], social networks [17-20], the World-Wide Web [11], the Internet [24], food webs [16], and chemical-reaction networks [14].

Another important characteristic of real-world networks is related to their nodal degree distribution. Unlike the bell-shaped Poisson distribution of random graphs, the degree distribution of many real-world networks have been documented to have power-law degree distribution,

$$p(k) \sim k^{-\gamma} \tag{5}$$

where $p(k)$ is the probability that a node has k edges. Networks with power-law distributions are often referred to as *scale-free networks* [29]. The power-law distribution implies that there are a few nodes with many edges; in other words, the distribution of nodal degrees has a long right tail (resulting in an extremely large variance) of values that are far above the mean (as opposed to the fast decaying tail of a Poisson distribution, which results in an small variance). Power-law distributions of both the in-degree and out-degree of a node have been also observed in a variety of directed real-world networks [9, 10] including the World-Wide Web, metabolic networks, networks of citations of scientific papers, and telephone call graphs. Although scale-free networks are prevalent, the power-law distribution is not universal. Empirical work shows that the total node degree distribution of a variety of real networks often has a scale-free regime with an exponential cutoff, i.e. $p(k) \sim k^{-\gamma} f(k/k^*)$ where k^* is the cutoff [8, 17]. The existence of a cutoff has been attributed to physical costs of adding links or limited capacity of a vertex [17]. In some networks, the power-law regime is not even present and the nodal degree distribution is characterized by a distribution with a fast decaying tail [8, 17]. It is also not clear that a scale-free network optimizes properties of network behavior, and alternatives have been proposed [32].

The goal of the present chapter is to investigate the statistical properties of large-scale distributed product development networks. We show that large-scale PD networks, although of a different nature, have general properties that are shared by other social, technological, and biological networks.

3. Data

We analyzed distributed product development data of different large-scale organizations in the United States and England involved in vehicle design [33], operating software design [34], pharmaceutical facility design [35], and a sixteen story hospital facility design [35]. A PD distributed network can be considered as a directed graph with N nodes and L arcs, where there is an arc from task v_i to task v_j if task v_i *feeds information to* task v_j. The documentation of the directed links between the tasks has been based on structured interviews with experienced engineers and design documentation data (design process models). In all cases, the repeated nature of the product development projects and the knowledgeable people involved in eliciting the information flow dependencies reduce the risk of error in the construction of the product development networks. More specifically, Cividantes [33] obtained the vehicle development network by questioning in person at least one engineer from each task "where do the inputs for the task come from (e.g., another task)?" and "where do the outputs generated by the task go to (e.g., another task)?" The answers to these questions were used by him to construct the network of information flows [33]. The operating software development network was obtained from module/subsystems dependency diagrams compiled by Denker [34]; and both the pharmaceutical facility development and the hospital facility development networks were compiled by Newton and Austin [35] from data flow diagrams and design-process model diagrams [36] deployed by the organizations. An example of a diagram from the pharmaceutical facility and sixteen-story hospital facility process models is shown in Figure 1.

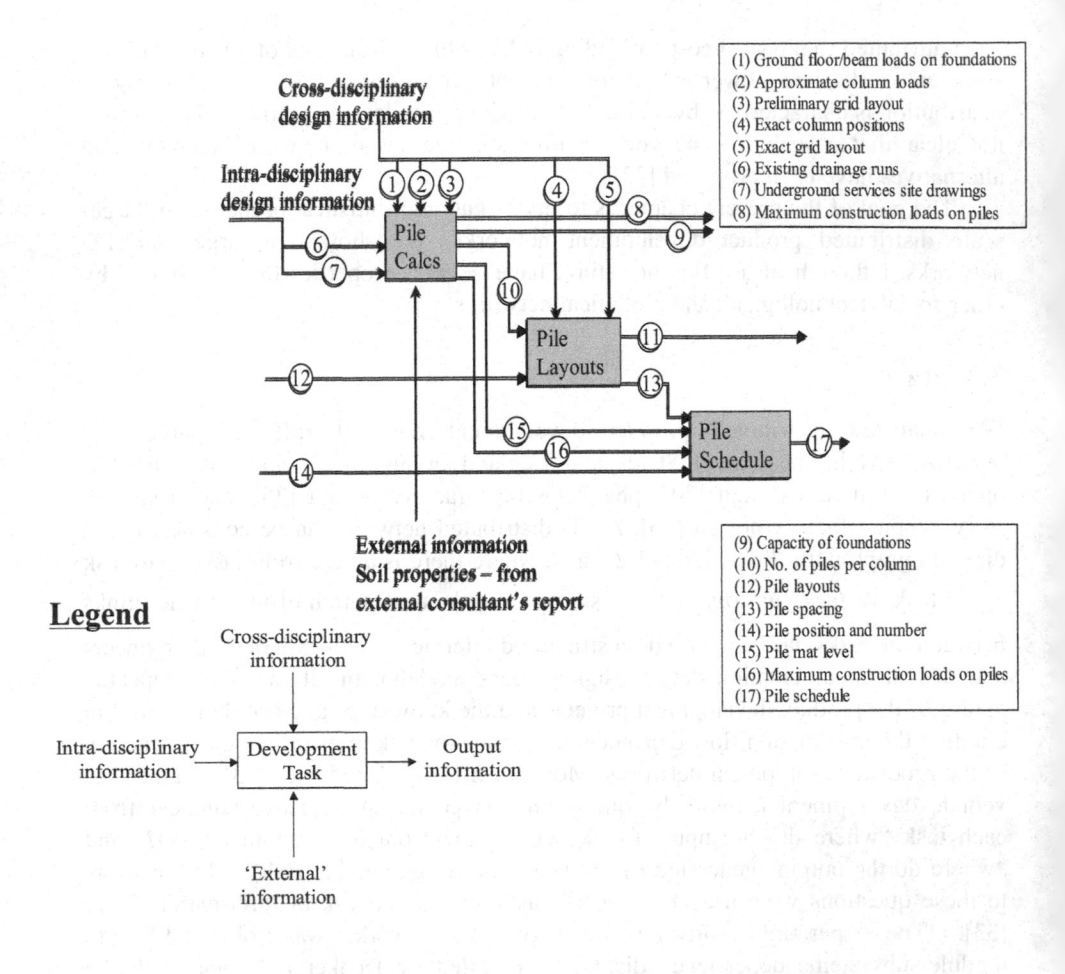

Figure 1 Example of a diagram from a design process model. Such diagrams were used to construct the pharmaceutical facility and the sixteen-story hospital facility networks (adapted from [37]).

4. Results

4.1 Small world properties

An example of one of these distributed PD networks (operating software development) is shown in Figure 2. Here we consider the undirected version of the network, where there is an edge between two tasks if they exchange information between them (not necessarily reciprocal communication). We see that this network is sparse ($2L/N(N-1) = 0.0114911$) with the average degree of each node only 5.34, which is small compared to the number of possible edges $N-1 = 465$. A clear deviation from a purely random graph is observed. We see that most of the nodes have low degree while a few nodes have a very large degree. This is in contrast to the nodal degree homogeneity of purely random graphs, where most of the nodal degrees are concentrated around the mean. The software development network also illustrates the 'small-world' property (see Section 2), which can be detected by measuring two basic statistical characteristics: 1) the average distance (geodesic) between two nodes; and 2) the clustering coefficient of the graph. Small-world networks are a class of graphs that are highly clustered like regular graphs ($C_{real} \gg C_{random}$), but with small characteristic path length like a random graph ($\ell_{real} \sim \ell_{random}$). For the software development network, the network is highly clustered as measured by the clustering coefficient of the graph ($C_{software} = 0.327$) compared to a random graph with the same number of nodes and edges ($C_{random} = 0.021$) but with small characteristic path length like a random graph ($\ell_{software} = 3.700 \approx \ell_{random} = 3.448$).

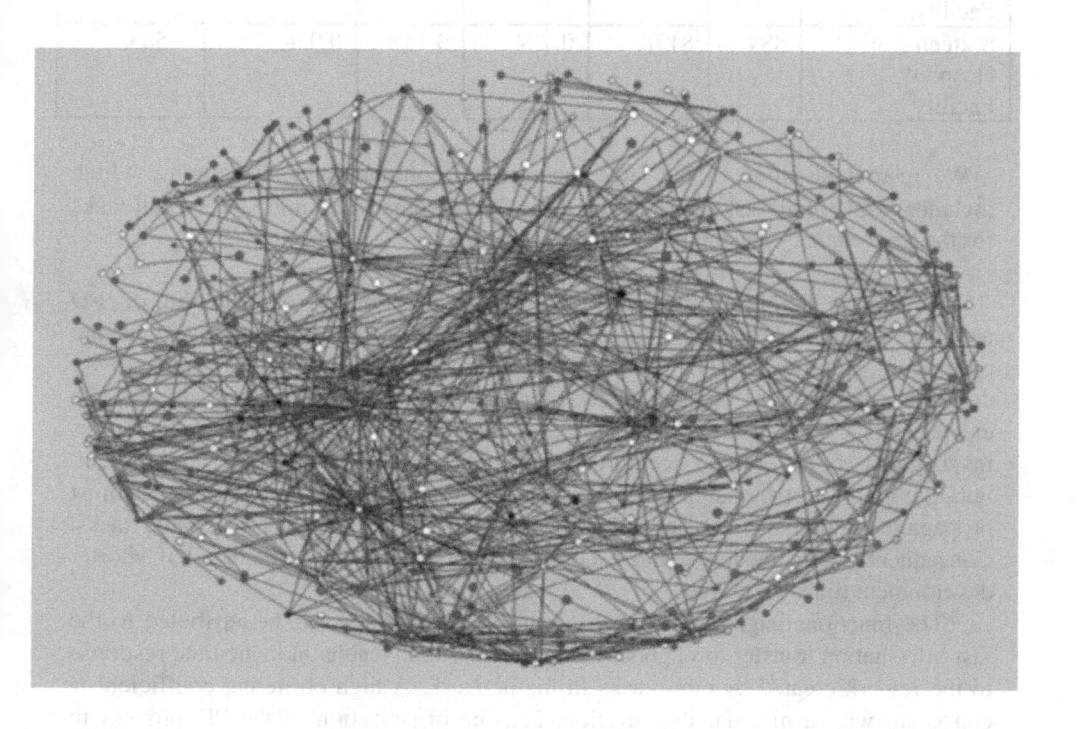

Figure. 2. Network of information flows between tasks of an operating system development process. This PD task network consists of 1245 directed information flows between 466 development tasks. Each task is assigned to one or more actors ("design teams" or "engineers") who are responsible for it. Nodes with the same degree are colored the same.

In Table 1, we present the characteristic path length and clustering coefficient for the four distributed PD networks examined in this chapter, and compare their values with random graphs having the same number of nodes and edges. In all cases, the empirical results display the small-world property ($C_{real} \gg C_{random}$ and $\ell_{real} \approx \ell_{random}$).

Table 1 Empirical Statistics of the four large-scale PD Networks

Network	N	L	C	ℓ	C_{random}	ℓ_{random}
Vehicle	120	417	0.205	2.878	0.070	2.698
Operating Software*	466	1245	0.327	3.700	0.021	3.448
Pharmaceutical Facility	582	4123	0.449	2.628	0.023	2.771
Sixteen story Hospital Facility*	889	8178	0.274	3.118	0.024	2.583

* We restrict attention to the largest connected component of the graphs, which includes ~82% of all tasks for the Operating Software network, and ~92% of all tasks for the Sixteen story Hospital Facility network.

An interpretation of the functional significance of the architecture of PD networks must be based upon a recognition of the factors that such systems are optimizing. Shorter development times, improved product quality, and lower development costs are the key factors for successful complex PD processes. The existence of cycles in the PD networks, readily noted in the network architectures investigated, points to the seemingly undeniable truth that there is an inherent, iterative nature to the design process [2]. Each iteration results in changes that must propagate through the PD network requiring the rework of other reachable tasks. Consequently, late feedback and excessive rework should be minimized if shorter development time is required.

The functional significance of the small-world property can be attributed to the fast information transfer throughout the network, which results in immediate response to the rework created by other tasks in the network. A high clustering coefficient is consistent with a modular organization; i.e., the organization of the PD process in clusters that contain most, if not all, of the interactions internally and the interactions or links between separate clusters is minimized [1-3, 49]. The dynamic model developed in [5] shows that a speed up of the PD convergence to the design solution is obtained by reducing or 'ignoring' some of the task dependencies (i.e., eliminating some of the arcs in the corresponding PD network). A modular architecture of the PD process is aligned with this strategy.

4.2 In-degree and out-degree distributions

We compared the cumulative probability distributions $P_{in}(k)$ and $P_{out}(k)$ that a task has more than k incoming and outgoing links, respectively (see Figure 3)[30]. For all four networks, we find that the in-degree and out-degree distributions can be described by power-laws (the "scale-free" property) with cutoffs introduced at some characteristic scale k^*; $k^{-\gamma} f(k/k^*)$ (typically the function f corresponds to an exponential or Gaussian). More specifically, we find scaling regimes (i.e., straight-line regimes in the figure) for both $P_{in}(k)$ and $P_{out}(k)$. We note however that the cutoff k^* is lower by more than a factor of two for $P_{in}(k)$ than for $P_{out}(k)$. This is a new observation that has not been found before [38].

The "scale-free" property suggests that complex PD task networks are dominated by a few *highly central* tasks. This is in contrast to the bell-shaped Poisson distribution of random graphs, which leads to a fairly homogeneous network where each node has approximately the same number of links (and thus *equally* affecting the network behavior). The 'failure' (e.g., excessive rework, lack of integration ability, or delays) of central PD tasks will likely affect the *vulnerability* of the *overall* PD process. Focusing engineering efforts and resources (e.g., funding and technology support) as well as developing appropriate control and management strategies for central PD tasks will likely maintain the sustainability and improve the performance of the PD process.

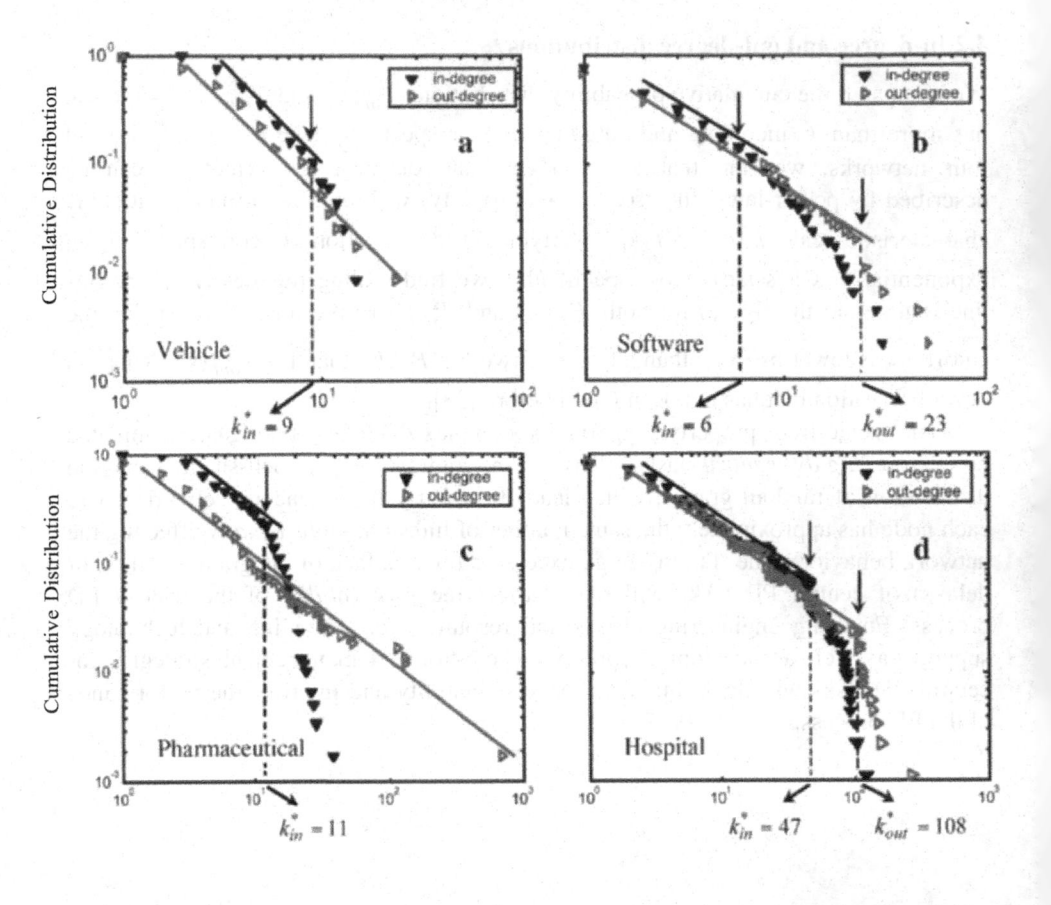

Figure 3 Degree distributions for four distributed problem solving networks. The log-log plots of the cumulative distributions of incoming and outgoing links show a power law regime (Pearson coefficient $R > 0.98$, $p < 0.001$) with or without a fast decaying tail in all cases. The in-degree distribution has a lower best visual fit cutoff k_{in}^* in each case. **a**, Vehicle development with 120 tasks and 417 arcs. The exponents of the cumulative distributions are $\gamma_{vehicle}^{in} - 1$ and $\gamma_{vehicle}^{out} - 1$, where $\gamma_{vehicle}^{in} \approx 2.82 \pm 0.25$ and $\gamma_{vehicle}^{out} = 2.97 \pm 0.24$ denote the exponents of the associated probability density functions. **b**, Software development with 466 tasks and 1245 arcs, where $\gamma_{software}^{in} \approx 2.08 \pm 0.13$ and $\gamma_{software}^{out} \approx 2.25 \pm 0.15$. **c**, Pharmaceutical facility development with 582 tasks and 4123 arcs, where $\gamma_{pharmaceutical}^{in} \approx 1.92 \pm 0.07$ and $\gamma_{pharmaceutical}^{out} \approx 1.96 \pm 0.07$. **d**, Hospital facility development with 889 tasks and 8178 arcs, where $\gamma_{hospital}^{in} \approx 1.8 \pm 0.03$ and $\gamma_{hospital}^{out} \approx 1.95 \pm 0.03$.

In order to analyze the structure of PD networks, it is important to study the relationships between the in-degree and out-degree of tasks. Thus, for example, we are interested in questions such as "Do tasks with high out-degree also have relatively high in-degree?" We address such questions by plotting the relationship between the in-degree and out-degree of tasks (Figures 4a-d). Interestingly enough it turns out that to a large extent, when considering the vehicle, pharmaceutical and hospital product development networks, the results reveal almost no correlation between the in-degrees of tasks and their out-degrees; i.e. there are tasks that have a small in-degree but yet have a large out-degree, and vice-versa. To illustrate this finding, we present in Table 2 the top 10 tasks of the vehicle development network at General Motors' Research & Development Center ranked according to their in-degree and out-degree centrality measures. We see that only 2 out of the 10 tasks (underlined in the table) appear both in the in-degree ranking and in the out-degree ranking. This finding implies that, generally, there is a clear distinction between large-scale generators of information (i.e. with high out-degree) and large-scale consumers (i.e. with high in-degree); a high generator of information could be a low consumer and vice versa. This further suggests that a distinction has to be made between in- and out-centrality as far as control and management strategies are concerned. Moreover, those tasks that have both high in- and out-centrality (e.g., 'track total vehicle issues' at General Motors' vehicle design in Figure 4) are likely to play a unique role during the product design process. The dynamical model presented in Section 5 shows that the nature of in- and out-degree correlations has profound and subtle effects on the behavior of PD processes defined in top of complex networks.

The presence of cutoffs in node degree distributions has been attributed to physical costs of adding links and limited capacity of a node [17]. Such networks may also arise if network formation occurs under conditions of preferential attachment with limited information [31]. As previously noted [17, 31], the limited capacity of a node, or limited information-processing capability of a node are similar to the so-called "bounded rationality" concept of Simon [28].

We find that there is an *asymmetry* between the distributions of incoming and outgoing information flows. The narrower power law regime for $P_{in}(k)$ suggests that the costs of adding incoming links and limited in-degree capacity of a task are higher than their counterpart out-degree links. We note that this is consistent with the realization that bounded rationality applies to incoming information, and to outgoing information only when it is different for each recipient, not when it is duplicated. This naturally leads to a weaker restriction on the out-degree distribution.

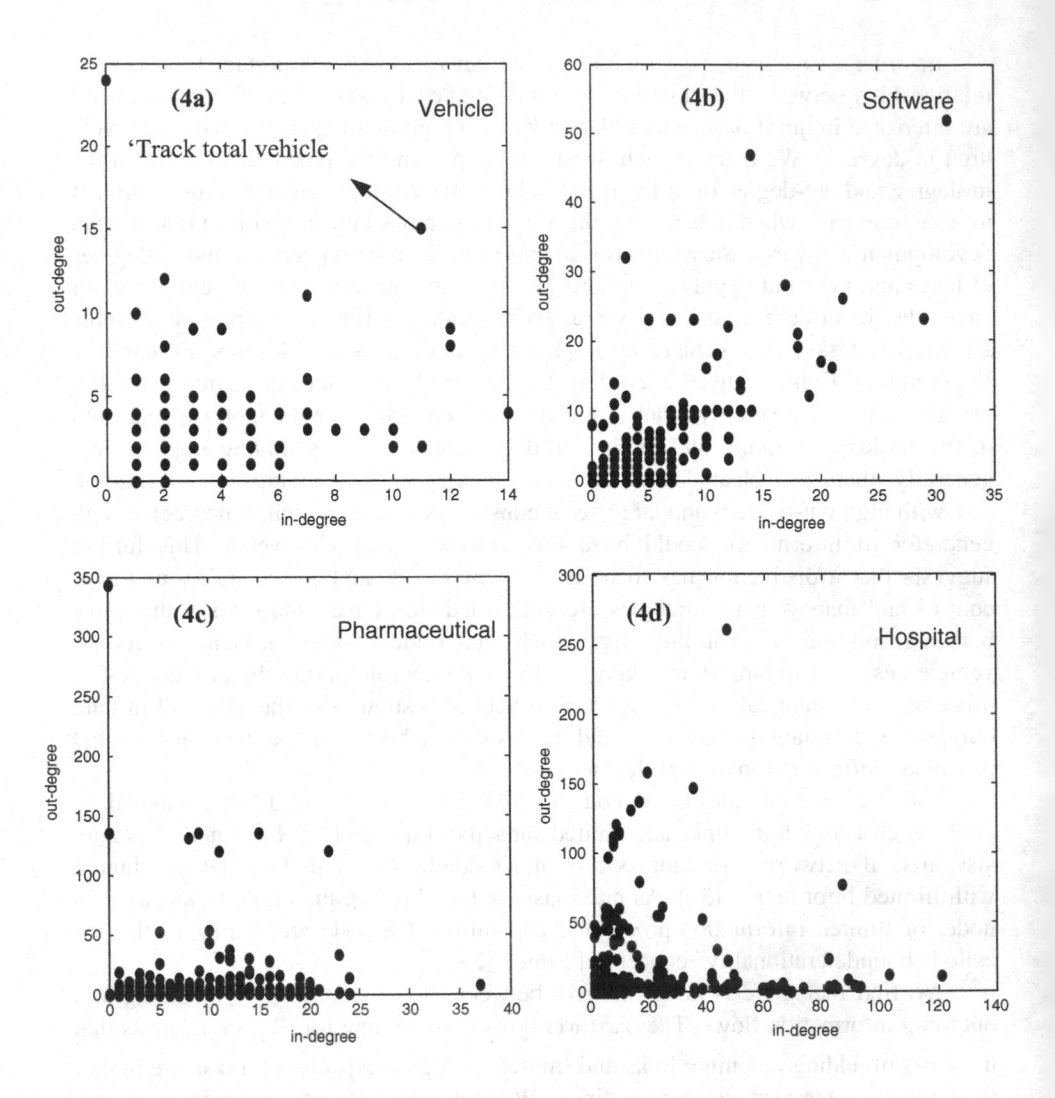

Figure 4 In-degree as a function of out-degree for four distributed PD networks. Figures 4a, 4c, and 4d show almost no correlation between the in-degrees of tasks and their out-degrees (Pearson coefficient of 0.17, 0.1, 0.11 respectively). Figure 4b shows a significant positive correlation (Pearson coefficient of 0.76).

Table 2 The top 10 tasks of the vehicle development network at General Motors' Research & Development Center ranked according to their in-degree and out-degree centrality measures.

Task	In-Degree	Task	Out-Degree
Develop Mainstream Integrated Concept Vehicle Model	14	Develop Nine Box Summary	24
Maintain Vehicle Mainstream Chart and Update Engineering Product Content Sheet	12	Track Total Vehicle Issues	15
Conduct Performance Synthesis and Analysis in Quick Study Phase	12	Set Engineering Target Parameters (Concept Technical Descriptors)	12
Track Total Vehicle Issues	11	Recommend Final Architecture	11
Review Quick Study Deliverables	11	Identify Target Architectures	10
Assess Risks in Performance Requirements	10	Develop Critical Product Characteristics / Key Voices	10
Prepare Program QRD Matrix	10	Develop Engineering Product Content Sheet	9
Follow up and Maintain Open Issues - Front Compartment	9	Maintain Vehicle Mainstream Chart and Update Engineering Product Content Sheet	9
Follow up and Maintain Open Issues - Passenger/Rear	8	Create Initial Visual Surfaces	9
Follow up and Maintain Open Issues - Chassis	8	Establish Body BOM Sharing Strategy	9

An additional functional significance of the asymmetric topology can be attributed to the distinct roles of incoming and outgoing links in distributed PD processes. The narrow scaling regime governing the information flowing into a task implies that tasks with large incoming connectivity are practically absent. This suggests that distributed PD networks limit conflicts by reducing the multiplicity of interactions that affect a single task, as reflected in the incoming links. Such architecture reduces the amount and range of potential revisions that occur in the dynamic PD process, and thus increases the likelihood of converging to a successful solution. Our empirical observation is found to be consistent with the dynamic PD model presented in the next section. There it is shown that additional rework might slow down the PD convergence or have a destabilizing effect on the system's behavior. As a general rule, the rate of problem solving has to be measured and controlled such that the total number of design problems being created is smaller than the total number of design problems being solved.

The scale-free nature of the outgoing communication links means that some tasks communicate their outcomes to many more tasks than others do, and may play the role of coordinators (or product integrators see [5]). Unlike the case of large numbers of incoming links, this may improve the integration and consistency of the problem solving process; thus reducing the number of potential conflicts. Product integrators put the separate development tasks together to ensure fit and functionality. Since late changes in product design are highly expensive, product integrators continuously check unfinished component designs and provide feedback to a large number of tasks accordingly.

5. A Dynamical Model on Complex PD Networks

This section introduces a deliberately simple model of product development on complex directed networks, which captures important features of PD dynamics (e.g., see [3]). We characterize the model's behavior by using analysis and simulations performed on the empirically heterogeneous directed network topologies examined in Section 4 (rather than on simplified fully connected or lattice topologies). In our model, there is a network of interconnected nodes (elemental tasks); each can be in a 'resolved' or 'unresolved' state. Each node could be affected by those nodes that directly reach it, and could affect those nodes that are directly reachable from it. The rule by which a resolved node becomes unresolved depends stochastically on the *number* of unresolved contiguous incoming nodes – the higher the number of unresolved neighbors, the higher the probability of becoming unresolved. This rule reflects the repetition (rework) of tasks due to the availability of new information and input changes generated by other contiguous tasks. An unresolved node may be fully resolved with probability that depends on both its self-completion rate (internal problem-solving rate), and on the number of unresolved neighboring nodes. Incorporating the effect of task j on task i (which possibly differs between each pair of tasks) as well as including non-binary states (e.g., the number of design problems or open issues associated with a task) can be readily done but does not offer additional understanding on the issues addressed here. Although the motivation is different, it is worthwhile to note that the model considered here is similar in spirit to dynamic models that have been studied in the context of collective action, percolation, majority-vote cellular automata, self-organized criticality, spin-flip Ising dynamics, and epidemic spreading [9, 10].

5.1 Model

We consider a network where each node (a task in the network) can be in one of two states, '0' or '1' representing unresolved or resolved states, respectively. We consider a dynamic process occurring at discrete times, $1, 2,..., t$. Node states are updated synchronously, indicating a parallel mode of product development[1]. Let $s_i(t)$ be the state of node i at time t. We consider two cases:

---

[1] Note that this assumption can be easily relaxed by randomly selecting, at each time point, a node for an update.

Case 1. Node i is resolved at time t (i.e., $s_i(t) = 1$):

Let k_i be the in-degree connectivity of node i, and let $k_i(t) = k_i - \displaystyle\sum_{j:<j,i>\in E} s_j(t)$ be the

number of neighboring unresolved nodes that are directly connected by directional links (arcs) to node i at time t. Node i changes its state according to the following stochastic rule:

$$s_i(t+1) = \begin{cases} 0 & \text{with probability } \tanh(\beta_i k_i(t)) \\ 1 & \text{with probability } 1 - \tanh(\beta_i k_i(t)) \end{cases} \qquad (6)$$

where β_i is a parameter that reflects the sensitivity of the node i state to its neighboring unresolved nodes, and $\tanh(x)$ is the hyperbolic tangent function defined by

$$\tanh(x) = \frac{e^x - e^{-x}}{e^x + e^{-x}} \qquad (7)$$

The stochastic dynamic rule allows for node state realizations to vary over time even if the node has the same *number* of unresolved neighbors at different times. The parameter β_i captures the tendency of a node to be affected by its neighbors. For $\beta_i = 0$, the node's behavior is completely decoupled from its neighbors. A low β_i corresponds to the case where a node's behavior is not influenced much by the states of its neighbors. For $\beta_i \to \infty$, each node's behavior is completely dependent on its neighbors: any non-zero number of unresolved neighbors will render the node unresolved at the next iteration.

Case 2. Node i is unresolved at time t (i.e., $s_i(t) = 0$):

In this case, node i changes its state according to the following stochastic rule:

$$s_i(t+1) = \begin{cases} 0 & \text{with probability } 1 - r_i(1 - \tanh(\beta_i k_i(t))) \\ 1 & \text{with probability } r_i(1 - \tanh(\beta_i k_i(t))) \end{cases} \qquad (8)$$

where r_i is a parameter that reflects the internal completion rate of task i ($0 \le r_i \le 1$). Here we assume that the node can be resolved if two events occur: (1) the node is not affected by its unresolved neighbors; and (2) the task is successfully completed internally in one unit of time with probability r_i.

5.2 Analytic Results for Random (Erdös-Rényi) Networks

The relaxation of the system to the uniformly resolved state (i.e., $s_i(t) = 1$ for all tasks) depends on the free parameters β_i, r_i, the initial state of the network, and the PD network topology. Although there is no theorem guaranteeing the relaxation of the network to the uniformly resolved state, we apply a mean-field approximation [47] to the stochastic model we have defined in order to gain insight about the convergent final state. We derive a rate equation for the density of unresolved tasks at time t,

$\alpha(t) = 1 - \dfrac{\sum_i s_i(t)}{N}$. We assume that for every task i, $r_i = r, \beta_i = \beta$. We also make

the following *homogeneity* condition, which holds particularly well for a completely random graph: for every task i, its number of unresolved neighbors is approximately $k_i(t) \cong \langle k_{in} \rangle \alpha(t)$, where $\langle k_{in} \rangle$ denotes the average in-degree of a task in the network. The global density of unresolved tasks $\alpha(t)$ evolves according to the rate equation,

$$\frac{d\alpha(t)}{dt} = (1-\alpha(t))\tanh(\beta\langle k_{in}\rangle\alpha(t)) - \alpha(t)r(1-\tanh(\beta\langle k_{in}\rangle\alpha(t))) \qquad (9)$$

After substitution of the hyperbolic tangent function and $\overline{\beta} = \beta\langle k_{in}\rangle$, we obtain

$$\frac{d\alpha(t)}{dt} = (1-\alpha(t))\left(\frac{e^{\overline{\beta}\alpha(t)} - e^{-\overline{\beta}\alpha(t)}}{e^{\overline{\beta}\alpha(t)} + e^{-\overline{\beta}\alpha(t)}}\right) - \alpha(t)r\frac{2e^{-\overline{\beta}\alpha(t)}}{e^{\overline{\beta}\alpha(t)} + e^{-\overline{\beta}\alpha(t)}}$$

At an equilibrium $\dfrac{d\alpha(t)}{dt} = 0$; thus, we obtain a single equation to be solved for $\alpha(t) = \alpha$,

$$\alpha = \frac{e^{\overline{\beta}\alpha} - e^{-\overline{\beta}\alpha}}{e^{\overline{\beta}\alpha} - e^{-\overline{\beta}\alpha} + 2re^{-\overline{\beta}\alpha}} = f(\alpha) \qquad (10)$$

We conclude that the only stable fixed point of this equation is $\alpha^* = 0$ if $f'(\alpha)|_{\alpha=0} < 1$, and has a non-zero solution if $f'(\alpha)|_{\alpha=0} > 1$. Thus, for $\overline{\beta} < r$, the global density of unresolved tasks at equilibrium is $\alpha^* = 0$ (i.e., all tasks are successfully completed). This result has a simple intuitive interpretation: if the task's internal completion rate r exceeds the average total sensitivity of the task to its unresolved incoming neighbors $\overline{\beta} = \beta\langle k_{in}\rangle$, the PD process will converge to the uniformly resolved state; otherwise, it is quite likely that the PD process will converge to a state where a non-zero fraction $\alpha^* > 0$ of the tasks remains unresolved[2].

When the fraction of unresolved tasks at equilibrium is very small ($\alpha^* \ll 1$), we can find a closed-form expression for α^* by expanding $f(\alpha)$ in Equation (10) to the second order in α and solving for α^* :

$$f(\alpha) = \frac{\overline{\beta}}{r}\alpha + \frac{\overline{\beta}^2(r-1)}{r^2}\alpha^2 + O(\alpha^3) \qquad (11)$$

---

[2] In other words, a threshold behavior occurs at $\beta\langle k_{in}\rangle / r = 1$.

Hence,

$$\alpha^* \approx \frac{r(\overline{\beta} - r)}{\overline{\beta}^2(1 - r)} \qquad (12)$$

In order to gain further insight regarding the rate of convergence to the fixed point of equation (9), we solve it approximately as follows. For small values of α, Taylor's expansion yields,

$$\tanh(\overline{\beta}\alpha) = \overline{\beta}\alpha + O(\alpha^3)$$

Thus, the differential equation (9) is approximated by

$$\frac{d\alpha}{dt} = (1 - \alpha)\overline{\beta}\alpha - \alpha r(1 - \overline{\beta}\alpha) \qquad (13)$$

The solution of Equation (13) is

$$\alpha(t) = \frac{\overline{\beta} - r}{re^{(r - \overline{\beta})t}(\overline{\beta} - 1) + \overline{\beta}(1 - r)} \qquad (14)$$

For $\overline{\beta} < r$, $\lim_{t \to \infty} \alpha(t) = 0$ and the system converges exponentially to the uniformly resolved state. For $\overline{\beta} > r$, the fraction of unresolved tasks decays at an exponential rate, and eventually saturates at a non-zero fraction of unresolved tasks $\alpha^* = \lim_{t \to \infty} \alpha(t) = \frac{\overline{\beta} - r}{\overline{\beta}(1 - r)}$ (notice that when α^* is small, as assumed, $\frac{\overline{\beta}}{r} \approx 1$ and thus the prediction of Equation (12) is consistent with the estimate above).

The deterministic analysis of the model has involved a number of assumptions, which can be tested by simulation. First, a directed random graph with a prescribed average in-degree of tasks (and same average out-degree) has been generated, and all tasks have been initially selected to be unresolved. The graph contains 10^5 tasks with connectivity $k_{in} = k_{out} = 12$. We have simulated the model using a synchronous discrete-event implementation. Figure 5 compares a typical simulation run to the corresponding deterministic solution (14). The simulation run has followed the deterministic solution quite well. Performing multiple independent simulation runs using the same parameters has shown that the variation in the equilibrium obtained across different simulation runs has been quite small.

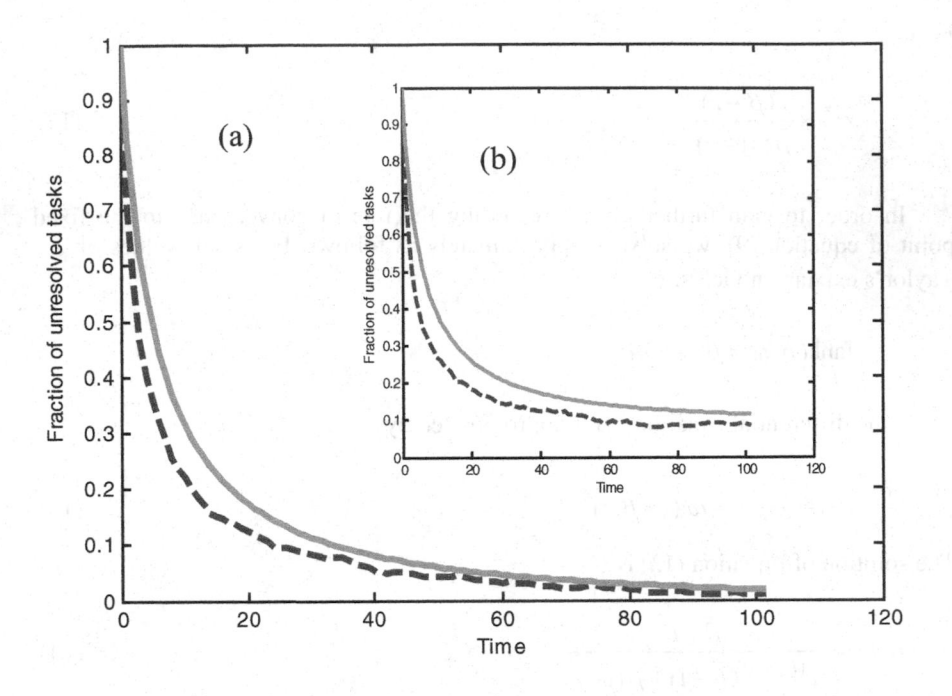

Figure 5. Comparison between average fraction of unresolved tasks versus time as predicted by deterministic theory (solid curve) and a typical simulation run (broken curve) on a randomly-generated graph with 10^5 nodes. The average number of incoming arcs connected to a node is $\langle k_{in} \rangle = 12.0192$. In Figure (**a**), the time evolution of $\alpha(t)$ when the sensitivity and the internal completion rates of tasks are $\beta = 0.061$ and $r = 0.75$, respectively. In this case, $\beta \langle k_{in} \rangle < r$ and the simulation run converges to the uniformly resolved state as predicted by theory. In Figure (**b**), in this case, the sensitivity and the internal completion rates are $\beta = 0.064$ and $r = 0.75$, respectively. In equilibrium, the average fraction of unresolved tasks between $t = 65$ and $t = 100$ (a stationary regime) is $\alpha^* = 0.087$, which agrees reasonably well with the prediction $\alpha^* = 0.09084$ given in Equation (12).

5.3 Analytic Results for General PD Networks

The extreme heterogeneity of the connectivity distributions of *undirected* scale-free networks significantly affects the dynamical processes that propagate through these networks. In particular, it was shown that the large fluctuations, $\langle k^2 \rangle$, of power-law connectivity distributions cannot be neglected as far as system dynamics is concerned, even for finite-size systems [10]. For *directed* PD networks, we show below that the first-order joint moment of the joint in-degree and out-degree distributions (i.e., $\langle k_{in}k_{out}\rangle$) plays an important role in determining the PD dynamics.

In order to take into account the extreme heterogeneity of real-world directed PD networks (see Section 4), we modify the mean-field analysis presented for random (Erdös-Rényi) networks by writing the rate equations governing the time evolution of $\alpha_k(t)$, where $\alpha_k(t)$ is the relative density of unresolved tasks with given in-degree connectivity k. In the analysis below, we also neglect the degree correlations among neighboring tasks (called "mixing by degree" in [10]). This assumption could be tested, for *undirected* networks, by measuring the linear (Pearson) correlation $r(k_i,k_j)$ between the sets $\{k_i\}$ and $\{k_j\}$ of total degrees for all tasks i and j at either ends of an edge in the network. This measure reflects the tendency of tasks of similar degrees to be connected to one another. For *directed* PD networks, however, there are four possible correlation coefficients: $r(k_{in}^i,k_{in}^j)$, $r(k_{in}^i,k_{out}^j)$, $r(k_{out}^i,k_{in}^j)$, and $r(k_{out}^i,k_{out}^j)$, where the index i indicates the source node of the directed edge, and j refers to the destination node. In [50], we show that the determining correlation as far as PD dynamics goes is related to the correlation coefficient $r(k_{in}^i,k_{in}^j)$. Thus, we have tested the correlation coefficients $r(k_{in}^i,k_{in}^j)$ for each of the PD networks studies. These have been found to be 0.0943 (vehicle), -0.0644 (software), 0.2452 (pharmaceutical), and -0.0750 (hospital); in support of our assumption. Still, for the sake of completeness, the general case where there are explicit correlations among the tasks' connectivities, is presented in [50].

The dynamical mean-field rate equations now become,

$$\frac{d\alpha_k(t)}{dt} = (1-\alpha_k(t))\tanh(\beta k\theta(t)) - \alpha_k(t)r(1-\tanh(\beta k\theta(t))) \qquad \forall k \qquad (15)$$

where $\theta(t)$ is the probability that any given incoming link (arc) to a task originates from an unresolved task. At an equilibrium $\frac{d\alpha_k(t)}{dt}=0$, $\theta(t)=\theta$ and thus we obtain a single equation to be solved for $\alpha_k(t) = \alpha_k$,

$$\alpha_k = \frac{e^{\beta k\theta} - e^{-\beta k\theta}}{e^{\beta k\theta} - e^{-\beta k\theta} + 2re^{-\beta k\theta}} \qquad (16)$$

In order to solve the equations in (16), we need to derive an expression for θ. First, we define the probability q_m that an incoming link to a task originates from another task with m outgoing links. Since it is more likely that a randomly chosen link originates from a node with high out-degree connectivity, the probability q_m is proportional to $mP_{out}(m)$ and the normalized distribution is given by,

$$q_m = \frac{m P_{out}(m)}{\sum_s s P_{out}(s)} = \frac{m P_{out}(m)}{\langle k_{out} \rangle} \tag{17}$$

We conclude that θ is given by,

$$\theta = \sum_m \sum_k \frac{m P_{out}(m) P(k_{in} = k | k_{out} = m) \alpha_k}{\langle k_{out} \rangle} =$$

$$= \sum_m \sum_k \frac{m P(m,k)}{\langle k_{out} \rangle} \left(\frac{e^{\beta k \theta} - e^{-\beta k \theta}}{e^{\beta k \theta} - e^{-\beta k \theta} + 2r e^{-\beta k \theta}} \right) = f(\theta) \tag{18}$$

where $P(k_{in}, k_{out})$ is the joint probability distribution of k_{in} and k_{out}. Consequently, Equation (18) yields a *consistency* equation for θ. After solving Equation (18) for θ, the average density of unresolved tasks in the system at equilibrium is evaluated by the relation $\alpha = \sum_k P_{in}(k) \alpha_k$.

Finding an exact solution of Equation (18) can be difficult, depending on the particular form of $P(k_{in}, k_{out})$. Fortunately, we don't have to solve Equation (18) explicitly to gain a qualitative understanding of the underlying PD dynamics. Following the same argument as in Equation (10), we obtain that,

$$f'(\theta)\big|_{\theta=0} = \sum_m \sum_k \frac{m P(m,k)}{\langle k_{out} \rangle} \left(\frac{\beta k}{r} \right) = \frac{\beta \langle k_{in} k_{out} \rangle}{r \langle k_{out} \rangle} = \frac{\beta}{r} \left(\frac{Cov(k_{in}, k_{out})}{\hat{k}} + \hat{k} \right) \tag{19}$$

where $\langle k_{in} k_{out} \rangle$ denotes the first-order joint moment of the joint probability distribution $P(k_{in}, k_{out})$, $Cov(k_{in}, k_{out})$ denotes the covariance[3] of the two random variables k_{in} and k_{out}, and $\hat{k} = \langle k_{in} \rangle = \langle k_{out} \rangle$. Thus, for the general case considered here, the model exhibits a threshold behavior at $f'(\theta)\big|_{\theta=0} = 1$, which implies that an initial seed of unresolved tasks would lead, at equilibrium, to the uniformly resolved state if $f'(\theta)\big|_{\theta=0} < 1$; i.e., for $\beta \langle k_{in} k_{out} \rangle < r \langle k_{out} \rangle$. We also note that if k_{in} and k_{out} are uncorrelated then $\langle k_{in} k_{out} \rangle = \langle k_{in} \rangle \langle k_{out} \rangle$, from which we recover the condition $\frac{\beta \langle k_{in} \rangle}{r} < 1$ obtained for homogeneous random networks (see Section 5.2).

The above analysis suggests that the dynamical model does exhibit a threshold behavior, as for homogeneous networks. However, for positive covariance, the range of $r - \beta$ values for which the system converges to the uniformly resolved state is more constrained compared with the corresponding homogeneous situation (see Figure 6). It is concluded that high covariance values exhibited in PD networks must be compensated for by either reducing the sensitivity of tasks to their neighbors

---

[3] We have shown elsewhere that some directed networks used to model the topology of the World Wide Web yield very large covariance values that go to infinity as the number of nodes in the network goes to infinity.

(reflected by the parameter β) or by increasing their internal problem-solving rates (reflected by the parameter r), if the project is expected to converge to the uniformly resolved state. As mentioned in Section 4.2, the observed low correlation between the in-degrees and out-degrees of nodes for some PD networks implies that $\dfrac{Cov(k_{in}, k_{out})}{\hat{k}} \ll 1$ in Equation (19), and thus the predicted threshold behavior at $\dfrac{\beta \langle k_{in} \rangle}{r} = 1$ (as for random homogeneous networks) could still be a good approximation (this is confirmed by simulations).

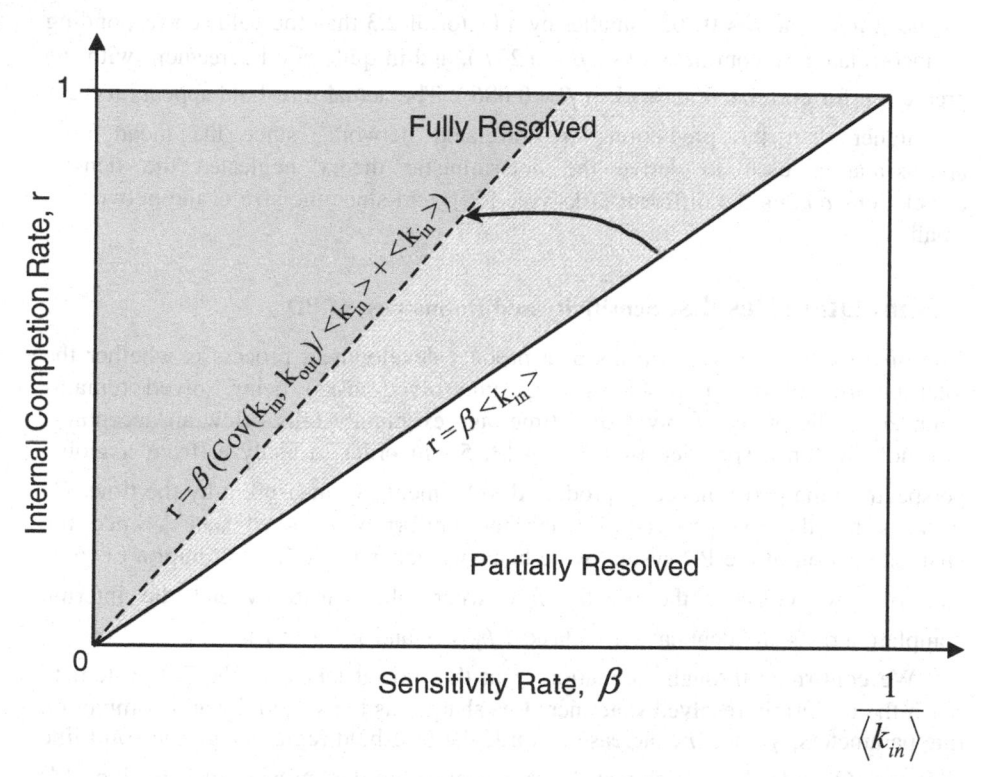

Figure 6. Dynamical behavior of a PD network with uncorrelated and correlated topologies. For a fixed network topology (i.e. $\langle k_{in} \rangle$ and $Cov(k_{in}, k_{out})$ are known values), the dynamics is characterized in terms of the parameters r and β. The parameter space $r - \beta$ is divided in two distinct regimes: fully resolved (gray) and partially resolved (white). The solid line represents the transition between these two regimes for a completely random (uncorrelated) network while the dashed line represents the transition between these two regimes for a positively correlated network (i.e., $Cov(k_{in}, k_{out}) \geq 0$).

The above deterministic analysis has been tested by simulating the model on the software network described in Section 3 with $\hat{k} = 3.163$. The software network indicates a high degree of interdependence or covariance between the two random variables k_{in} and k_{out} (i.e. $Cov(k_{in}, k_{out}) \approx 5.59$ and correlation ≈ 0.76). For example, for internal completion rates $r_i = 0.75$ $\forall i$, a threshold behavior is predicted at a value of $\beta \approx 0.086$ for which $\dfrac{\beta}{r}\left(\dfrac{Cov(k_{in}, k_{out})}{\hat{k}} + \hat{k}\right) = 1$. It is instructive to compare the threshold thus obtained with the prediction for an uncorrelated random network, $\beta \approx 0.237$. The actual measurement of the threshold has been found at $\beta \approx 0.103$, smaller by a factor of 2.3 than the value corresponding to uncorrelated random networks ($\beta \approx 0.237$), and in quite good agreement with the prediction for correlated networks ($\beta \approx 0.086$). The actual threshold appears to be a bit higher than the prediction for correlated networks since the mean field approximation used to derive the deterministic theory neglected the density correlations *among* the different tasks (see [50]) and since the size of the network is small.

6. Simulation Results: Sensitivity and Robustness of PD

One of the most practical aspects of a product development process is whether the total number of design problems (e.g., unresolved tasks) being solved remains bounded as the project evolves over time, and eventually falls below an acceptable threshold within a specified time frame [3, 5]. In order to analyze (from a global perspective) the performance of product development, we measure it by the time T^* it takes for the PD process to reach the uniformly resolved state[4]. Since the characterization of the PD system-wide behavior depends on the distribution of β_i's and r_i's, we consider the special case where the sensitivity and the internal completion rates are identical across tasks ($\beta_i = \beta$ and $r_i = r$ $\forall i$).

We confirmed through simulations that the time it takes for the PD system to reach the uniformly resolved state increases sharply as the sensitivity and completion rate parameters, β and r, increase towards the threshold regime (e.g., the solid line in Figure 6). Indeed, we note that the exponent in the denominator of Equation (14) begins to dominate the other factors for values of t for which $(r - \overline{\beta})t \approx 1$. Thus, we expect the inverse of the mean convergence time, $1/T^*$, to grow linearly with the scaled parameter $\delta = r - \overline{\beta} = r - \beta\langle k_{in}\rangle$ – the "threshold gap." This is illustrated in Figure 7 for the pharmaceutical network, where the inverse of the convergence time

---

[4] Here we guarantee convergence by proper selection of parameters. In general, we could use a performance measure T_p, where T_p is the earliest time at which the fraction of resolved tasks is greater or equal to p.

$1/T^*$ is plotted against the threshold gap δ, which verifies the predicted linear dependence of $1/T^*$ on δ over a large range of threshold gap values.

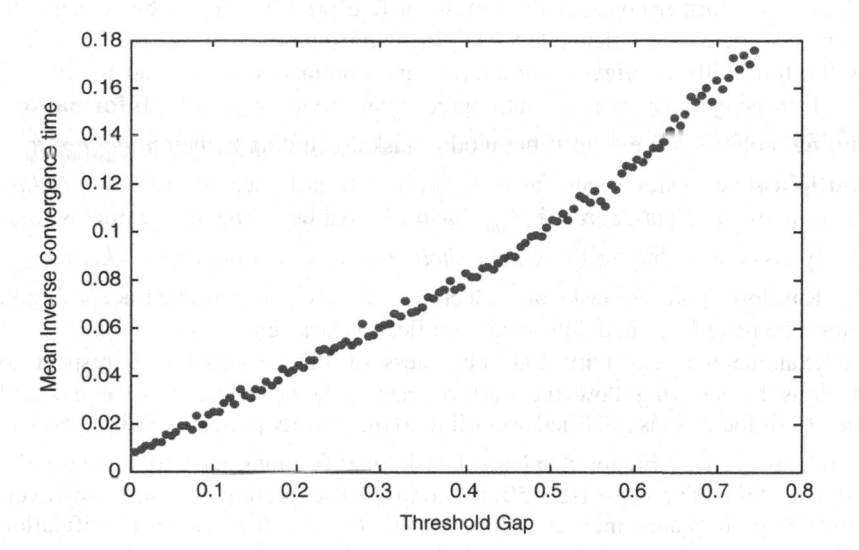

Figure 7. The inverse of the mean convergence time, $1/T^*$, for different values of the threshold gap δ. The average number of in-coming arcs connected to a node is $\langle k_{in} \rangle = 6.37$, and the internal completion rates are $r_i = 0.75 \; \forall i$. The values of β go from 0 to 0.117 in increments of 0.001, and convergence times were averaged over 100 independent simulation runs. The plot shows a linear relationship (Pearson coefficient $R > 0.98$, $p < 0.001$).

We further examine the dynamics of the PD process by analyzing the sensitivity as well as robustness (also known as error tolerance) of the PD network topology with respect to internal and external perturbations such as planned and unplanned design changes. We demonstrate two important properties of complex PD networks: (1) their dynamic behavior is highly *insensitive* (error tolerant) to random perturbations, yet highly *sensitive* (responsive) to perturbations that are targeted at specific tasks, and (2) if wisely exploited, the sensitivity of PD complex networks to targeted perturbations can yield great benefits with minimal effort, yet the sensitivity characteristic may also result in detrimental effects if not properly controlled.

In the following, perturbations are considered as either "planned" or "unplanned" task modifications that could affect the performance[5] of the PD process. Planned task modifications are defined as deliberate improvements of task parameters, and

---

[5] As before, the performance is measured as the time it takes for the PD process to reach the uniformly resolved state.

include[6]: (1) decreasing the value of sensitivity rates β_i , or (2) increasing the value of internal completion rates r_i. Effective improvement of tasks, however, will not select tasks randomly, but rather will preferentially direct resources to the most "important" tasks. Hence, we further consider the following five priority policies by which task improvements could be implemented: (1) '**Information-Generating**' policy – first modify the task with the *highest out-degree*, and continue selecting and modifying tasks in decreasing order of their out-degree connectivity k_{out}; (2) '**Information-Consuming**' policy – same as in 1, but modify tasks according to their *in-degree* k_{in} ; (3) '**Multiplicative**' policy – same as in 1, but modify tasks according to the *product of their in-degree* and *out-degree* $k_{in}k_{out}$; and (4) '**Additive**' policy – same as in 1, but modify tasks according to the *sum of their in-degree and out-degree* $k_{in} + k_{out}$; and (5) '**Random**' policy – tasks are selected randomly, and modified accordingly. The latter scheme reflects an uninformed modification strategy.

We examine the sensitivity and robustness of PD networks with respect to perturbations by studying how the performance is being affected when a small fraction, f , of the tasks is modified according to the priority policies specified above. In general, as seen in Figure 8, planned task modifications tend to increase the performance of the PD process (see [50] for additional supporting material). However, while the PD performance increases slowly with f when the random modification scheme is applied, a drastically different behavior is observed when the deliberate modification schemes are utilized. When tasks are modified preferentially (by either one of the above modification policies), the performance of the PD network increases rapidly, becoming about twice larger as its original value even if only 6% of the tasks are modified. This sensitivity to deliberate perturbations is deeply-ingrained in the inhomogeneity property of the in-degree and out-degree connectivity distributions of PD networks as indicated by their long right tails and extremely *large* variances (see Section 4). More specifically, the inhomogeneity property related to the out-degree connectivity means that the PD network is dominated by a few tasks that generate information to a large number of other neighboring tasks. Similarly, the inhomogeneity associated with the in-degree distribution implies that the PD network is dominated by a few tasks that consume information from a large number of neighboring tasks. Consequently, improvement efforts that are channeled towards these dominating tasks (e.g., increasing their internal completion rates) are expected to *drastically* alter the overall network's performance.

The aggregate-based policies (Multiplicative and Additive) seem to generally outperform the single-based policies ('Information-Generating' and 'Information-Consuming'). This result is rooted in the nature of the directed information flows forming the links among the tasks. While not uniquely affected by either the in-degree or out-degree connectivity distributions alone, both distributions are needed to understand the dynamics of the PD process. Tasks with large in- and out-degrees have

---

[6] Restructuring (redesigning) the task connectivities is another means for improving performance. Here, we assume a fixed network topology.

both significant internal complexity associated with assembling the information of several other tasks and significant external dependability upon which others rely. Thus, it is plausible to expect that tasks with large in- and out-degrees could hamper the PD process.

Figure 8 (see also the supporting material presented in [50]) shows that for the vehicle, pharmaceutical and hospital networks the performance of the 'Information-Consuming' priority policy (based on in-degree connectivities) is poor relative to the other policies. As observed in Section 4, these networks have the following proporties: (1) the correlation between the in-degree and out-degree of tasks is small, and (2) the in-degree distribution has a cutoff that is significantly lower than the corresponding out-degree cutoff. This suggests that other networks that satisfy these properties and utilize the 'Information-Consuming' priority policy might also perform less effectively. Indeed, an early cutoff of the in-degree distribution (relative to the out-degree cutoff) implies that tasks with large incoming connectivities are practically absent. Also, a lack of degree correlation implies that it is unlikely that a highly information-generating task (i.e. with large out-degree connectivity) is also highly information-consuming (i.e. with high in-degree connectivity). Consequently, the PD dynamics is generally expected to be more responsive to modifications that include high out-degree connectivity of tasks.

Finally, we observe that, for the software network, all the "non-random" priority policies perform similarly. This might be expected for networks for which the in-degree and out-degree connectivities are highly correlated (e.g., Pearson correlation of 0.76). Next we analyze the effect of *unplanned* changes of tasks on the PD performance (see Figures 10 and 11 in [50]). To simulate unplanned changes, we modified tasks by impairing their sensitivity or completion rate parameters. As seen in Figures 10 and 11 in [50], the PD performance decreases slowly with f when tasks are changed randomly. On the other hand, a drastically different behavior is observed if unplanned changes are targeted at central tasks. When tasks are modified preferentially (by either one of the above modification policies), the performance of the PD network decreases rapidly, becoming about twice lower as its original value even if only 6% of the tasks are modified.

Overall, Figures 8-11 illustrate the double-faceted characteristic of PD sensitivity – if wisely planned, the sensitivity of PD complex networks to targeted perturbations can yield great benefits with minimal effort (Figures 8 and 9), yet the sensitivity characteristic may also result in detrimental effects if not properly controlled (Figures 10 and 11).

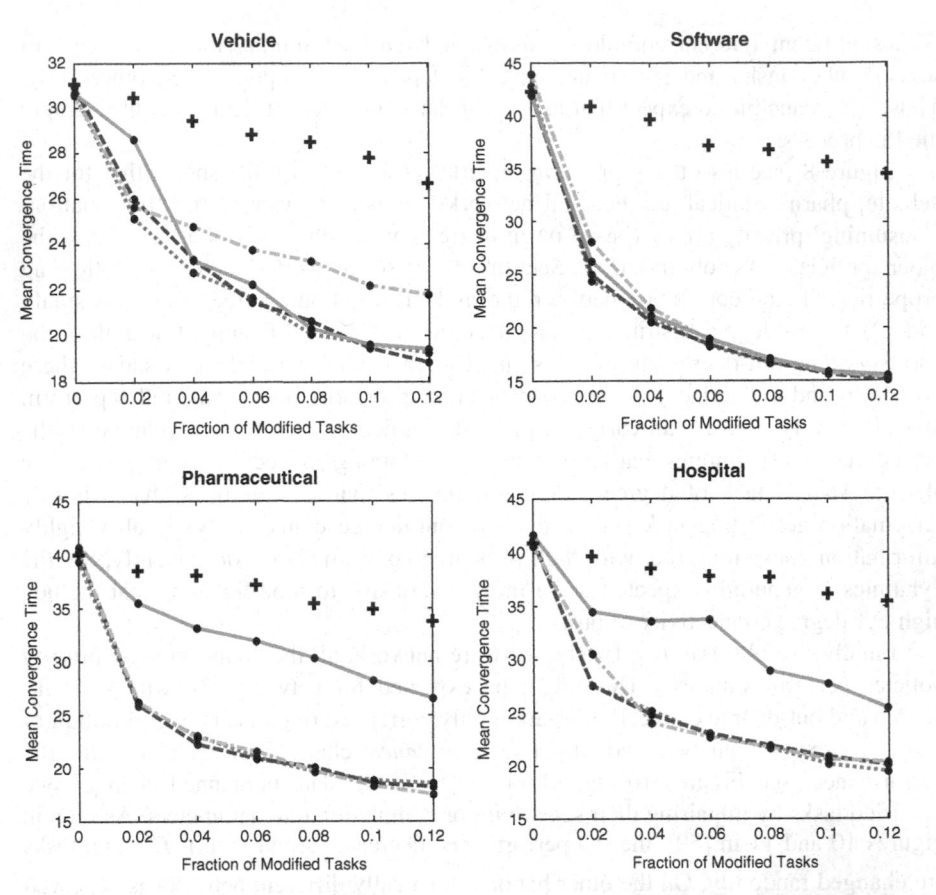

Figure 8. Comparison between five priority policies: Multiplicative (red dotted line), Additive (blue dashed line), Information-Generating (green dash-dot line), Information-Consuming (magenta solid line), and Random (+). The figure presents the PD performance versus the fraction of modified tasks for which the completion rates are *improved*. For the non-random priority policies, each data point is the average of 1000 realizations. For the Random priority policy, each point is the average of 30 different modified task selections, performed for 100 independent runs. The average in-degree, sensitivity rate, internal completion rate *prior to modification*, and modified internal completion rate are, respectively, as follows: **Vehicle.** $\langle k_{in} \rangle = 3.475$, $\beta = 0.135$, $r = 0.5$, $r^+ = 1$; **Software.** $\langle k_{in} \rangle = 3.163$, $\beta = 0.06$, $r = 0.5$, $r^+ = 1$; **Pharmaceutical.** $\langle k_{in} \rangle = 6.371$, $\beta = 0.065$, $r = 0.5$, $r^+ = 1$; **Hospital.** $\langle k_{in} \rangle = 9.741$, $\beta = 0.045$, $r = 0.5$, $r^+ = 1$.

7. Summary and Conclusions

In the last few years, the study of complex network topologies has become a rapidly advancing area of research across many fields of science and technology [8-10]. One of the key areas of research is understanding the network properties that are optimized by specific network architectures [17, 23, 27, 31, 32]. Here we have analyzed the statistical properties of real-world networks of people engaged in product development activities. We have shown that complex PD networks display similar statistical patterns to other real-world networks of different origins, and have shown how the underlying network topologies provide direct information about the characteristics of PD dynamics. In particular:

- PD complex networks exhibit the "small-world" property, which means that they react rapidly to changes in design status;
- PD complex networks are characterized by inhomogeneous distributions of incoming and outgoing information flows of tasks. Consequently, PD task networks are dominated by a few highly central 'information-consuming' and 'information-generating' tasks;
- PD networks exhibit a noticeable *asymmetry* (related to the cut-offs) between the distributions of incoming and outgoing information flows, suggesting that the incoming capacities of tasks are much more limited than their counterpart outgoing capacities. The cut-offs observed in the in-degree and out-degree distributions might reflect Herbert Simon's notion of bounded rationality [28], and its extension to *group-level* information processing.
- Focusing engineering and management efforts on central 'information-consuming' and 'information-generating' PD tasks will likely improve the performance of the overall PD process;
- 'Failure' of central PD tasks affects the vulnerability of the overall PD process;
- Positive correlation between the in-degree and out-degree of a task tends to limit the range of the parameters' values for which the system converges to the uniformly resolved state.
- PD dynamics is highly error tolerant, yet highly responsive to perturbations that are targeted at specific tasks.

In the context of product development, what is the meaning of these patterns? How do they come to be what they are? We propose several explanations for these patterns. Successful PD processes in competitive environments are often characterized by short time-to-market, high product performance, and low development costs [7]. In many high technology industries, an important tradeoff exists between minimizing time-to-market and development costs and maximizing the product performance. In PD task networks, accelerating the PD process can be achieved by "cutting out" some of the links between the tasks [5]. Although the elimination of some arcs should result in more rapid PD convergence, this might worsen the performance of the end system. Consequently, a tradeoff exists between the elimination of task dependencies (speeding up the process) and the desire to improve the system's performance through the incorporation of additional task dependencies. PD networks are likely to be highly optimized when both PD completion time and product performance are accounted for. Recent studies have shown that an evolutionary algorithm involving *minimization of link density and average distance* between any pair of nodes can lead to non-trivial types of networks including truncated scale-free networks; i.e. $p(k) = k^{-\gamma} f(k/k^*)$ [23, 27]. This might suggest that an evolutionary process that incorporates similar generic optimization mechanisms (e.g., minimizing a weighted sum of development time and product quality losses) might lead to the formation of a PD network structure with the small-world and truncated scale-free properties.

Another explanation for the characteristic patterns of PD networks might be related to the close interplay between the design structure (product architecture) and the related organization of tasks involved in the design process. It has been observed that in many technical systems design tasks are commonly organized around the architecture of the product [25]. Consequently, there is a strong association between the information flows underlying the PD task network and the design network composed of the physical (or logical) components of the product and the interfaces between them. If the task network is a "mirror image" of the related design network, it is reasonable that their large-scale statistical properties might be similar. Evidence for this can be found in recent empirical studies that show some design networks (electronic circuits [22] and software architectures [23]) exhibit small-world and scaling properties. The scale-free structure of design networks, in turn, might reflect the strategy adopted by many firms of reusing existing modules together with newly developed modules in future product architectures [2]. Thus, the highly connected nodes of the scale-free design network tend to be the most reusable modules. Reusing modules at the product architecture level has also a direct effect on the task level of product development; it allows firms to reduce the complexity and scope of the product development project by exploiting the knowledge embedded in reused modules, and thus significantly reduce the product development time.

Of greatest significance for the analysis of generic network architectures, we demonstrated a previously unreported difference between the distribution of incoming and outgoing links in a complex network. Specifically, we find that the distribution of incoming communication links always has a cutoff, while outgoing communication links is scale-free with or without a cutoff. In the cases studied, when both distributions have cutoffs, the incoming distribution has a cutoff that is significantly lower by more than a factor of two. From a product development viewpoint, the functional significance of this asymmetric topology has been explained by considering a bounded-rationality argument originally put forward by Simon in the context of human interactions [28]. Accordingly, this asymmetry could be interpreted as indicating a limitation on the actor's capacity to process information provided by others rather than the ability to transmit information over the network. In the latter case, boundedness is less apparent since the capacity required to transmit information over a network is often less constrained, especially when it is replicated (e.g., many actors can receive the same information from a single actor by broadcast). In light of this observation, we expect a distinct cut-off distribution for in-degree as opposed to out-degree distributions when the network reflects communication of information between human beings as a natural and direct outcome of Simon's bounded rationality argument. It would be interesting to see whether this property can be found more generally in other directed human or non-human networks. It seems reasonable to propose that the asymmetric link distribution is likely to hold for such networks when nodes represent information processing elements.

The chapter analyzes an *intra-organizational* network where PD tasks are nodes. It would be interesting to see if the statistical patterns uncovered for intra-organizational networks remain invariant when moving to the *inter-organizational* level where enterprises form the nodes (e.g., supply chain networks, see [43, 44]). We conjecture that the level of abstraction will not significantly change the qualitative structure of the network's topology; but may change the embedded parameters underlying the network's characteristics (e.g., coefficients and cut-offs of the power-law distributions). We have identified two generic categories of network nodes: "information-consuming" and "information-generating." We believe that this categorization could be expanded by at least three methods: 1) considering other unit centrality measures (e.g., closeness and betweenness centrality, see [42]); 2) analyzing the structure of sub-graphs ("building blocks") embedded in the networks; and 3) assigning richer data structures that more naturally describe a product development; e.g., adding characteristics to each task or adding information bandwidth (weights) to links. Finally, it would be interesting to see (by direct observations) if the group-level information-processing capacity reflected by the distributions' cutoffs can be extended; e.g., by redesigning the structure or topology of the network or by incorporating sophisticated information technologies and transaction protocols.

References

[1]   C. Alexander, *Notes on the Synthesis of Form* (Harvard University Press, Cambridge, MA, 1964).

[2]   D. Braha and O. Maimon, *A Mathematical Theory of Design: Foundations, Algorithms, and Applications* (Kluwer Academic Publishers, Boston, MA, 1998).

[3]   A. Yassine, and D. Braha, "Complex Concurrent Engineering and the Design Structure Matrix Method," Concurrent Engineering, September 2003, vol. 11, no. 3, pp. 165-176.

[4]   M. Klein, H. Sayama, P. Faratin and Y. Bar-Yam, "The Dynamics of Collaborative Design: Insights from Complex Systems and Negotiation Research," Concurrent Engineering, September 2003, vol. 11, no. 3, pp. 201-209.

[5]   A. Yassine, N. Joglekar, D. Braha, S. Eppinger and D. Whitney, Information Hiding in Product Development: The Design Churn Effect. Research in Engineering Design. Vol. 14 (3). 131-144.

[6]   S. M. Osborne, Product Development Cycle Time Characterization Through Modeling of Process Iteration . MSc. Thesis, Massachusetts Institute of Technology, 1993.

[7]   K. B. Clark, "Project scope and project performance: the effect of parts strategy and supplier involvement on product development," *Management Science* 35 (10), 1247–1263 (1989).

[8]   S. H. Strogatz, "Exploring Complex Networks," *Nature* **410**, 268-276 (2001).

[9]   R. Albert and Barabási, A.-L., "Statistical Mechanics of Complex Networks," Reviews of Modern Physics 74, 47-97 (2002).

[10] [M. E. J. Newman, "The Structure and Function of Complex Networks," SIAM Review **45**, 167-256 (2003).

[11] R. Albert, H. Jeong, and A.-L. Barabási, "Diameter of the World Wide Web," Nature 401, 130-131 (1999).

[12] M. Faloutsos, P. Faloutsos, and C. Faloutsos, "On Power-Law Relationships of the Internet Topology," *Comp. Comm. Rev.* **29**, 251-262 (1999).

[13] D. J. Watts, and S.H. Strogatz, "Collective dynamics of 'small-world' networks," *Nature* **393**, 440-442 (1998).

[14] H. Jeong, B. Tombor, R. Albert, Z. N. Oltavi, and A.-L. Barabási, "The Large-Scale Organization of Metabolic Networks," Nature 407, 651-654 (2000).

[15] H. Jeong, S. Mason, A.-L. Barabási, and Z. N. Oltvai, "Lethality and Centrality in Protein Networks," Nature 411, 41-42 (2001).

[16] J. M. Montoya and R. V. Solé, "Small World patterns in Food Webs," J. Theor. Bio. **214**, 405-412 (2002).

[17] L. A. N. Amaral, A. Scala, M. Barthélémy and H. E. Stanley, "Classes of Small-World Networks," *Proc. Nat. Ac. Sci USA* **97**, 11149-11152 (2000).

[18] M. E. J. Newman, "The Structure of Scientific Collaboration Networks," *Proc. Nat. Ac. Sci USA* **98**, 404-409 (2001).

[19] M. E. J. Newman, "Scientific Collaboration Networks. I. Network Construction and Fundamental Results," *Phys. Rev. E* **64**, 016131 (2001).

[20] M. E. J. Newman, "Scientific Collaboration Networks. II. Shortest Paths, Weighted Networks, and Centrality," *Phys. Rev. E* **64**, 016132 (2001).

[21] D. J. de S. Price, "Networks of Scientific Papers," *Science* **149**, 510-515 (1965).

[22] R. Ferrer, C. Janssen, and R. V. Solé, "Topology of Technology Graphs: Small World Patterns in Electronic Circuits," *Phys. Rev. E* **63**, 32767 (2001).

[23] S. Valverde, R. F. Cancho, and R. V. Solé, "Scale Free Networks from Optimal Design," Europhys. Lett. 60, 512-517 (2002).

[24] R. Albert, H. Jeong, and A.-L. Barabási, "Error and Attack Tolerance in Complex Networks," Nature **406** , 378-382 (2000).

[25] S.D. Eppinger, D.E. Whitney, R.P. Smith, and D.A. Gebala, "A Model-Based Method for Organizing Tasks in Product Development," Research in Engineering Design **6** (1), 1-13 (1994).

[26] D.V. Steward, "The Design Structure System: A Method for Managing the Design of Complex Systems," IEEE Transactions on Engineering Management **28**, 71-74 (1981).

[27] R. F. Cancho, and R. V. Solé, SFI Working Paper 01-11-068 (2001).

[28] H. A. Simon, *The Sciences of the Artificial* (MIT Press, Cambridge, MA, 1998).

[29] A.-L. Barabási, and R. Albert, "Emergence of Scaling in Random Networks," Science **286**, 509-512 (1999).

[30] Note that a power-law distribution of the in-degree distribution (respectively, the out-degree distribution) $p_{in}(k) \sim k^{-\gamma_{in}}$ with exponent γ_{in} translates into a power-law distribution of the cumulative probability distribution $P_{in}(k) \sim \sum_{k'=k}^{\infty} k'^{-\gamma_{in}} \sim k^{-(\gamma_{in}-1)}$ with exponent $\gamma_{in}-1$.

[31] [31] S. Mossa, M. Barthélémy, H. E. Stanley, and L. A. N. Amaral, "Truncation of Power Law Behavior in "Scale-Free" Network Models due to Information Filtering," *Phys. Rev. Lett.* **88**, 138701 (2002).

[32] B. Shargel, H. Sayama, I. R. Epstein and Y. Bar-Yam, "Optimization of Robustness and Connectivity in Complex Networks," *Phys. Rev. Lett.* **90** (6), 068701 (2003).

[33]   A. Cividanes, private communication. See also A. Cividanes, MSc. Thesis, Mechanical Engineering Department, Massachusetts Institute of Technology, 2002. A complete description of the tasks, the list of interviewees, and the result of the survey are available at   http://necsi.org/projects/braha/largescaleengineering.html.   For   further   details regarding the data collection process at GM's Research & Development Center see Cividanes's thesis.

[34]   S. Denker, private communication; Available at http://necsi.org/projects/braha/largescal-eengineering.html

[35]   A. Newton and S. Austin, private communication; Available at http://necsi.org/projects/braha/argescaleengineering.html

[36]   For a detailed description of data flow and design-process model diagrams see S. Austin, A. Baldwin, B. Li and P. Waskett, "Analytical Design Planning Technique: A Model of the Detailed Building Design Process," *Design Studies* 20 (3), 279–296 (1999).

[37]   S. Austin, A. Baldwin, B. Li and P. Waskett, "Integrating Design in the Project Process," *Proceedings of the Institution of Civil Engineers,* 138 (4), 177–182 (2000).

[38]   Braha, D. and Bar-Yam Y. "Topology of Large-Scale Engineering Problem-Solving Networks." Physical Review E. Vol. 69, 016113, January 2004.

[39]   Cross, R., Borgatti, S. P., and Parker A. "Making Invisible Work Visible: Using Social Network Analysis to Support Strategic Collaboration," California Management Review, 44/2 (Winter 2002): 25-46.

[40]   M. Granovetter, "The Strength of Weak Ties," *American Journal of Sociology,* 78, 1360-1380 (1973).

[41]   D. Krackhardt, and J.R. Hanson, "Informal Networks: The Company behind the Chart," *Harvard Business Review,* 71 (4), 104-111 (1993).

[42]   S. Wasserman, and K. Faust, *Social Network Analysis* (Cambridge University Press, Cambridge, 1999).

[43]   Reitman, V. "Toyota's Fast Rebound", *Wall Street Journal,* May 8, 1997.

[44]   Nishiguchi, T and Beaudet, A. The Toyota Group and the Aisin Fire. Sloan Management Review, Fall, 1998.

[45]   J. Mihm, C. H. Loch, A. Huchzermeier, "Problem -Solving Oscillations in Complex Projects" *Management Science* 49 (6), 733-750 (2003).

[46]   Erdös, P. and Rényi, A. "On random graphs," Publicationes Mathematicae 6, 290–297 (1959).

[47]   J. Marro, and R. Dickman, *Nonequilibrium Phase Transitions in Lattice Models* (Cambridge University Press, Cambridge, 1999).

[48]   Bar-Yam, Y. and Epstein, I. R. "Response of complex networks to stimuli," *Proceedings of the National Academy of Sciences* 101, pp. 4341-4345 (2004).

[49]   Y. Bar-Yam, Dynamics of Complex Systems (Perseus Books, Reading MA, 1997).

[50]   Braha, D. and Bar-Yam Y. "The Statistical Mechanics of Complex Product Development: Empirical and Analytical Results," NECSI Technical Report 2004-09-01, September 2004.

Chapter 4

On the Nature of Design

Sergi Valverde and Ricard V. Solé
Complex Systems Lab
ICREA-Universitat Pompeu Fabra, Barcelona
svalverde@imim.es
ricard.sole@upf.edu

1 Introduction

At the beginning of the industrial revolution, an extraordinary event attracted the attention of scientist, philosophers and layman alike. It was so extraordinary in fact that even today we are fascinated by it and by the no less uncommon people who got involved. The subject of this story was an amazing machine, more precisely an automaton. Known as the Turk, it was a mechanical chess player, made of wood and dressed in a Turkish-like costume (see Fig. 1). It played chess with Napoleon, inspired Charles Babbage and moved the great Edgar Allan Poe to write a critical essay about the nature of the automaton [1].

Although mechanical automata were not new at the time the Turk appeard into the scene in 1770, it was certainly a far-sighted invention. From the available accounts of these times, it had to be a rather impressive rival. Kempelen's automaton was life-size, and was able to move its head and eyes and move the chessmen forward. It was also able to say a few words such as "Check".

The expectation and doubts raised by the Turk were the result of its life-like, intelligent behavior. The machinery inside the Turk did not look complicated enough to explain the virtually astronomical repertoire of movements observed. So to speak, the hardware was impressive but the software was missing. Previous automata achieved fame by displaying a given repertoire of mechanical actions that were repeated again and again with the same sequence. Vaucanson's duck, for example, imitated a real bird and was able to quack, flap its wings and even simulate digesting food. In spite of the complexity of these actions, the internal mechanism was a clock-like system with wheels and levers. All these mechanisms and the wires connecting them with the different parts of the automaton were included inside a large pedestal. But the chess player faced a great challenge: to be able to play a game with an enormous potential combinatorics. How it could it be possible? Although Charles Babbage considered the possibility of building an intelligent machine after playing with the Turk in 1819, it was apparent to him that the automaton was probably hiding a human inside it. That was of course the case.

Figure 1: Von Kempelen's chess player automaton, the Turk. Here a front view is shown, with the cabinet doors open showing the internal mechanisms which were claimed to power the automaton's abilities. In reality, a man was actually hidden inside the cabinet and manipulated the automaton.

An interesting point here is how life-like (or "human") the Chess Player was. In a more general context, the Turk raises questions on the boundaries between life and the artificial. What features of living systems can be captured by man-made designs? Engineered structures seem to be closer to the physical than the biological world. But this might actually be a misleading conclusion. The key difference that distinguishes biology from physics is that biological systems perform computations. The origin of such difference stems from the role that information plays in the first, which is not shared by the second [2]: there is an evolutionary payoff placed on being able to predict the future. More complex organisms are better able to cope with environmental uncertainty because they can compute and can also make calculations that determine the appropriate behavior using what they sense from the outside world. Such computing systems emerge through evolution as a consequence of different (non exclusive) mechanisms [3, 4, 5, 6, 7].

Perhaps the earliest exploration of a theoretical basis for life-like structures is Von Neumann's study of self-reproducing automata. While looking at the minimal, formal conditions required for a given system to replicate itself, Von Neumann found that these automata should include two key ingredients in their architecture: *hardware* and *software* [8]. As noted by several authors, there is a surprisingly good mapping between Von Neumann's finding and the actual structure of cellular organization. Although this work was formulated several years before Watson and Crick's discovery of DNA, it already presented a formal picture of *what should be expected* to be observed.

Computer science has been evolving over the last 50 years in many directions, but in some fundamental sense it has been frozen into a well-defined view of computing based on von Neumann's ideas of computers (see below). Although there was an early fascination in getting inspired by nature while thinking on how machines should compute, such initial fascination rapidly faded out. Powerful designs rapidly emerged and became real. Fast computers were built and biological metaphors became unnecessary.

When looking at the architecture of cells, three basic components can be properly identified (together with a membrane structure separating the inside from the outside):

- The genome, and the regulation pathways defined by interactions among genes;

- The proteome, defined by the set of proteins and their interactions; and

- The metabolome (or metabolic network) also under the control of proteins that operate as enzymes.

The last two components define the hardware of cells, while the first is the software. Roughly speaking, the instructions written in the DNA sequence (the genetic material) are executed provided that the appropriate hardware is present. Perhaps not surprisingly the jargon of molecular cell biology is full of terms suggesting that computations are taking place, such as *transcription, translation* or *genetic code* [9].

Technological design and evolution reveal a number of traits in common with natural evolution [3]. On the other hand, many patterns found in nature seem to result from a combination of optimal designs together with strong structural constraints [3][4][5]. Such similarities are made more apparent while looking at the overall pattern of interactions among components both in cells and artifacts [10].

Understanding the origins and implications of computation in biology as well as in technology requires understanding the system level behavior of both hardware and software. Although hardware has received great attention, software has been less appreciated in spite of its fundamental relevance, complexity and plasticity. This chapter constitutes a very early attempt to explore the main features of the architecture and functional organization of software, from the microscopic to the macroscopic level. In particular, we will report different network patterns observed in software structures. By understanding how these patterns originate, we might be able to provide tentative answers to some fundamental questions, such as:

1. Are there constraints to optimality in technological designs?

2. What type of emergent patterns result from engineering?

3. Is there tinkering at some level in the organization of artifacts?

4. Are emergent patterns similar to those observed in natural structures?

5. Are there fundamental differences in the global structures observed in natural and artificial structures?

6. Can we get inspiration from biological patterns in order to obtain new types of designs?

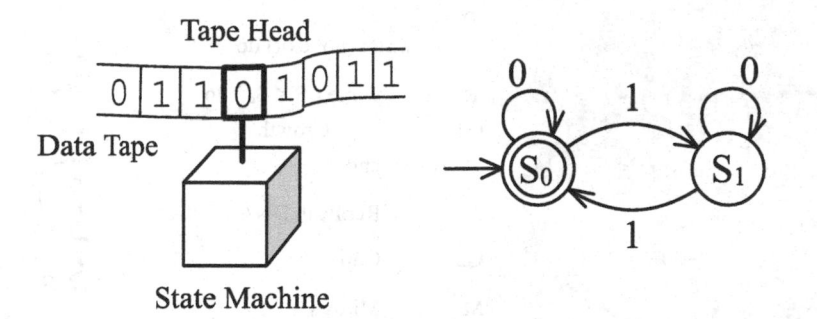

Figure 2: Schematic representation of a state machine (left). The state transition diagram for a simple finite state machine that accepts binary words with an even number of ones (right). Accepting states are depicted by double circles (see text).

2 Computing Machines

In order to start our exploration, we need to approach the problem in a way that is largely independent of specific computing machine features. In this context, computation theory [14] allows us to explore key features of complex systems by integrating architecture and function at different levels. Any computing device can be modelled with an abstract state machine. A state machine consists of a (possibly infinite) data tape of symbol cells and a computing device (see Fig. 2). The computing device handles the data tape according to a predefined set of rules (or transition functions). The behavior of this machine can be described graphically by means of a state transition graph, where nodes depict machine states and directed links represent the possible transitions between states. At every step, a transition rule is selected depending on the current device state and the symbol read by the tape head. The rule instructs the device if a new symbol must be written on the tape, what is the next state and the direction the data tape shifts (i.e: move to the left or to the right).

A diversity of computing models can be obtained by disabling certain features of the most powerful model of computation, the Turing machine. This computer can write symbols in the data tape and also shifts the tape to the right or to the left, as desired. It can be shown that the Turing machine is powerful enough to simulate any computer [24]. On the other hand, the simplest model of computation is the finite state machine (FSM). The finite automaton has a read-only and unidirectional data tape. Another limitation of this machine is the finiteness of input words placed in its tape. This type of machine is only able to emulate a very constrained family of computers. In spite of these limitations, this class of machines really deserves some analysis because many dynamical features of cell biology, such as the cell cycle, can be described in terms of the FSM (see Fig. 3).

a
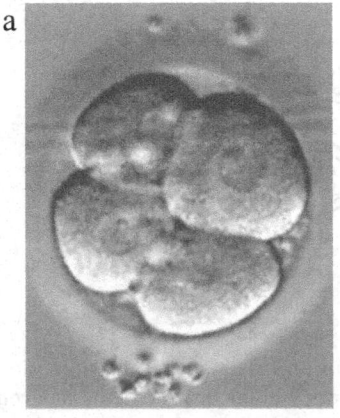

b

while (not exit) do

R   while ($S < S_c$) do
G1   Growth
             end

S   ReplicateDNA
G2   Gap
M   Mitosis

             end

c
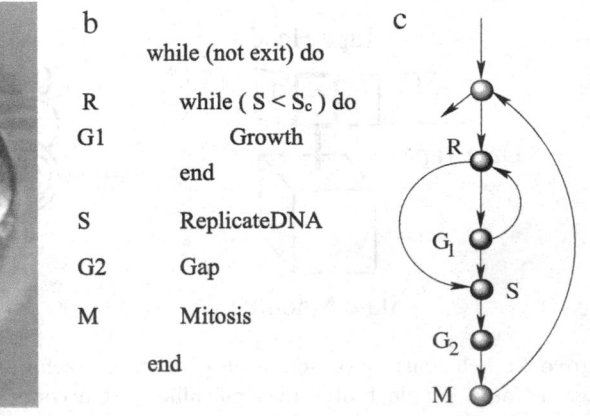

Figure 3: Many processes taking place inside living cells can be understood in terms of a computation. Some particular situations can be easily mapped into a discrete, finite state machine (see text). An example is the cell cycle. In (a) a micrograph of a dividing mammalian cell is shown. In (b) the corresponding basic algorithm for the cell cycle is given and the associated finite automaton is displayed in (c).

We can illustrate the inner workings of the FSM with an example. Fig. 2 shows the state transition diagram for a simple finite automaton. This simple computing device is intended to detect if the input word has an even number of ones. To perform this function, only two states are required. Every link is labelled with a symbol used to match the next state transition. Initially, the machine is configured in the starting state, which is denoted by a pointing arrow (i.e: the state S_0 in the above figure). The tape is loaded with an input word (say "011") and the head is pointed to the first symbol (that is, "0"). The computing device performs the transition $\langle S_0, 0 \rangle \rightarrow S_0$ and the tape head is shifted one position to the next symbol. In the following transition, the machine reads "1" and the new state becomes S_1. Finally, the symbol "1" is read and the state changed again to S_0. There are no more symbols to read so the machine checks if the current state is an ending state. In this case the word is accepted but not every word presented to the machine leaves the machine at an ending state (i.e: try "111").

In order to recognize more complex words it is necessary to extend the finite state machine capabilities. For instance, no finite state machine is able to accept words having an arbitrary number n of ones and zeros like $0^n 1^n$. The machine should be able to process a potentially infinite number of states. Counting requires the device to remember the number of symbols past read by using a memory storage, which is precisely the ability possessed by the Turing machine. The physical realization of the Turing machine is the Von Neumann architecture.

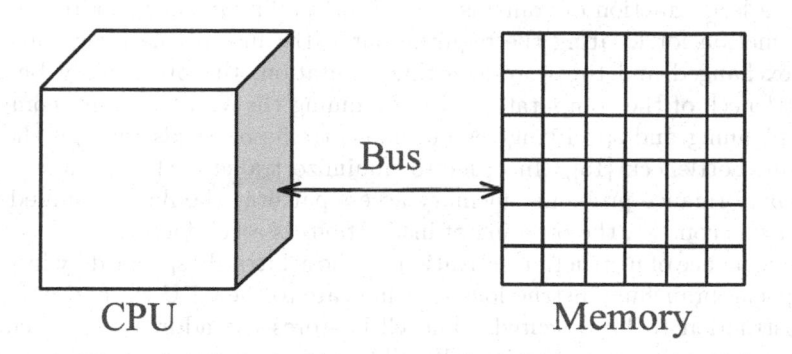

Figure 4: The Von Neumann architecture has three different parts: the CPU, the memory and the bus connecting both (see text for detailed description). The program is stored in the memory along with data, parameters and temporary calculations. The CPU traverses the memory recognizing program instructions and performing their associated actions. A little memory (registers) is included in the CPU and used during normal operation like for locating the current instruction.

The Von Neumann computer consists of three differentiated components: a central processing unit or CPU, a data store (or memory) and a wires connecting both components (or bus) (see Fig. 4). The CPU is a state machine which is able to recognize a number of special words (also called "program instructions"). The memory is a finite grid of cells analogous to the data tape used by the Turing machine. But unlike the Turing machine, the CPU is capable of direct access to any particular memory cell simply by referencing its position in the data store (or "memory address").

3   The Memory Stored Program

The signature of the universal computer is the memory stored program [24]. Any complex system requires this component in order to properly react and adapt to a changing environment. The program is a sequence of instructions that describe the computer's behavior. The CPU scans this sequence and activates different actions by accepted instructions. It can be shown that a very small instruction repertoire is enough to implement any program: arithmetic operations between two memory locations, store and/or retrieval of memory cells, and branching instructions. Complex behavior is achieved by the interpretation of the stored program, which yields a particular interleaving of calculations and memory accesses.

Assignment is the key instruction of the Von Neumann computer. The assignment stores a word or a number (often the value returned by evaluating a numeric expression) in a given memory cell. This places a very important restriction because the computer is only able to handle a single word-at-a-time [15]. Moreover, this computation model requires large amounts of data traffic exchanges through the bus in order to do useful work. The situation is worsened because a large fraction of traffic is wasted for sending memory addresses, that is, information for locating the required data. Because of the large volume of words exchanged and the word-at-a-time limitation, the bus rapidly becomes the bottleneck of the computation. Programming the Von Neumann computer means planning and specifying the enormous traffic of words through the Von Neumann bottleneck [15]. In order to minimize traffic exchanges and partly reduce performance problems, memory access patterns should be planned with care. In this context, the ordering of instructions is a key factor.

The sequence of instruction activation can be arbitrarily specified by inserting some special branching instructions that indicate to the CPU the address of the next instruction to be executed. The CPU stores the address of the current program instruction in a special cell called "program counter" register (PC). When the current operation finishes, the PC is automatically incremented in order to point to the following instruction. Branching instructions can change the state of the program counter in several ways and are very important in defining program behavior. For instance, there are unconditional branching instructions that allow the CPU to move to a distant location after (or before) the current instruction. Sometimes the jump is executed depending on the success of some arithmetic test or condition. Loops (executing the same instructions several times) can be implemented by combining unconditional and conditional instructions.

3.1   Flow Graphs

Program dynamics is defined at the interplay between memory contents and the instruction branching process. In this context, graphs are a useful tool for expressing interaction between different program parts. As static structures, they provide the skeleton on top of which function takes place. Still, this flow graph is an incomplete characterization of program complexity because (besides other reasons) the interaction between instructions and memory cells is not depicted. However, as will be shown below, such static structures can be highly constrained in terms of the possible range of graphs that can be found for a given purpose. The flow graph describes the instruction processing order, encoding all possible branchings between program instructions[13]. Every node $v \in V$ in the flow graph (also named "basic block") represents a continuous sequence of instructions. The last instruction of a node is always a branching instruction or decision point. Directed links $(v, u) \in E$ signal the transferring of control from the last instruction of the source node v to the first instruction at the destination node u. Execution flows unidirectionally from the entry to the exit node, which are two special nodes present in any flow graph.

We call *in-degree* the number of links entering a node and *out-degree* the number of links exiting a node. Nodes with out-degree two denote conditional branching. With conditional branching, the transfer of flow depends on the evaluation of a conditional clause (link taken/not taken). On the other hand, unconditional branching is represented by nodes with out-degree one. Additional performance information may be attached to both nodes and links. For example, useful performance profiles are frequency of visits to a node or the frequency of traversing a link.

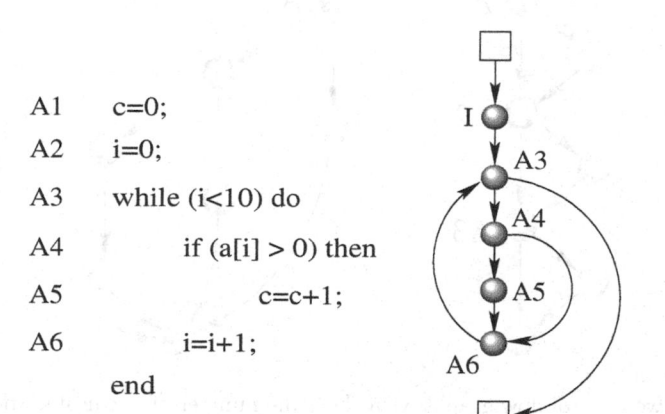

A1   c=0;

A2   i=0;

A3   while (i<10) do

A4   if (a[i] > 0) then

A5   c=c+1;

A6   i=i+1;

      end

Figure 5: A simple program and its flow graph. Entry and exit points are denoted by empty boxes. Several instructions (A1,A2) are grouped within the initialization node (I). The main feature of the control flow graph is a loop, which is expressed with the back link (A6,A3). The looping condition is tested at node A3. When $i \geq 10$ the entire loop is skipped. At this moment, variable c equals the number of non-zero entries in vector a_j. Note also the branching point at A4 where the link (A4, A5) or the link (A4, A6) is chosen depending on the value of a memory cell. A multiplicity of different programs can be mapped onto the same control flow graph.

3.2   Program Predictability

Designers must pay attention to the evaluation order of instructions, that is, to the structure of the flow graph. An innocent permutation of a random pair of program instructions may yield very different program semantics. For example, let us swap instruction A6 and A1 in Fig. 5. We obtain an unresponsive program that never finishes and gets caught in an infinite loop. Unfortunately, it was shown by Turing that no automatic procedure is able to detect if a program stops for any given input configuration [24]. This so-called halting problem is deeply related to the inability to predict the future behavior of a computer program. In spite of the apparent simplicity of some programs, it turns out that we *cannot* predict what microscopic states result from the composition of simple instructions.

Any useful artificial object must be predictable. Software engineering should not be an exception to this rule. Take planetary missions for example. This is the kind of system that requires autonomy and strong tolerance under highly stressing environments. In the current state of art, reasonably predictable software is obtained by doing a lot of tests in many different scenarios. Unfortunately, the engineers can not plan in advance every situation faced by the software controlling the robot deployed on an unknown planet.

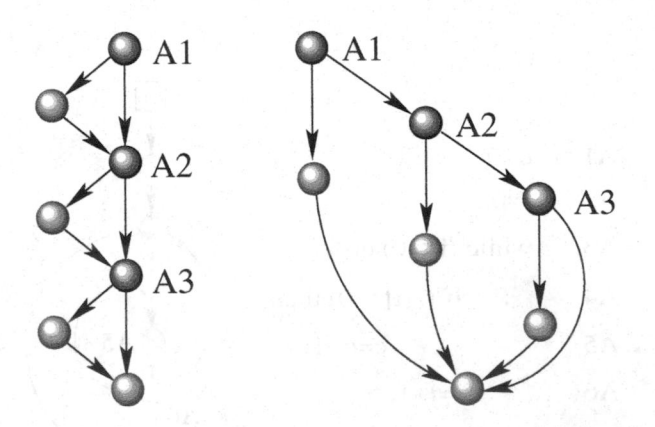

Figure 6: Two control flow graphs with the same number of predicates and branches, but very different number of (acyclic) paths. The control flow graph on the left defines 2^3 possible paths to be compared with the 4 possible paths of the right control flow graph. In this sense, the left directed graph is less predictable than the right one. Basically, this is reflected in the in-degree of some nodes on the left. Those nodes will be the crossroad of more than one path and thus increasing the uncertainty of the expected program behavior.

Predictability is also deeply linked to performance. For example, modern computers exploit regularities found in programs to yield better performance. A good example is provided by data and instruction caching. In order to avoid slow memory accesses, frequently referenced memory portions are copied into a fast memory (or cache). This scheme results in speed-up only if program execution displays a certain predictability, that is, if the processor accesses the same memory region more than once.

Dynamic software behavior can be understood in terms of a walking through the flow graph. However, even the simplest flow graphs display an enormous number of potential execution paths. Walk length is theoretically unbounded because it is possible to visit some links many times (such as the link (A6,A3) in Fig. 5). We can also restrict the discussion to acyclic walks, also called paths. A path is a finite sequence of links $(u_1, u_2), (u_2, u_3), (u_3, u_4), ..., (u_{n-1}, u_n)$ in the flow graph where the destination of each link equals the source of the following link. No single link is reported more than once. The number of links in the path is the path length.

Unfortunately, flow graphs of large software systems still have an enormous number of potential paths. For instance, the popular application Microsoft(R) Word contains more than 2^{64} potential paths [23]. Again, the serious limitations deduced from the halting problem prevents us from differentiating between potential paths and actually visited paths from the static program description. In fact, this is equivalent to the problem of determining if a related program halts or not, which we know to be undecidable in general [24].

Surprisingly, the statistical analysis of real flow graph indicates that some programs are more predictable than others (see Fig. 6). Empirical studies have revealed that programs as a whole traverse a tiny fraction of the millions of potential program paths. Software performance profiles typically reveal that 90% of execution time is spent in a very small number of so-called "hot paths". This subset of traversed paths is largely independent of parameter variability, suggesting that software dynamics is not a purely driven process. In addition, real programs display a non-negligible amount of local correlation. Both global and local empirical regularities have been exploited in modern computer hardware in order to predict the next instruction to be executed by storing past branching outcomes (a mechanism analogous to data caching). In addition, these regularities enable the programmer to focus on a few program paths. This offers advantages when detecting performance bottlenecks and/or computation errors.

4   Programming and Separation of Concerns

Branching allows the reuse of instructions without the need for code duplication. The branching instruction allow us to partition the program into disjoint code pieces that alternate execution flow. Such a simple technique minimizes the number of program instructions and saves scarce memory resources, an important constraint in old computers. Beyond performance requirements, there is also a more important reason for structuring the program. The partition promotes the view that different functionalities must be provided by different program components [11]. Ideally, the mapping between functionality and components should be one-to-one but it was realized early on that, for complex programs, it is really difficult to make a clear division of labor.

Now imagine that our program achieves clear separation of responsibilities. In this case, the program state can be partitioned into disjoint pieces, each with a well-defined function. Then the whole state transition function can be expressed as the concatenation of simpler state transition functions. This program structure extends to memory organization because each piece computes a well-defined part of the global program state, without overlaps. In this ordered system, the whole coincides with the sum of its parts. Unfortunately, real software practices show us that the above clear separation is not an easy objective to reach. We find it very difficult to decompose the global program state in a linear combination of simple pieces. Instead, global program behavior is often defined by the interaction between more than one component (i.e, one memory cell accessed by two distant program instructions). In this case, interaction involves complex temporal correlations. Because of the constraints, our systems tend to exhibit complex dynamics that are more than the sum of the parts.

4.1 Object-Oriented Programming

Another important factor influencing programming is the language used for expressing the program. Using the same reduced set of machine instructions turns out to be very inconvenient for human designers. Different artificial languages have been considered for this task. The requirements imposed by a programming language are numerous and, to a certain extent, contradictory. Fortunately, the listener (the computer) is so constrained that ambiguity must be removed from the artificial language. This greatly simplifies the syntax of programming languages.

A broad characterization splits the world of programming languages into two big groups: declarative and procedural. When using the former type of language, the program constitutes a formal specification of what is wanted to be computed. Declarative languages (such as Prolog) only tell the computer what is desired and not how to achieve it. Conversely, a program written with a procedural language (like C, C++ or Basic) is a step-by-step detailed recipe of how to perform the computation. All previously presented code samples are instances of programs written in a procedural language. They are detailed plans that, when interpreted by the computer, yield the desired behavior. Declarative languages are very desirable from the user point of view but suffer from severe performance problems.

Modern programming practices (i.e: object-oriented languages like C++ or Java) are procedural and extensible. These languages are so powerful that they enable the programmer to create new software entities that represent real-world concepts. At this level, the program is understood in terms of high-level processes that manipulate abstract entities. The programmer looks at the domain of software application trying to localize the relevant entities and the relationships between them. For example, if a business application deals with customers and selling orders, the programmer should explicitly define the 'customer' and 'order' entities as part of the program description. In this example, 'customer' and 'order' are related to each other because a particular order is placed by a given customer. In addition, attributes and processes are attached to entities. For example, with every customer the program stores her name and a process that enables the customer to issue a new order.

The complexity associated to software requires an evolutionary approach. The number of requirements is so large that we cannot develop the full application in a single step. In addition, some requirements are just unknown when we start to develop the system. This iterative process requires the programmer to switch between local and the global views of the software system. The programmer tends to focus her efforts on a single part of the complete design, which later is integrated with the whole system. Moreover, several developers can work on different parts simultaneously. Unfortunately, experience tells us that one cannot simply add more and more human developers and expect that the development process will be greatly shortened. As the number of programmers increases, so do the chances of unwanted interaction. There will be conflicting design decisions among programmers, which may result in project delays and overruns. Soon, the cost of communication outweighs the benefits of having many programmers working in parallel. In this context, an adequate global architecture of the software system (i.e., the set of components and their relationships) can be of great help.

4.2   Class Structure

Every advantage offered by object-oriented programming is based on the concept of class. The class recognizes that software development must be incrementally performed. When a new functionality must be added to the system, the engineer addresses a little computing device encoded by a class. The class machinery is nothing more than a grouping of variables and the code operating on them. The class variables are also known as attributes and each block of instructions is called a method. In order to avoid redundancy, the class might be allowed to cross its boundaries by accessing attributes and methods from other classes. The entire system is viewed like a network of interacting and simpler computing devices.

Inside the class, method interaction can be direct and indirect. The former type of interaction takes place when a method transfers the execution flow to another method, like directed edges in the flow graph. An indirect (but somewhat stronger) method interaction is through shared variables (like the methods f and g in Fig. 7). The only way that a method can alter the behavior of another method is by exchanging a variable. The execution path in a method depends on the values of accessed variables.

The notion of class cohesion is one of the most important object-oriented features. Good class design displays strong cohesion, that is, "a class should not be a collection of unrelated members, but all the members of a class should work together to provide some behaviors of the corresponding objects" [27]. Designs with strong class cohesion are believed to be more maintainable and reusable.

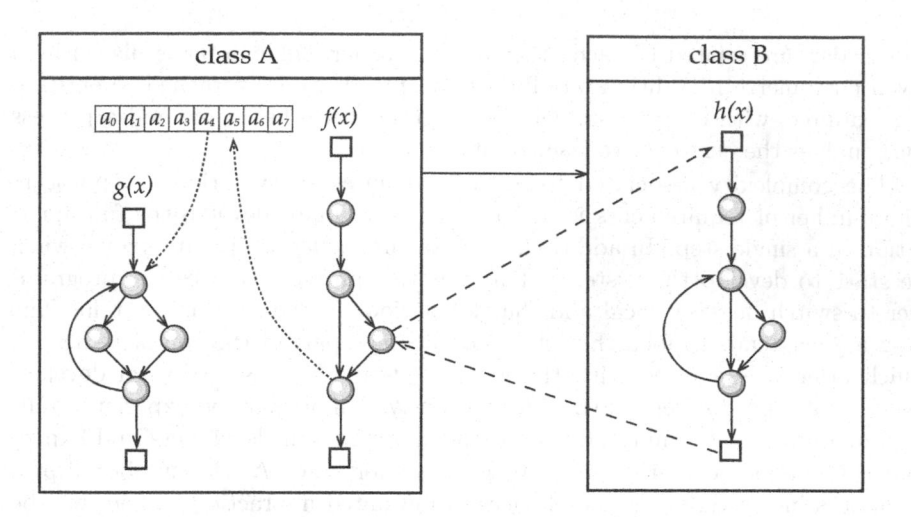

Figure 7: Mixed graphical representation for a simple software system consisting of two classes (also components) A and B. The diagram shows the following elements. Class A defines a single attribute (a vector of eight elements $a_0, a_1, a_2, ..a_7$) and two methods f and g. Every method is described by its control flow graph. Methods f and g indirectly interact by means of the data vector. The class B encapsulates the method h. Eventually, f transfers the control to h. The dashed line indicates that control flow is crossing class boundaries.

When the class methods are loosely interconnected, this is a sign of poor design and the class must split into several classes. Conversely, a class with strong cohesion will be difficult to split into isolated parts [28].

In this case, methods are closely related to each other by shared attributes. A bipartite graph representation is well-suited for measuring class cohesion. The bipartite graph $G = (F, V, E)$ consists of two disjoint sets F and V of nodes representing methods and attributes, respectively. Only interaction between two nodes of unlike sets is displayed. An edge belongs to the graph $\{f, v\} \in E$ if the method $f \in F$ references the variable $v \in V$. Two methods f and g will be indirectly related only if they share the same variable, that is, only if the $\{f, v\} \in E$ and $\{g, v\} \in E$. This bipartite graph is called an attribute-method reference graph. For instance, the attribute-method reference graph for class A in Fig. 7 consists of only three nodes: one for attribute a_i and another two for the methods f and g. The graph has only two edges connecting the attribute with the two methods. The attribute-method reference graph for class B in the same figure will have only a single node for the method h.

Once the reference graph is defined, the cohesion is measured as a function of the fraction of methods that should be removed in order to disconnect it [28]. Fig. 8 illustrates the notion of cohesion captured by this measurement.

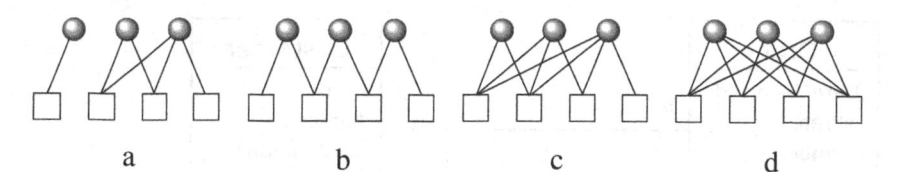

Figure 8: From left to right, attribute-method reference graphs for several classes displaying increasing levels of cohesion (see text). Attributes are displayed with circles and methods with boxes. Note that every box encloses a control flow graph. (a) shows an already disconnected class with weak cohesion. This is a symptom of poor design suggesting that two unrelated functionalities were enclosed within a single class. A better design will split the class in two different classes. Complete bipartite graph (d) always displays very strong cohesion. All methods must be removed in order to disconnect the graph. (Adapted from [28]).

4.3   Class Diagrams

Complex software systems rarely consist of a single class. The typical software system performs several functionalities distributed (more or less) evenly among a collection of interrelated classes. Software engineers are aware of this and they explicitly depict this large-scale organization through class diagrams. For this purpose, graphical languages like UML have been devised to communicate software designs in a standard way[36]. A simple UML class diagram is displayed in Fig. 9.

Class diagrams represent a number of interesting features about software designs. A quick look at this diagram gives a global idea of the internal software structure. The class diagram is an abstraction of the domain of software actuation, the entities and the nature of their relationships (i.e: in the previous sample commercial application, its class diagram should reflect the interaction between customers and orders). In order to introduce new software functionalities or to fix unwanted software behavior (also known as bug fixing), programmers navigate the information space defined by the class diagram. It is believed that some class diagrams enable fast identification of the relevant software pieces that must be changed or modified, thus reducing the total amount of effort spent by the programmer to accomplish her task. Can we identify and measure common structural patterns of class graphs? How is software quality reflected in its structure?

An early answer to the previous questions was given in their seminal book by Gamma, Helm, Johnson and Vlissides [29]. In their book, the authors proposed to assess the quality of an object-oriented system by looking at the patterns of collaborations between classes. It turns out that for some particular design problems there is some preferred solution which is more frequently selected among other candidate solutions. [29] presents a full catalogue of common solutions (or design patterns) observed in object-oriented programming. Every single design pattern has a name and is described by its intent, motivation and structure.

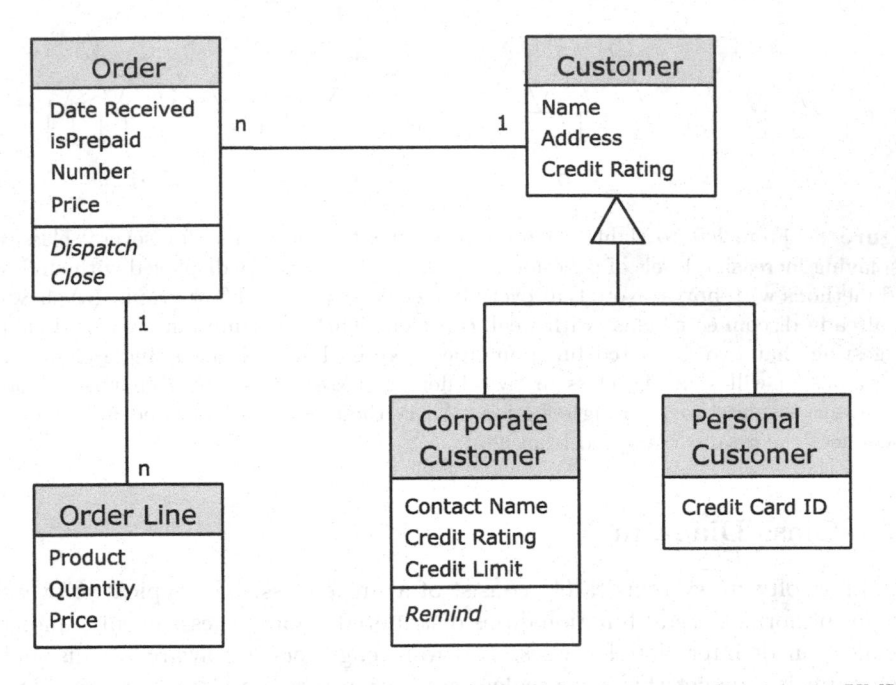

Figure 9: A simple class diagram for a commercial software application, in UML notation. The diagram shows five classes: Customer, Corporate Customer, Personal Customer, Order and Order Line. Every class is divided into three sections: name (shaded), attributes and methods (in cursive). Classes might relate to each other in three ways: composition, inheritance and use. These relationships are denoted by decorated links connecting two classes. For instance, the fact that the one customer can place more than one order is represented by a single relationship between 'customer' and 'order'. The numbers at the end-points of the link are the multiplicity of the relationship, telling how many objects will participate in the relationship. In the example, the customer is related to 'n' orders, but every order is only related to a single customer. Another typical relationship in UML diagrams is inheritance. This applies when two classes are similar but have different features. In the figure, both corporate customer and personal customer are related to the customer class by an inheritance relationship, indicating that they are able to place orders but in different ways.

In addition, the book classifies patterns in several families depending on their purpose (what a pattern does) and their scope (specifies whether the pattern applies primarily to classes or to objects). Are these patterns the signature of universal laws followed by high-quality software structures?

5   The Small World of Software Architecture

It is clear that the so-called "design patterns" approach is a qualitative catalogue of software knowledge. Here, we propose a new approach to document software knowledge, which is based on the quantitative study of structural patterns in object-oriented systems. The first requirement of this new approach is to represent software structure with a network. The software graph is defined by a pair $\Omega_s = (W_s, E_s)$, where $W_s = \{s_i\}, (i = 1, ..., N)$ is the set of $N = |\Omega|$ classes and $E_s = \{\{s_i, s_j\}\}$ is the set of edges/connections between classes. The *adjacency matrix* ξ_{ij} indicates that a static interaction exists between classes $s_i, s_j \in \Omega_s$ ($\xi_{ij} = 1$) or that the interaction is absent ($\xi_{ij} = 0$). Nodes in the software graph are black boxes hiding internal class complexity. Similarly, links in the class graph hide the specific meaning of an underlying static collaboration between classes. There are two ways to recover the software graph: (1) from the class diagram described in UML or (2) from the source code itself.

When there is no explicit UML class diagram available, the analysis of source code is the best method for recovering the software graph. In [18] we have described a simple algorithm that recovers the software graph from C++ or Java source code. Automatic documentation tools implement similar algorithms [22]. The reconstruction process is implemented by a finite state machine, which looks for class definitions by finding all instances of the keyword "class" in the source code. The analysis of the class declaration body provides the links connecting classes. In this case, we look for the so-called "inheritance" and "uses" relationships found within the class. Every time the parsing process detects a class attribute, an edge is set from the owner class to the referenced class. Our definition of the software graph does not make any distinction between different types of static collaborations, which are always represented with a plain link. The reason for not attaching semantic information to nodes and links is that here we are only interested in modelling and characterizing structural patterns. Fig. 10 shows a software graph recovered from a real software application.

We define the average path length l as $l = \langle l_{min}(i,j) \rangle$ over all pairs $s_i, s_j \in \Omega_s$, where $l_{min}(i,j)$ indicates the length of the shortest path between two nodes. The clustering coefficient is defined as the probability that two classes that are neighbors of a given class are neighbors of each other. Poissonian graphs with an average degree $\bar{k}$ are such that $C \approx \bar{k}/N$ and the path length follows:

$$l \approx \frac{\log N}{\log(\bar{k})} \tag{1}$$

C is easily defined from the adjacency matrix, and is given by:

$$C = \left\langle \frac{2}{k_i(k_i - 1)} \sum_{j=1}^{N} \xi_{ij} \left[\sum_{k \in \Gamma_i} \xi_{jk} \right] \right\rangle_{\Omega_s} \tag{2}$$

This provides a measure of the average fraction of pairs of neighbors of a node that are also neighbors of each other. In a remarkable paper [17], Watts and Strogatz observed that many social, biological and technological networks, while very different in purpose and nature, share a number of common traits. All these systems are instances of what is known as small-world. Small-world

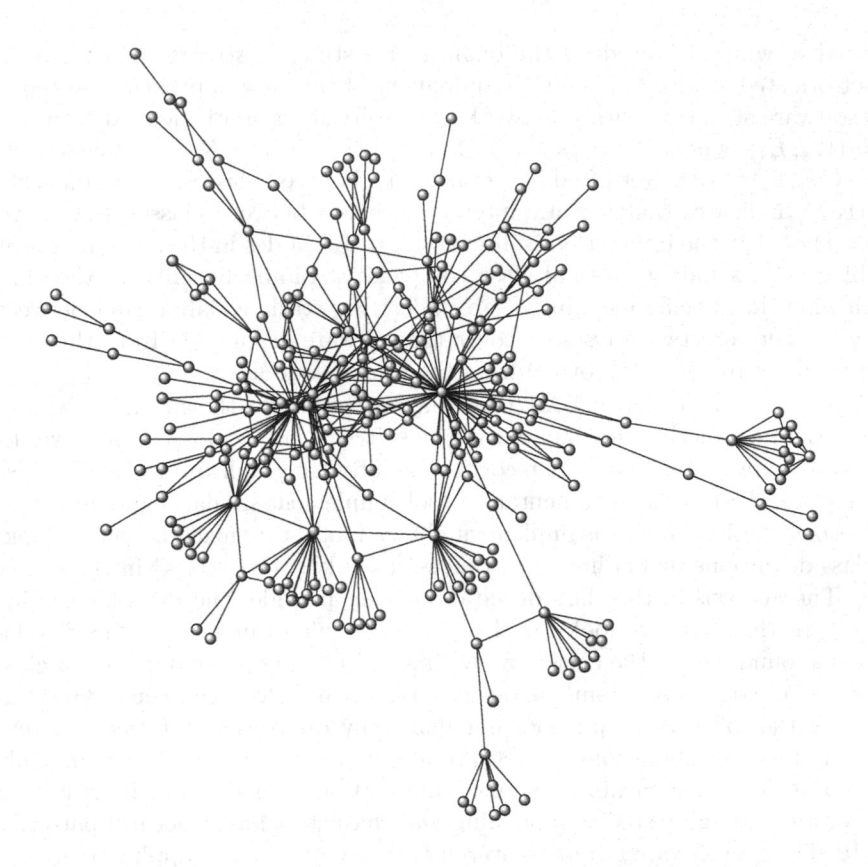

Figure 10: The largest connected component of the software graph Ω_s reconstructed from the source code of the 3D tool Aztec (http://aztec.sourceforge.net).

networks displays high clustering C, that is, nodes are connected in local neighborhoods. The surprising thing about small-world networks is that, in spite of the limited scope of nodes, the average path length l is very low. That is, any node is reachable within a small number of hops. This is achieved by means of a small number of key links or "shortcuts" that act like bridges connecting distant network regions.

We have analyzed the class diagrams for 29 different software systems (see [18] for a detailed analysis). These diagrams are examples of highly optimized structures, where design principles call for diagram comprehensibility, grouping components into modules, flexibility and reusability (i.e. avoiding the same task to be performed by different components). Although the entire plan is controlled by software engineers, no design principle explicitly introduces small-worldness. The resulting software graphs, however, turn out to be small worlds (see Fig. 11).

The small-world structure has important effects on the resulting network dynamics. A small-world communication network like the Internet propagates

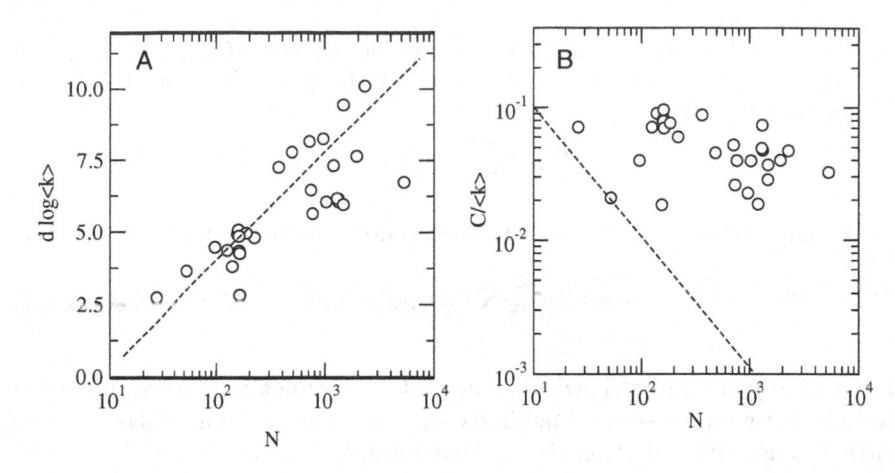

Figure 11: Average path length against network size for 29 different software systems. (a) Normalized distance grows with the logarithm of the number of classes, as expected in small world networks (see text). (b) Normalized clustering strongly departs from the predicted relation followed by random graphs (dashed line).

messages very quickly because of the low average path length. Moreover, its clustered nature makes the network very resilient to the loss of single elements. That is, there is enough redundancy in the number of paths connecting two nodes because their neighborhood is densely connected. Studies of synchronization in networks has also shown how small-world properties can be exploited in order to reach a globally synchronized state in a decentralized and robust manner [19]. This phenomenon is deeply related to computation. Indeed, we have determined that a large number of software applications are small-worlds. It might be that our software graphs arrange in small-world settings for similar reasons. But, how does the system reach the small-world architecture? How is the small-world exploited by computation processes?

5.1   A Simple Explanation for the Small World

The microscopic software structure is captured by the relationship between classes and methods (see Fig. 8), which accepts a bipartite graph $G = (W_s, W_m, D)$ representation (also known as class-method reference graph). The set W_s of classes and the set W_m or methods are disjoint, that is, $D = \{\{s_i, m_j\}\}$ where $s_i \in W_s$ and $m_j \in W_m$. The adjacency matrix ψ_{ij} for the bipartite graph encodes these connections:

$$\psi_{ij} = \begin{cases} 1 & \{s_i, m_j\} \in D \\ 0 & otherwise \end{cases} \tag{3}$$

This is an $N \times M$ binary matrix where $N = |W_s|$ (the number of classes) and $M = |W_m|$ (the number of methods). Relationships between nodes of the same kind can be recovered by means of one-mode projection of the bipartite graph

G (see Fig. 12). The projections yield two one-mode graphs $G_s = (W_s, D_s)$ and $G_m = (W_m, D_m)$. The adjacency matrix ψ^s for the graph G_s is related to the adjacency matrix ψ by

$$\psi^s_{ij} = \sum_k \psi_{ik}\psi_{jk} \tag{4}$$

and a similar relation holds between the adjacency matrix ψ^m for the graph G_m and ψ,

$$\psi^m_{ij} = \sum_k \psi_{ki}\psi_{kj} \tag{5}$$

Interestingly, it can be shown that projections are not random graphs even if the links between classes and methods are chosen at random. There are two important constraints affecting the projected graphs.

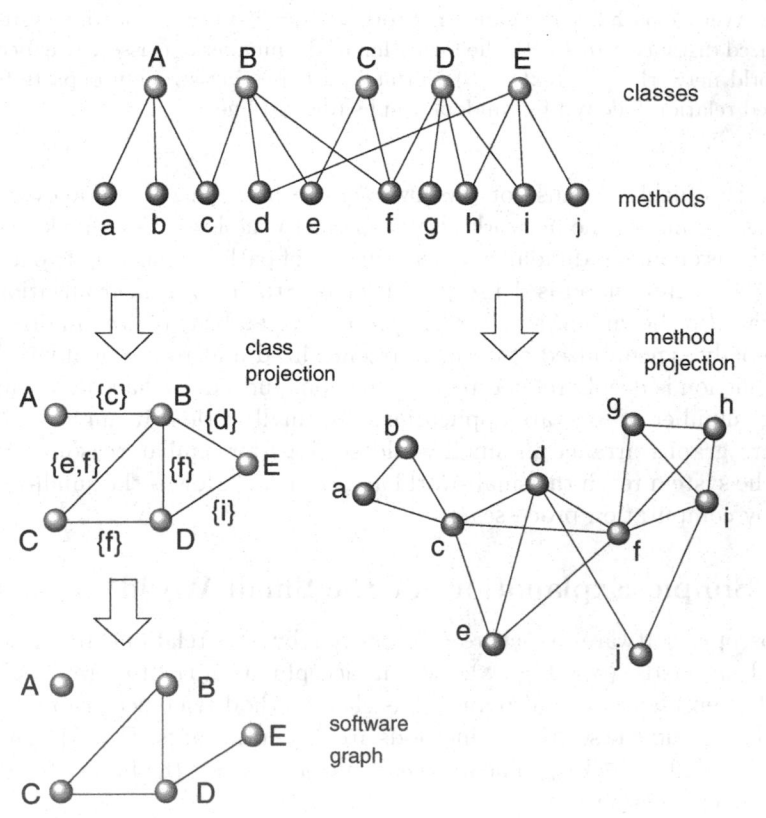

Figure 12: The reference graph relating classes and methods (top graph) can be projected in two one-mode graphs (middle graphs). The software graph is a subgraph of the class projection. Random bipartite graphs yield highly clustered one-mode graphs for free, that is, the software graph is constrained to follow certain non-random topological properties (see text).

The bipartite structure induces high clustering and low average path length. Let us consider the method projection G_m. In this projection, two methods m_i and m_j will be related if they both access the same class s_k. Projection predicts that all methods owned by a class will be related to each other, thus yielding highly clustered G_m graph (even if the bipartite graph is sparse) (figure 12 right). In terms of software engineering practices, the above means that classes tend to display strong cohesion (see figure 12 (c) and (d)). Actually, this hypothesis is in agreement with empirical studies of object-oriented software[28]. It can be shown that the projection of a random bipartite graph is always a small-world. Newman et al. [16] derived the equations for the clustering C and averaged path distance l when the underlying bipartite graph has Poisson distributed connections. The clustering coefficient $C(G_s)$ for the class projection of a random bipartite graph is:

$$C(G_s) = \frac{1}{\mu + 1} \tag{6}$$

where μ is the average number of methods referencing a class. There is a similar equation for the clustering coefficient $C(G_m)$ of the method projection:

$$C(G_m) = \frac{1}{\nu + 1} \tag{7}$$

where ν is the average number of classes referenced by a method. Note that $M\nu = N\mu$. The average distance l for the one-mode class projection of a Poisson bipartite graph is also very small:

$$l(G_s) = \frac{logN}{logz} \tag{8}$$

and

$$l(G_m) = \frac{logM}{logz} \tag{9}$$

where $z = \mu\nu$ is the expected average degree for the one-mode projection.

The bipartite network approach assumes that the class projection will explain most of the topological properties of the software graph. The software graph $\Omega_s = (W_s, E_s)$ should be a subset of the class projection $G_s = (W_s, D_s)$:

$$|E_s| = p\,|D_s|$$

where p is the fraction of edges lacking in the software graph with respect to the class projection. The differences between the Ω_s and G_s could be due to errors in the approximate reconstruction process of Ω_s described in the previous section or simply because the bipartite approach is not correct. We have performed a comparison between the topological properties of the software graph Ω_s and the same measures taken from the class projection G_s (see table 1). There is considerable agreement between the average degree, the clustering coefficient and average path length, thus suggesting that macroscopic software graph properties derive from the microscopic bipartite network (see Fig. 13).

Net	$<k>$	C	l
Ω_s	4.29	0.16	5.52
G_s	4.24	0.19	5.21
Random	13.04	0.08	2.71

Table 1: Comparison between software graph Ω_s and the projected graph G_s from the software bipartite graph G. Both Ω_s and G networks were obtained from a large software system analyzed in [18]. A Poisson bipartite graph is provided for comparison. Parameters: $N = 1071$ classes, $M = 9218$ methods, $\mu = 10.59$ and $\nu = 1.23$.

Still, we can appreciate how the real software graph deviates from the random bipartite network (having the same parameters as the real software system). The clustering coefficient is about one order of magnitude larger than the random counterpart (see table 1). Moreover, the average path length is two times larger than random. The random bipartite explanation tells us that the projection will be naturally correlated (clustered) and will be a small-world even if the underlying graph is uncorrelated (which is not the case), that is, the small-world property is achieved for free. The software graph will display a small-world architecture because of the encapsulation mechanism associated with object-oriented languages. The consequences of this observation still remain to be completely uncovered but we can safely conclude that the small-world behavior of software structures is not an additional requirement selected by human designers during software development.

5.2   Scale-Free Networks

Dijkstra was the first to claim that software engineers must be not only concerned with function but also with software structure [11]. He recognized the importance of having a well-organized software system that enables easy changes and modifications. He guessed that such an ideal software structure will be represented by a hierarchical tree. In these systems, function is provided by assembling the behavior of simple components which have clearly defined responsibilities. The coarse layout of these systems will resemble a planar graph, where nodes are software modules and edges depict existing collaborations. In this context, the aesthetics of the graph are associated with clever design. The planarity of the graph is interpreted as a signature of the clarity and clear separation of concerns achieved by its human designer.

The ordered diagram is also very homogeneous. It is easy to detect a repeating pattern in the way nodes connect to each other. In a tree, every node has only one predecessor and a slowly varying number of successors. This regular pattern can be detected by looking at the degree distribution $P(k)$ or the probability of a node having k connections. For a homogenous network like the tree or the random graph, this distribution follows the exponential distribution (equation 1.2). The main feature of this distribution is the small variance around

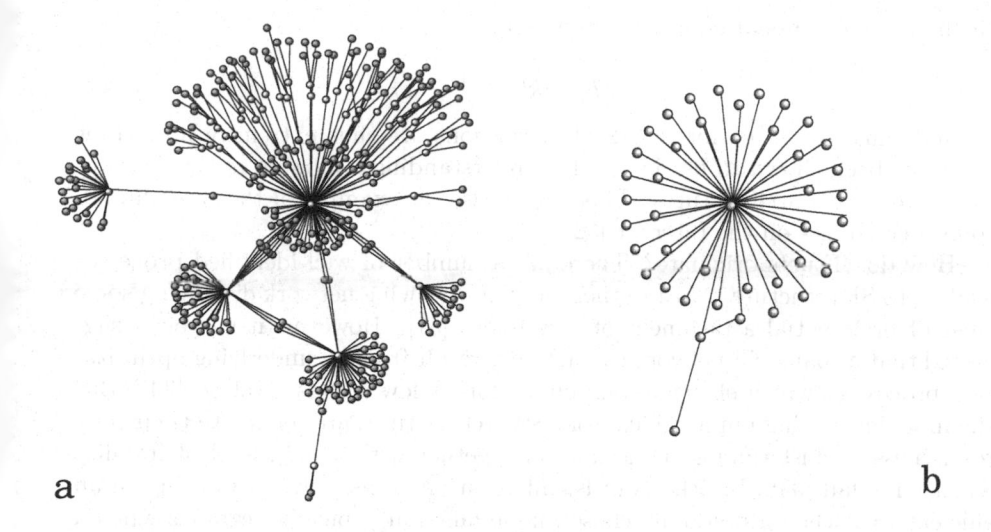

a   b

Figure 13: Looking at different granularities of the same software system. (a) A large subgraph of the class/method bipartite graph for the software application poEdit v1.2.5 (http://poedit.sourceforge.net/). Class nodes and method nodes are displayed with empty balls and gray balls, respectively. A noticeable feature is the asymmetry between average class degree ($\mu = 8.65$) and average method degree ($\nu = 2.38$). (b) The class projection for the previous bipartite graph coincides with the software graph obtained from the source code. The one-mode graph is star-shaped with a single triangle. The class at the center is the so-called hub.

the average degree. The graph has a well-defined scale.

An additional, widespread feature of many complex networks is the scale-free behavior of their degree distributions. Specifically, we have

$$P(k) = Ak^{-\gamma}\exp(k/k_c) \tag{10}$$

where A is a normalization constant, k_c is a cut-off degree and the scaling exponent γ is typically constrained to a range $\gamma \in (2,3)$. As k_c increases, the tails of the distribution become larger and the graph will display a majority of nodes having few links and a small number of nodes (the hubs) having a large number of connections [31][33]. These graphs are called 'scale free' (SF) and are found in many different contexts, from natural to technological systems [37]. Their ubiquity seems to stem from shared organizing principles [30]. SF networks are known to display some unexpected statistical features. In particular, looking at the moments of the degree distribution, i. e.

$$M_\mu = \int_1^\infty k^\mu P(k)dk \tag{11}$$

(with $\mu = 1, 2, ...$) and assuming that $P(k) \approx Ak^{-\gamma}$, it is easy to show that the average degree is well defined, leading to $< k > = (\gamma - 1)/(\gamma - 2)$, whereas the higher moments are not, since they scale as

$$M_\mu = k^{\mu-\gamma+1} \tag{12}$$

and thus $M_\mu \to \infty$ for $\mu \geq 2$. Fluctuations are thus extremely important and have been shown to be the key for understanding a number of key features exhibited by SF architectures. This is the case for example of the spreading of computer viruses on the Internet [32].

How do SF nets originate? There are a number of well-identified processes leading to SF structure. Most of them rely in a growing network displaying some rules of preferential attachment of new nodes [31]. However, it has been suggested that a sparse SF network can actually result from an underlying optimization process in which efficient communication at low cost is involved [34]. But the most interesting implications from SF architectures are related to their high robustness against random node failure, together with a high level of fragility when hubs fail [35]. In other words, information transfer keeps working in an efficient way when a randomly chosen node fails but typically degrades when a highly connected node fails. These observations have been shown to have immediate implications for reliable network architecture. Since a system's sensitivity to component failure is a fundamental problem in any area of engineering, it is important to recognize how network topology influences system performance.

Interestingly, all the software networks studied here are scale-free [18], that is, the degree distribution in software graphs scales with degree, $P(k) \sim k^{-\gamma}$. In order to properly estimate the scaling exponent γ, we have used the cumulative distribution $P_>(k)$, defined as follows:

$$P_>(k) = \sum_{k'>k} P(k') \tag{13}$$

so if $P(k) \sim k^{-\gamma}$, then we have

$$P_>(k) \sim \int P(k')dk' \sim k^{-\gamma+1} \tag{14}$$

A clear regularity is that the exponents obtained from the directed software network differ from the undirected one. Typically, we observe $\gamma \sim 2.5$, with $\gamma_{in} < \gamma$ and $\gamma_{out} > \gamma$. In other words, if we look at the number of outgoing and incoming links, the resulting degree distributions are different (see figure Fig. 14). They are more heavy tailed for the in-degree and more rapidly decaying for the out-degree distribution. Classes with high in-degree typically result from broad reuse [21]. The reasons for such an asymmetry might be rooted in the economization of development effort and related costs [18]. In principle, maximum in-degree is unbounded because linking to a class imposes no cost on the reused class. On the other hand, the complexity of the class increases with the number of used classes. The benefit of reusing some externally provided functionality is overwhelmed by the additional machinery required, which limits the maximum out-degree.

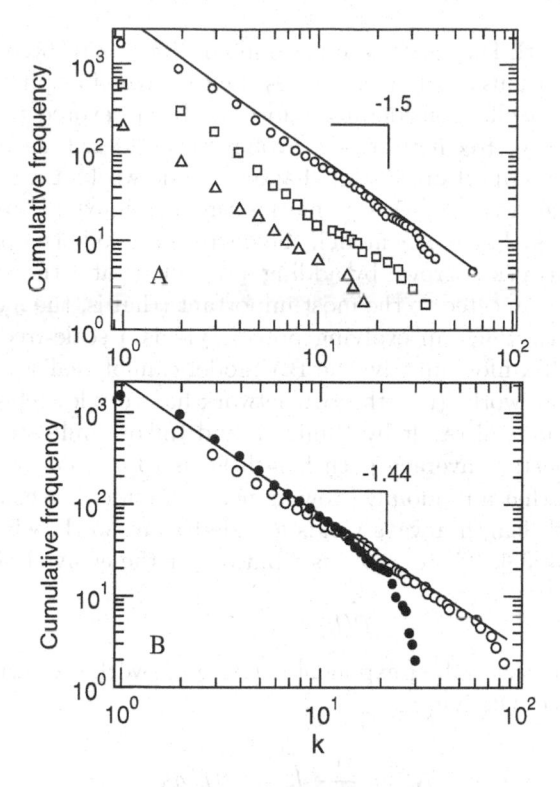

Figure 14: (a) Cumulative degree distributions for different software graphs varying in size: N=129 (triangles), N=495 (squares) and N=1488. All distributions have an exponent about -2.5 in spite of the obvious differences in size and functionality. (b) Asymmetry of in-degree (open circles) and out-degree (black circles) distributions for ProRally 2002 system. The in-degree distribution is the probability that a given component is reused by k_{in} other components. Conversely, the out-degree distribution is the probability that a component uses k_{out} other components.

Our recent studies suggest that the scale-free pattern is the fingerprint of some universal pattern of software development. The SF pattern is an emergent property of software evolution: the overall architecture is not specified within the design principles and yet it seems to be the universal result of software development. The fact that all the systems analyzed display SF structure, in spite of the obvious differences in size, functionality and other features, indicates that strong constraints are at work during software evolution.

5.3   Software Evolution yields Scale-Free Networks

The small-world behavior of software appears to be an unavoidable consequence of encapsulation, that is, simpler functions are grouped into components. However, the random bipartite model does not explain the slightly higher-than-random average path length. The unaccounted differences force us to look for alternative explanations about their causes. One source of inspiration is the evolution of software itself. Considerable effort has been devoted to studies about the generation of scale-free networks by evolving processes. It seems that contingency plays a very important role in shaping the network. Barabsi and Albert (BA) proposed the "rich-gets-richer" mechanism as a universal process yielding a scale-free network. In the BA model, the degree of a node is a proxy of its importance. The network is grown by adding a new node at a time whose m links are preferentially connected to the most important (that is, the more connected) existing nodes [31]. Such an evolving process yields a scale-free network with an exponent of -3. Unfortunately, the BA model cannot realistically reproduce other features of networks (i.e., the BA network has very low clustering).

A related theoretical result by Puniyani and Lukose indicates that growing networks with constant average path length leads to a scale-free network [38]. They have shown that a randomly growing network under the constraint of constant average path length always yields a scale-free network, with an exponent between -2 and -3 [38]. The degree distribution for the evolved SF network is:

$$P(k) \approx k^{3-\frac{\alpha}{\beta}} \tag{15}$$

where $\alpha < 1$ is the scaling exponent relating network size with the fluctuations in network connectivity:

$$N^{\alpha} = \frac{1}{\langle k \rangle} \int^{k} k^2 P(k) dk \tag{16}$$

and β is the scaling exponent linking the degree distribution cutoff k_c with size, i.e:

$$k_c \approx N^{\beta} \tag{17}$$

For the systems analysed in [18], we get $\beta = 0.62 \pm 0.09$ and $\alpha = 0.42 \pm 0.08$, which gives a predicted scaling exponent $\gamma = 2.59$. This is in very good agreement with the averaged exponent for the studied systems $< \gamma >= 2.57 \pm 0.07$.

In order to check if the SF software network is related to a pattern of constrained growth, we have analyzed the evolution of the computer game Prorally 2002, which is a large video game (about 2000 classes in its final release) developed by Ubisoft during two years. We have collected a large sample of class graph snapshots taken at different moments of Prorallys evolution. From this data set we have computed the time series of its average path length and we have observed that, after an initial and sudden jump, the evolution of average path length appears to be constant during Prorallys evolution. Fig. 15 displays the evolution of the average path length for the ProRally 2002 system [20][18].

Interestingly, the sudden jump in average path length does not correspond to a sharp increase in development activity due to some external deadlines or other external pressures. The evolution of number of nodes and links is almost

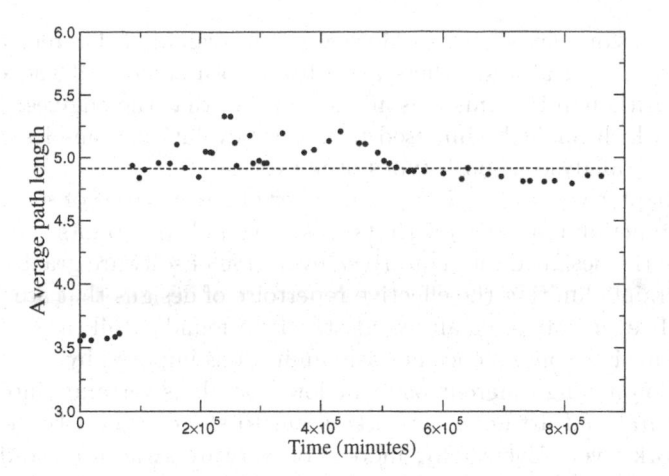

Figure 15: The evolution of average path length for the ProRally 2002 system. In spite that system size grows in a linear way, the average path length is kept constant during the project lifetime (see text).

linear (not shown) and thus, the average amount of work spent at each moment is more or less constant. Other projects we have analyzed show a similar growth pattern. We have shown that this pattern of constrained growth predicts the observed exponent of the degree distribution, thus confirming the result by Puniyani and Lukose. However, the origin of this constrained growth pattern remains to be explained. This is an apparently difficult question because building a model of software evolution appears to be a very complicated task. It is widely acknowledged that software is probably one of the most intricate human inventions. In principle, any useful model of software structure should take into account the many different mechanisms involved in computer programming. Unfortunately, some of the principles underlying computer programming are relatively unknown. For instance, it seems important to consider cognitive skills of computer programmers involved in this task. In this context, we are developing stochastic models of network growth that imitate common programming practices (like code duplication). The comparison of synthetic networks generated with these models with real software networks can shred some light into the mechanisms responsible of software growth and ultimately, the models will enable us to understand software development in a quantitative and unambiguous manner.

6 Conclusions

Understanding the origins of natural and artificial complexity requires the consideration of both their function and architecture. In order to perform a given function in an efficient way, not all topological patterns of interactions among units are allowed. Cost and proper communication are two essential requirements for most complex systems. Moreover, constraints of different nature exist: some are historical and others arise from both structural and dynamical limitations. Although it seems reasonable to think that the engineer overcomes the barriers which might be imposed to natural evolution, some lessons can be learnt from the analysis of both types of networks.

In this chapter we have reviewed a number of key features of software architecture and function in relation with its evolution and constraints. We have seen that, in spite the designed, human-driven evolution of software graphs, there are strong constraints limiting the effective repertoire of designs that are ultimately reachable. The scale-free, small world structure found in all, large-scale computer programs is a consequence of basic limitations imposed by an appropriate communication among different parts at low cost. It is certainly interesting to see that natural and artificial networks seem to share several key regularities at the network level. Eventually, models of software structure should provide insights into how internal and external forces constrains processes of artificial design.

Beyond the exact evolutionary rules shaping both types of structures, common principles might be at work. In this context, it is interesting to see that both software maps and electronic circuits [39] share some common features (such as their small world structure and their heterogeneity) with cellular networks, in spite of their differences in robustness. Such observations suggest that the plasticity implicit in cellular webs, as far as related to network topology, might inspire future developments of reliable technological systems by exploiting the common architectural patterns.

References

[1] T. Standage, *The Mechanical Turk*. Penguin Press, London (2002)

[2] J. J. Hopfield "Physics, computation, and why biology looks so different", J. Theor. Biol. 171, 53-60 (1994).

[3] S. J. Gould, *The Structure of Evolutionary Theory*, Belknap Press, Cambridge MA, (2002).

[4] S. A. Kauffman, *The Origins of Order: Self-Organization and Selection in Evolution*, Oxford University Press, 1993.

[5] F. Jacob, "Evolution as Tinkering", Science, vol. 196, 1161-1166, (1976).

[6] R. V. Solé and B. Goodwin, *Signs of Life: How Complexity Pervades Biology*

[7] B. Goodwin, *How the Leopard Changed Its Spots: Evolution of Complexity*, Charles Scribner's Sons, New York, (1994).

[8] J. von Neumann, "Theory of Self-Reproducing Automata", University of Illinois Press (edited and completed by A.W. Burke (1966).

[9] D. R. Hofstadter, *Gödel, Escher, Bach: an eternal golden braid*. Basic Books, New York (1979).

[10] R. V. Solé, R. Ferrer-Cancho, J. M. Montoya and S. Valverde, "Selection, Tinkering and Emergence in Complex Networks", Complexity, vol. 8(1), 20-33, (2002).

[11] E. W. Dijkstra, "The Structure of the 'T.H.E.' multiprogramming system", Comm. of the ACM, vol. 11, no. 5, pp. 453-457, (1968).

[12] D. L. Parnas, "On the criteria to be used in decomposing systems into modules", Comm. of the ACM, vol. 15, 1053-1058, (1972).

[13] A. V. Aho, R. Sethi and J. D. Ullman, *Compilers: Principles, Techniques and Tools*, Addison-Wesley, Reading, Mass (1988)

[14] J. E. Hopcroft, R. Motwani and J. D. Ullman, *Introduction to Automata Theory, Languages, and Computation*, Addison-Wesley, Boston, 2nd Ed. (2000)

[15] J. Backus, "Can programming be liberated from the von Neumann style?: a functional style and its algebra of programs", Comm. of the ACM, vol. 21, 8, 613-641, (1978)

[16] M.E.J Newman, S.H. Strogatz and D. J. Watts, "Random Graphs with arbitrary degree distributions and their applications", Santa Fe Institute working paper, SFI/00-07-042, (2000).

[17] D.J. Watts and S.H. Strogatz, "Collective Dynamics of Small-World Networks", Nature, vol. 393, no. 440, (1998).

[18] S. Valverde and R. V. Solé, "Hierarchical Small-Worlds in Software Architecture", Santa Fe Institute, working paper SFI/03-07-044, (2003).

[19] S. Strogatz, *Sync: The Emerging Science of Spontaneous Order*, Hyperion Books, (2003).

[20] S. Valverde, R. Ferrer-Cancho and R. V. Solé, "Scale-Free Networks from Optimal Design", Europhysics Letters, 60, pp. 512-517, (2002).

[21] C. R. Myers, "Software Systems as Complex Networks: the Emergent Structure of Software Collaboration Graphs", Phys. Rev. E, vol. 68, 046116, (2003).

[22] A similar lexical reconstruction method is also performed by the automatic documentation system Doxygen by Dimitri van Heesch (http://www.doxygen.org).

[23] T. Ball and J. R. Larus, "Programs follow paths", Microsoft Research, Redmond, WA, Tech. Rep. MSR-TR-99-01, January, (1999).

[24] A. M. Turing, "On Computable Numbers", Proc. of the London Math. Soc., 2-42, pp. 230-265, (1936).

[25] J. von Neumann, *The Computer and the Brain*, New Haven, Yale University Press (1958).

[26] T. McCabe, "A Complexity Measure", IEEE Trans. on Soft. Eng, SE-2, 4, pp. 308-320, (1976)

[27] J. Rumbaugh et al., *Object-Oriented Modeling and Design*, Prentice-Hall, (1991).

[28] H-S. Chae, Y-S. Kwon and D-H. Bae, "A Cohesion measure for object-oriented classes", Soft. Pract. and Exp., vol. 30, 12, pp. 1405-1431, (2000).

[29] E. Gamma, R. Helm, R. Johnson and J. Vlissides, *Design Patterns*, Reading, MA, Addison-Wesley, (1994).

[30] A.-L. Barabási and E. Bonabeau, "Scale Free Networks", Sci. Am., pp. 60-69, May (2003).

[31] A.-L. Barabási and R. Albert, "Emergence of Scaling in Random Networks", Science, vol. 286, pp. 509-512, (1999)

[32] R. Pastor-Satorras and A. Vespignani, "Epidemic Spreading in Scale-Free Networks", Phys. Rev. Letters, vol. 86, 3200-3, (2001).

[33] S.N. Dorogovtsev and J. F.F. Mendes, *Evolution of Networks: From Biological Nets to the Internet and the WWW*, Oxford, New York, (2003).

[34] R. Ferrer-Cancho and R. V. Solé, "Optimization in Complex Networks", *Statistical Physics in Complex Networks*, Lecture Notes in Physics, Springer, Berlin, (2003).

[35] R. Albert, H. Jeong and A.-L. Barabási, "Error and Attack Tolerance of Complex Networks", Nature, 406, pp. 378-382, (2000)

[36] M. Page-Jones, *Fundamentals of Object-Oriented Design in UML*, Addison-Wesley, New York (1999).

[37] D. Braha and Y. Bar-Yam, "Topology of Large-Scale Engineering Problem-Solving Networks", Phys. Rev. E, vol. 69, 016113, (2004).

[38] A. R. Puniyani and R. M. Lukose, "Growing Random Networks under Constraints", cond-mat/0107391, (2001).

[39] R. Ferrer-Cancho, C. Janssen and R. V. Solé, Topology of technology graphs: small world patterns in electronic circuits. Phys. Rev. E. 63, 32767 (2001).

Chapter 5

Creation of desirable complexity: strategies for designing self-organized systems

Carl Anderson
School of Industrial and Systems Engineering
Georgia Institute of Technology, Atlanta
Currently: Qbit, LLC, 6905 Rockledge Drive, 3rd Floor
Bethesda, MD 20817
canderson@qbit.com

1. Introduction

Thirty years ago, Barry Malzberg described what may have been the first vision of a self-organized[1] application: "surely ant and man could coexist peacefully…we might even be able to voyage to the stars together, the ants developing a communications network that would implement our vast technological resources" [53, p. 118]. Although we have indeed developed ant-based routing for telecommunications [17, 32, 63], more generally progress has been disappointing. To date, self-organization research has focused mostly on dissecting and understanding these complex systems, elucidating the links between the micro and macro levels, and identifying their general characteristics and ingredients. We have lingered in the proof-of-concept stage, demonstrating that simple agents with local interactions and feedbacks can lead to complex, adaptive group-level properties. However, despite various authors waxing lyrical about the enormous potential for problem-solving self-organized systems [15, 33, 46, 47], there are surprisingly few practical applications in use.

---

[1] Defined as "a process in which pattern at the global level of a system emerges solely from numerous interactions among the lower-level components of the system. Moreover, the rules specifying interactions among the system's components are executed using only local information, without reference to the global pattern" [20, p. 8]. See [1] for a review of other definitions.

The problem is, of course, that we are amateurs in the process of *designing* self-organized systems, or as Karl Sims [67, p. 22] puts it, in "the creation of desirable complexity." In many ways, this is not too surprising. After all, the cards seem stacked against us. We know that small changes in individual level rules can lead to enormous changes in global behavior [2, 13, 19, 20, 73]. We also know that positive feedback may amplify small perturbations in such a way that a phase transition occurs, shifting our whole system to a new, very different, and perhaps undesirable regime [*ibid.* 11, 76]. And, we know that the moderate number of agents in these systems implies unimaginably large state spaces, ones in which the region of interest, the ones possessing our particular desired system level properties, are miniscule. Finding needles in haystacks might actually be easier. The above features are, however, part of the draw of self-organized systems: their potential as a powerful, flexible, robust, and decentralized problem solving architecture.

How then does one design a self-organized system with a given set of system-level properties? Which possible design strategies could be used and what particular features and factors of the problem might favor one approach over another? In this chapter, I consider these questions at a broad level, discussing some of the general pros and cons of approaches such as bottom-up simulation, top-down engineering, analogy and mimicry, and interactive evolution. The aim is to extract some of the key criteria, decisions, and constraints that *might* help pinpoint an initial useful approach to tackling the design problem at hand. Later, I develop a key, similar to those used by biologists to identify species, with a few questions that may aid this process. I also offer a few thoughts on coping with pathological system behavior. I should stress that this is just a preliminary survey and a cursory overview. It does not offer any system-specific guidelines or magical recipes. Instead, its objective is more as a starting point for discussion about the general process of designing self-organized systems and, hopefully, might initiate some useful discussion for those groups tackling the design of specific self-organized systems.

2. Analogy and mimicry

I start with the simplest approach: copying. Rather than designing a new self-organized system from scratch, one might be lucky and know of, or find, an existing, analogous system that can act as a model. This strategy is the one most likely to succeed: if the original model works, then it is only a matter of implementing it correctly.

Philosophically, this strategy should be the least common of all the major design strategies considered in this chapter—what are the chances of knowing a self-organized system that exactly matches ones' needs? Practically, however, our ability to dream up new applications, and to get those projects funded, will be limited and shaped by current knowledge and technology; in other words, mimicry is a safe bet and may occur relatively frequently.

The use of such a model may take one of several major forms:

1) *Archetypal*: we may find a completely analogous model that is to be replicated. As there are always going to be some differences, even if slight, such situations are probably rare. This first category, therefore, probably falls mostly under *modeling* the original system, perhaps to understand it better.

2) *Prototypal*: more realistically, the model may not be exactly as required but can be tweaked to produce the desired result.

3) *Inspirational*: the model serves as a starting point, a basic source of inspiration, which is then modified in whatever way necessary to produce the desired result or to enhance system performance. This pragmatic, goal-oriented approach may, however, result in a system that has little resemblance to the original system. Further, the new system will require testing and validation, most likely using bottom-up simulation (Section 1.4).

In the first two forms, by implementing the same agent level behavior, including agent-agent and agent-environment interactions, one hopes to obtain the same system-level properties. Implicit are two important assumptions that must be met: first, that one has sufficiently detailed knowledge of the original system to demonstrate that it does indeed possess the system level properties required in the designed system, that is, to prove that one is copying the *correct thing*; second, that all the "necessary and sufficient" [20] agent-level proximate mechanisms are known in detail; that is, that one has all the necessary *correct information*. In most cases, this is a tall order, but there are many self-organized systems that have served as inspiration, metaphors and models for designed self-organized systems.

In many cases, these models come from the natural world, but why is this? One possible reason that biological metaphor is perhaps the most pervasive in designed multi-agent systems is inherent intra-agent behavioral flexibility. It is true that physical and chemical self-organized systems such as stone stripes, Bénard convection cells, and the swirls and spirals of the Beloussov-Zhabotinsky reaction [e.g., 8, 48, 70, 75] have been well studied, are often easy to manipulate in the laboratory, and give us enormous insight about the general mechanisms, properties and characteristics of self-organization. However, they could be regarded as rather static systems in that the agents have (relatively) fixed individual level behavior: physical laws of interaction and bonding with little room for agent evolution. Adaptive agents, and especially systems of adaptive agents, possess a far richer suite of complex group-level behavior. A second element in biological systems', and also social sciences', favor is that many of these systems have existed for long periods—in the case of ant colonies, around 100 million years—and have likely been subject to various selection pressures, thereby resulting in diverse, efficient and sometimes surprising and counterintuitive behavioral algorithms and mechanisms.

Collective robotics is an excellent example of a field that has drawn heavily from biology, especially sociobiology. Many roboticists recognize the potential benefits of adopting the behavioral algorithms of social insects, especially ants. Although individually dumb, ants form complex adaptive systems (colonies) that are decentralized, robust, flexible, and self-organized [12, 17, 64]. Moreover, the majority of the complexity resides not in the individual ants themselves but in the *network* of interactions [46, 47]. This has a powerful implication: by networking them, roboticists can get away with cheap, simple and therefore easily designed robots.

Although one is exploiting algorithms known to work in other systems, use of analogy and mimicry does not guarantee success. Let us suppose a researcher has been inspired by the collective decision making of ants and decides to mimic them in his system. He has complete knowledge of the ants' rules, feedbacks and interactions and copies these *exactly* in his set of agents. This still does not guarantee that his system will have the same, desired properties. Why might this be so?

First, differences in environment may shape the system level behavior differently. This is an aspect that I feel is underemphasized in the self-organization literature. For example, ground slope determines the orientation of stone or vegetation stripes [50 and references therein, 75], raid patterns may differ among army ant species solely due to differing food dispersion [29, 34], and substrate absorption properties affect the collective decision making abilities of ants in double bridge experiments [31] (See also Figure 1).

Second, agents may be relatively unreliable in the new system. For instance, many robotics studies that have tried to mimic collective behavior in ants [e.g., 22, 30, 42, 74]. While, to the uninitiated, this may not seem particularly innovative—after all, biologists and mathematicians can show exactly how and why these systems work—it is necessary, and often a significant challenge, to demonstrate that these potentially useful algorithms work in real instantiated robots. A robot system that works perfectly in simulation may fail spectacularly in reality because of sensor noise and component failure.

Third, and finally, bifurcations may exist. As mentioned above, many self-organized systems exhibit phase transitions in which the emergent properties only arise above some critical number or density of agents [11, 17, 20, 37, 70]. Most collective robotics studies involve 20 agents or less, very rarely 100 [36]. They are not harnessing the full potential of self-organization; the networks are far too small, an issue taken up in more detail later (Section 1.7).

3. Top-down design

Writers on self-organization tend to emphasize *emergence* and the need for a *synthetic* reductionist methodology to explain "high level behavior from low level causes" [16, p. 304]. Except in the very simplest cases, complete knowledge of the individual agents, their environment, and the coupling among them [60] does not provide sufficient insight or predictive ability as to the system-level behavior [e.g., 13, 23]. We must build these systems from the bottom-up (Section 1.4). It might, therefore, be a little surprising to some readers that I include a section on top-down design.

The classic self-organized examples (e.g., those commonly cited in [8, 17, 20, 70]) are seemingly devoid of any intermediate hierarchical levels between the lower-level agents and the upper-level system [*contra* 23]: agents contribute directly to the system level behavior, which, in turn, directly affects the agents (Figure 1). For example, a turning fish contributes to the school-level behavior, and this schooling, e.g., splitting

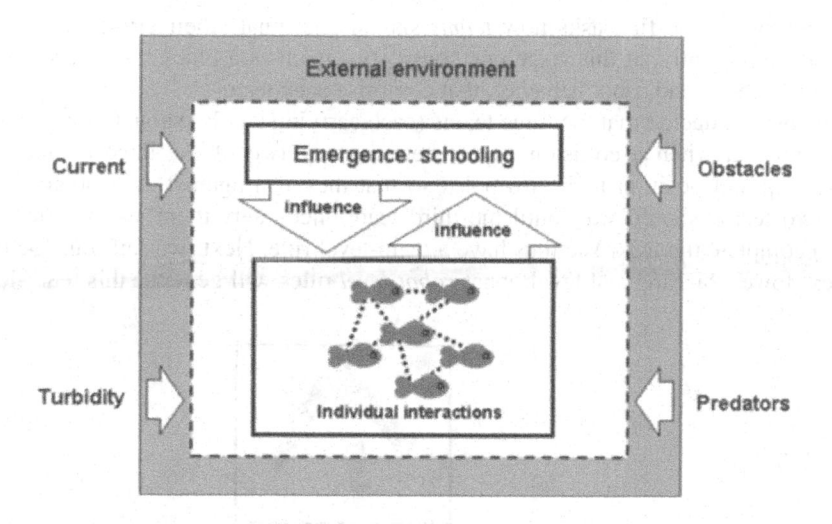

Figure 1. In a fish school, there are no intermediate hierarchical levels between the individual level, that is, the fish, and the system level, the self-organized school. Fish directly influence the school and the school directly affects the fish. Note that the system may be shaped by the environment; that is, from external factors such as water current and turbidity, predators, and obstacles. (Based upon [51] and a long lost webpage.)

to avoid an obstacle, directly affects the turning motion of individual fish. It is difficult, therefore, to imagine how one would design this from anything other than a bottom-up approach—one hypothesizes the individual level rules, implements them in an agent-based simulation model, studies the system-level behavior and if the outcome does not match that expected then modifies the hypothesis, reruns the model, and so on. It is an iterative process between the macro and micro levels [see 20].

There may be some situations, however, in which one or more intermediate hierarchical levels exist between the agent-level and the system-level. Hierarchy is often associated with top-down command and control, for example, directives from a company's CEO constraining and shaping the individual workers' behavior and the company's culture. However, this need not be the case; multiple hierarchical levels may exist in a self-organized system [e.g., 4]. Given such a system, one could start at the system level and question "what must happen at the next-lowest hierarchical level to produce this behavior," then design this level of behavior, repeat the question at the next lowest level, and so on.

This approach is perhaps best explained with a simple example. Suppose we wish to design a system in which a group of self-organized robots must construct a tripod structure. Without leaders, and with only local interactions and decisions, three teams of robots must coordinate their efforts to rest three poles against each other [cf. 5 and 41]. (Implicit is the assumption that a single robot cannot lift and move one of the poles by itself.) Thus, there are three hierarchical levels (Figure 2). Adopting this top down approach, one first asks how *teams* should coordinate their efforts to achieve the goal. Importantly, at this stage one considers a team as a black box, ignoring the individual robots, and robot behavior, that comprise each team.

Suppose we decide that the three teams (each carrying a pole) should move around at random and when a collision occurs between any two of the three teams, they should stop and position their two poles so that they rest against each other. Then, these two teams should wait until the third team encounters them and positions its pole to complete the task. We thus have a team-level rule. Next, we shift our focus to the next lowest hierarchical level: what *robot-level* rules will generate this team-level

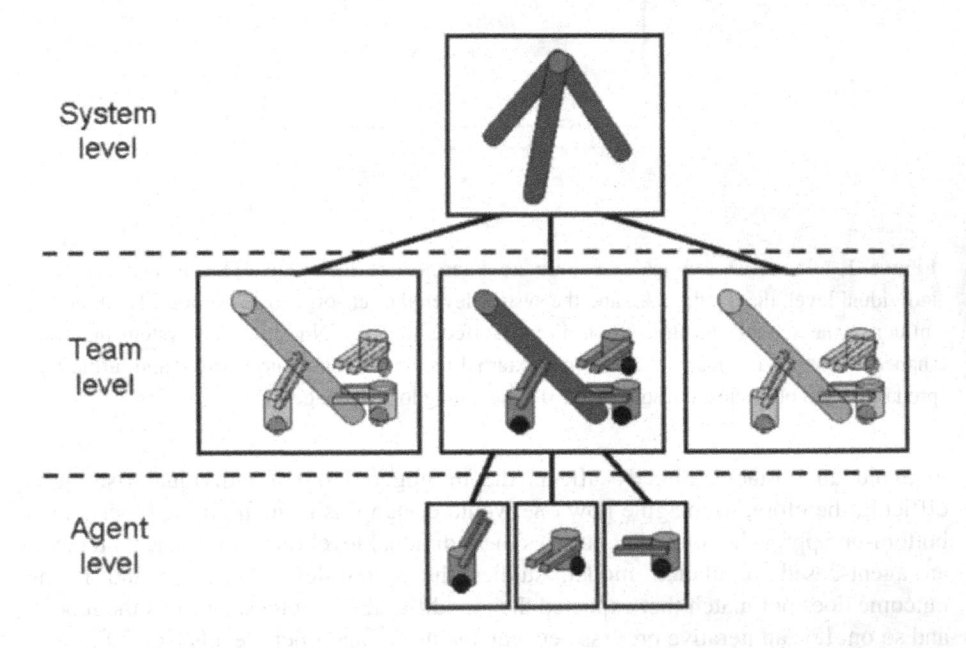

Figure 2. A top-down approach for designing a system of autonomous robots that must collaborate to construct a tripod. First, one starts at the system level, the tripod, and considers what is necessary at the team level to achieve this goal. Next, one focuses on a team and what is necessary at the individual level to achieve this subgoal.

behavior? Unfortunately, I don't have the complete answer, although it likely will involve the process of team formation (which may be relatively easy to program; individuals move around randomly until they encounter one of the three poles), collective movement (rather more tricky and depends upon how the pole is carried, e.g. dragged or carried aloft), collision detection (easy), and pole positioning (undoubtedly the hardest subtask).

The important point here is that self-organized systems may involve hierarchy and modularity and that we might prefer to start the process from the top down rather than from the bottom up, each stage moving to smaller and smaller, and possibly more well-defined, sub-problems. In other words, hierarchy and modularity can result in problem reduction [65]. While a bottom-up may appear to be the general, primary approach for self-organized systems, there may be cases in which a top-down approach may be favored. Further, I tentatively speculate that a hierarchical structure may be a factor that favors top-down. In addition, a combined top-down and bottom-up approach might be even better in some cases.

4. Bottom-up design

4.1. Why bottom up?

In the majority of cases, the most natural and perhaps only approach to design, test, and analyze self-organized systems is to model them from the bottom up. There are two main reasons for this.

First and foremost, in most self-organized systems the system level properties are emergent, a result of a complex set of interdependent (often non-linear) interactions, feedbacks, and perturbations among a set of agents that may differ in space, time and state. Unless simplifying assumptions can be made without detriment, too high a level of abstraction (meaning other modeling techniques such as mean field and differential equations) will miss these crucial aspects. Bottom up modeling, in most cases *agent-based* models, have the greatest fidelity to the process being the modeled or emulated. It is the only way to capture the crucial low level processes and interactions that produce the actual system level pattern [see 13, 60].

Second, modeling from the top down or with some (unconstrained) evolutionary approach may yield wonderfully simple algorithms that work incredibly well, but which may somehow be incompatible with the set of agents one plans to implement them. In self-organized systems many different rules may lead to the same system level behavior (therefore making the design task complex *sensu* [21]). This may be a blessing or a curse. While some systems may have broad "basins of attraction" [e.g., 77], potentially providing multiple design solutions for one's system [e.g., 35], a top-down approach may be led down just a single path, one that simply cannot work with the planned agents. For instance, the required rules may involve unfeasible sensory range or cognitive abilities. It may be easier therefore to start with suitably constrained agents and build the complexity upwards (possibly using evolutionary computation to parameterize the system).

4.2. Agent design strategies

Much has been written in a *descriptive* manner about agent qualities and characteristics in self-organized systems—the general characteristics that these agents and systems possess. However, there is an embarrassingly small literature that is *prescriptive*—"how to" guides or rules. "*Designing Self-Organized Systems FOR DUMMIES*" has yet to appear in bookshops. Building upon Kelly's [46, 47] important work with his "Nine rules of God," the best prescriptive review to date is probably that of Parunak [60].

Focusing on agent software development, although with more widely applicable insights, he derives "a set of general principles that artificial multi-agent systems can use to support overall system behavior significantly more complex than the behavior or the individual agents" [60, p. 69]; in other words, a set of general design principles for bottom up generation of self-organized systems. I shall simply state these principles, without additional comment, and encourage the interested reader to examine Parunak's more detailed and thorough treatment:

1. Agents should correspond to things in the problem domain rather than to abstract functions;

2. Agents should be small in mass (a small faction of the total system), time (able to forget), and scope (avoiding global knowledge and action);

3. The agent community should be decentralized, without a single point of control or failure;

4. Agents should be neither homogeneous nor incompatible, but diverse;

5. Agent communities should include a dissipative mechanism to whose flow they cannot orient themselves, thus leaking entropy away from the macro level at which they do useful work;

6. Agents should have ways of caching and sharing what they learn about their environment, whether at the level of the individual, the generational chain, or the overall community organization;

7. Agents should plan and execute concurrently rather than sequentially.

These guiding principles are both sensible and justified. Unfortunately, however, at this preliminary stage we are probably akin to a student with a poor grasp of statistics: armed with a partial list of tests, he can find a recipe detailing how to apply a particular test but doesn't have a deep understanding of precisely when and why to apply it. Like the student, we will only increase our depth of knowledge of system design by both hands on practice and a concerted effort to grasp the fundamentals (see Section 1.7).

Our relative lack of experience and understanding, and also the huge state spaces involved, contribute to the prevalence of evolutionary computation—including evolutionary strategies, genetic programming and genetic algorithms—to parameterize the systems. That is, to evolve the parameters of rules in a bottom-up simulation or even to randomly generate completely new rules in the hope that they will generate the desired system level behavior. While these techniques are relatively common and require little additional comment, an aspect that receives far less attention but, importantly, is a viable design strategy for self-organized systems *in some situations* is interactive evolution.

5. Interactive evolution

Pioneered by Dawkins [26], interactive evolution is a method designed to mimic a natural evolutionary process but in which the selective agent is the user (reviewed by [9,71]). He or she is presented with a number of variants, "mutant offspring," of the object being evolved and chooses one based upon some criterion, often simply esthetic value. That chosen object becomes the new parent from which a set of mutated offspring are generated (usually asexually) and displayed in the panel for the user to choose again. (In some cases, two or more offspring are chosen and bred with recombination, that is, sexually [66], or they are simply averaged [26].) In this manner, the user can select for a particular desired trait or simply explore parameter space in an interactive manner.

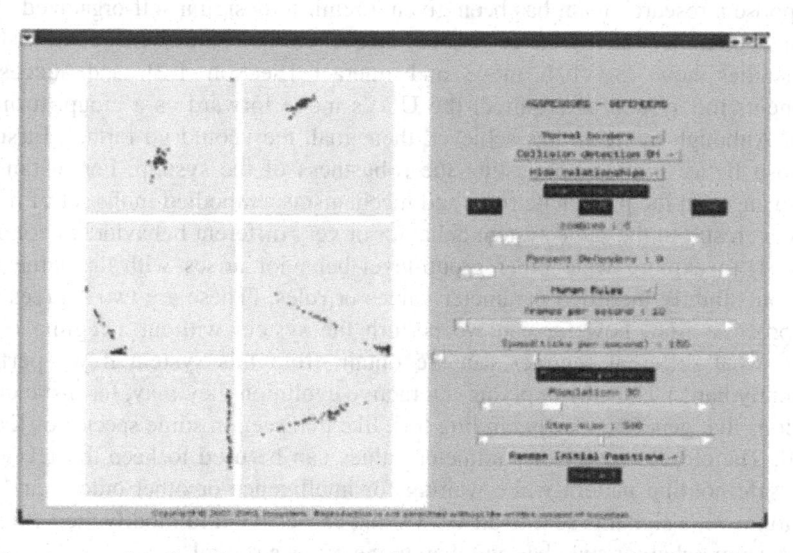

Figure 3: an example of an interactive evolution interface. To the left are six panels, each containing a separate self-organized system, in this case a set of agents playing the aggressor-defender game (from [18, 35]; reprinted with permission). The user selects a panel containing an "interesting" group-level behavior from which are bred six new "offspring" systems.

Dawkins [26, 27] developed his system to evolve simple stick-like creatures called biomorphs, but this methodology can also be used to select and evolve whole self-organized systems [2] [18, Figure 1.3; 35]. That is, the user is presented with a panel of variant *systems*, each with slightly different individual level rules, number of agents, threshold values etc., and he or she selects the one whose group-level behavior most

---

[2] Notice that selection, natural or otherwise, acts upon the whole system, which just happens to be self-organized; self-organization is simply a dynamic process that gives rise to the system's form, that is, its phenotype, based upon its individual level rules, that is, its genotype, and is not an alternative or rival to Darwinism, as some authors have suggested (e.g., [38, 45]; see also [20, pp. 88–89]).

closely matches the desired properties; this, in turn, is bred from to form a new panel of variants.

There are a number of features that determine the utility of this approach for designing self-organized systems. Interactive evolution (IE) is often associated with esthetic selection [e.g., 67], meaning that, in most cases, the user is not entirely sure what it is they are after—they select what looks good or feels right. (As such, IE has been used to develop novel, never-before-imagined art, including pictures [e.g., 66, 68] and sounds [28, 57], as well as providing an advanced computer aided design tool for items such as bridges, dams, and cars [14, 69, 71].) In this chapter, however, I am assuming that the user does know what it is they are after and that they do have a set of desired system-level properties. Nevertheless, IE can play a useful role in exploring the set of possible systems or system dynamics.

Suppose a research team has been given a remit to design a self-organized system of unmanned underwater vehicles (UUVs) that school in a similar manner to fish. The team studies how real fish move and interact [Section 1.2], and successfully implements the system as required: the UUVs move forward as a group, turning in unison. Although the team has achieved their goal, they could go further. First, they could use IE as *one* way of testing the robustness of the system. For instance, do slight changes in the proximate rules and mechanisms, embodied in the set of mutants on screen, result in the same system behavior or very different behavior? Second, they may wish to explore what other group-level behavior arises with the same set of agents but slightly modified parameter values or rules. (These are two degrees of the same process: first, how far can we perturb the system without a regime change; second, what different regimes can we obtain from this system if we perturb it sufficiently hard.) Thus, by applying interactive evolution they may, for instance, find conditions that generate slowly circling tori, like that seen in some species of fish [24, 58, 59]. Therefore, one set of parameter values can be used to keep the UUVs in a stable, safe holding pattern while waiting for intelligence or other orders, and when ready to move, a signal is sent to all UUVs that causes them to modify their rules, and so the system behavior switches and they move off as a school.

Generally, IE is most useful when it is not possible to write an explicit, closed form analytic expression for the fitness or objective function (as in esthetic selection, above). That is, evolutionary computation, such as genetic algorithms, may work well to search parameter space but one has to specify *precisely* what is to be optimized (the fitness function). In self-organized systems this may not always be possible. First, the system specifications may be qualitative or vague—e.g., "develop a self-organized system of robots that acts like a herd of wildebeest." Second, one may develop a set of relevant metrics but it is unclear quite how to weight those into a final fitness function. For example, rotating tori of fish are characterized by a low degree of polarization and high group angular momentum [24], but what precise relative values should one incorporate into a fitness function? Third, it may be perfectly possible to develop the metrics and define a fitness function but would be much easier, cheaper, and faster to evaluate the system by eye ([26], see [52] for an example). Fourth and finally, you may wish non-experts to design the required self-organized system; that is, experts set up the IE application but others who do not need, wish, or have the ability to understand the underlying mechanisms actually use it [e.g., 54].

Interactive evolution has a number of important constraints and limitations that affect its usefulness in designing self-organized systems:

1.   System behavior has to be *realized quickly* [28, 52, 54, 66, 69]. The user is likely to become disinterested if they must wait hours between rounds of generation and selection—*interactive* evolution implies real time generation [69]. This may be a particular problem with self-organization as each of the multiple systems on screen may contain a large number of agents.

2.   System behavior must be *easily evaluated*. The production of a self-organized structure, e.g. a set of self-assembling robots that form a bridge, may be simple to evaluate. Similarly, certain types of behavior, e.g. UUVs that form a torus, may be easily evaluated. However, for many other systems, this will not be the case. For instance, the key aspect may be the ability of the group of agents to form a complex adaptive system, able to tolerate environmental uncertainty and challenges, agent failure, and so on. Imagine trying to evaluate a new ant-based Internet routing protocol with IE. It would be impossible. In addition, IE is most suited to evaluation of all the screen's sub-panels simultaneously. For visuals this may not be too difficult but for other forms, such as sounds, this could be extremely tricky ([57], but see [28]).

3.   *Low population size*. There should only be a small number of mutant offspring systems on screen. Whereas evolutionary computation, such as genetic algorithms, may involve large population sizes (100 to 1000 or more), IE will involve far fewer offspring to be evaluated each iteration (e.g., 20–40: [66]; 9: [28, 57]; 16: [54]; 6: Figure 1.3). This is a product of limited computational power, cognitive ability to compare multiple offspring simultaneously, and screen size, a particular problem for self-organized systems if one needs to be able to distinguish each of the many agents in *each* system. Thus, exploration of the state space will be very limited each iteration and slow overall.

4.   *Fast convergence*. Ideally, a satisfactory solution should be arrived at in relatively few generations as users can easily fatigue [e.g., 39, 72]. While convergence rate may be difficult to control in many situations, some aspects may help; for instance, lowering the numbers of levels of a discretized parameter, increasing the population size (point 3 above), and use of visualized interactive evolutionary computation [39], a form of interactive evolution where a 2D representation of n-dimensional search space, as sampled from previous solutions, is presented alongside the panels of offspring.

In summary, interactive evolution will involve a slow stroll through state space. Not only is user selection a relatively slow process (compared to automatic evaluation), but each iteration will also likely have a high computational load: multiple agents in multiple systems. It is most useful when the systems can be computed and evaluated quickly and easily and/or where the development of useful metrics and fitness functions may be slow or difficult, or at least quicker and easier by human eye.

6. Avoiding pathological behavior

One of the problems with self-organizing systems is that they can be extremely myopic and the system may develop pathological behavior. Each agent's view of the system-level is so limited that no single individual, or other functional unit such as a team, has a global view. This can result in highly undesirable, inefficient, and even ludicrous global patterns of behavior. The classic illustrative example is that of circular milling in army ants (Figure 4). Ants of many army ant species are blind and entirely reliant on chemical trails to direct their movement. If a group of ants becomes isolated from the rest of the colony, each individual has no choice other than to follow the trail of the ant in front. However, in turn, the ant they are following has no choice other than to follow the ant in front of it, and so on. If the trail crosses itself, a circular trail or "mill" may develop in which the ants circle around and around until they drop dead from exhaustion. Beebe [10], for instance, reports a circular trail 1200 feet (365 m) in circumference with a circuit time of 2½ hours. This is obviously an extreme illustration but Anderson and Bartholdi [3] report similar pathological behavior in industry and human societies.

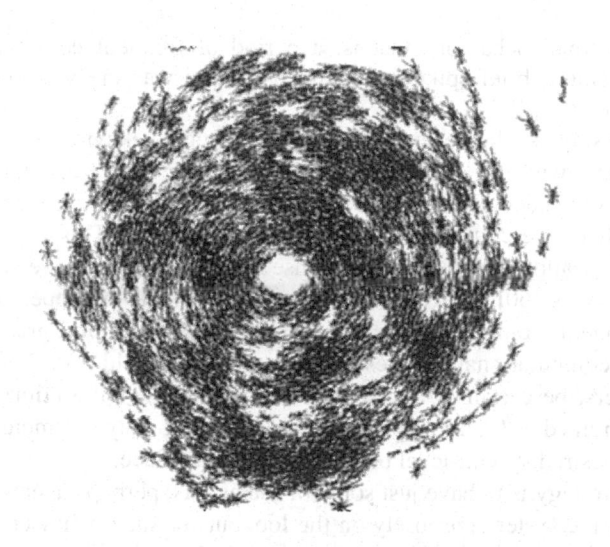

Figure 4. A circular mill in army ants. Being blind, each ant has no choice other than to follow the trail from the ant in front. If a group of ants is somehow isolated from the rest of the colony a circular trail, or "mill" may form in which the ants circle around and around (counterclockwise in this case) until starvation and death, or they accidentally break out of this vicious cycle. [From: ARMY ANTS by T. C. Schneirla, edited by H. R. Topoff © 1971 by W. H. Freeman and Company. Reprinted with permission.]

In the ants, such milling behavior is rare, caused by a significant colony perturbation, for example rain washing away portion of the trail [10] or an experimenter deliberately isolating a group of ants [62]. Each ant's "world view" is simply too limited to see the maladaptive circular trail. However, for us, when standing over the mill with our bird's eye view of the system, it is easy to identify such behavior. What potential strategies, therefore, are available 1) to design the system to recognize such pathological behavior *itself*, and 2) to deal with the behavior and move the system to a more desirable configuration?

First, it is perhaps worth recognizing that such undesirable behavior may only be temporary, perhaps a result of *drift* of the system through state space and not the result of a major *shift* or perturbation. Thus, in time, the system may drift back to a more efficient state. Consequently, if the sub-optimal behavior is not mission-critical, dangerous, or too costly, it may, in fact, not be a major issue or concern. An example is that of Morley and Ekberg's ([55], see also [56, 49]) decentralized scheme for allocating trucks rolling off an assembly line to booths for painting. The goal is to allocate trucks, which each have a customer-desired color, to booths currently painting that color, thus saving time and paint costs of the booths having to switch colors between trucks. They found that their new, self-organized scheme (where booths bid for trucks) was sometimes slower and less efficient that the previous, more traditional scheduling scheme (where trucks are allocated in advance to booths). While these inefficient periods sound unfavorable, the new scheme generated a 10 % reduction in paint use and $3 million dollars savings per annum. The key feature is that the sub-optimal behavior—that is, a period of frequent color switching—was *rare*, only causing a brief spike in wasted paint and was fairly insignificant in the grander scale of things.

Let us now suppose, however, that such pathological behavior really is significant. How can we deal with it? There are several potential levels of detection and control. First, we can engender agents with the ability to assess or recognize the global behavior. In short, we could make the agents less myopic. This comes at a cost of course. Agents would need to be able to sense a larger portion of the system, which is contra to Parunak's [60] advice of keeping agents "small in scope." It may require greater inter-agent communication, which would also imply greater intra-agent cognitive and computational abilities. And, even if we could do all of this, how can we, as designers, be certain to foresee all of the possible sub-optimal behavior and know all the remedies? This approach seems as if it may be more difficult than designing the desired system-level behavior in the first place.

A second strategy is to have just some of the agents, perhaps agents of a particular class, to patrol the system purposely on the lookout for such behavior—Parunak [60] terms these "watchdog agents." This may be a little easier to achieve. Only a few "well-endowed" agents would be necessary (rather than all agents, above), and their motion through the system may be sufficient to gain a reasonable global view without a significant increase in inter-agent communication. One example is the use of parent agents in one of the first agent-based telecommunication routing algorithms [7]. In this system, a set of load management agents route telephone calls across a network by updating routing tables to a source destination. A second, more global level of control is provided by a smaller set of parent agents. These agents roam the network monitoring congestion and manage the population level of load agents by launching new load agents in congested areas [7]. This is obviously a very simple example, but it is not too hard to imagine bestowing the ability in parent agents to detect circular routes, a truly pathological behavior in routing.

Paranuk [60], without going into detail, advises that "It is best not rely on watchdogs at all, but if they are used, they should sense conditions and raise signals but not plan or take action." In general, this is true. Having a few key individuals, such as these, makes the system more centralized and therefore less robust. However, in practice, as discussed above, a truly decentralized detection and error correction scheme may be very difficult to design.

A final strategy is the use of an observer with a truly global perspective. Implicit is the assumption that this observer has far greater cognitive and computational abilities than the individual agents within the system and so can detect or foresee problematic behavior. Humans are an obvious example. Hubanks [43, p. 54] describes a self-organized military logistics program and hints at a human-level monitoring: "if the solution is not palatable, we would input policy directives and the system would recompute." Similarly, in the truck-painting example above, supervisors easily overrode the system behavior (e.g., to route the vice president's new truck through the best performing booth) by manipulating the parameters determining a particular booth's bid (G. Ekberg, pers. comm.). With the rapid advance in global positioning (GPS) and other technologies, it is now possible to monitor and control self-organized systems remotely and in real time—teleoperating. Thus, we have the potential for a system of semi-autonomous, fish school mimicking robots or missiles, with an operator thousands of miles away following their progress, ready to intervene and manipulate global behavior when necessary. This could take the form of explicit orders to *specific* agents or adjusting the parameters determining individual level behavior for *all* agents. For instance, detecting that the set of robots or missiles are about to crash, the operator could send a signal telling each agent to increase its "personal space" or zone of repulsion [24, 61] thus reducing agent density and avoiding potential disaster. Of course, such a labor-intensive strategy is only realistic for relatively few multi-agent systems but, again, the key point is that such pathological behavior should be rare.

If the system is well designed so that these problems seldom occur, then the overheads of a global observer could be low. For instance, it may not be necessary for the human observer to constantly monitor the system. Appropriate global metrics, such as group angular momentum [24], where a high value would be a reliable indicator of circular milling, can probably be devised in many cases. It only remains, therefore, for a system to collect and analyze the data automatically, alerting the human observer, or perhaps an expert system, if these metrics exceed certain thresholds. The bottom line is that these are self-organized systems and so have the potential to develop pathological behavior. There is no cure-all and a decision whether to tolerate such behavior depends on a host of factors including the severity and cost of the behavior, how cheap and disposable the agents are, the rarity of the behavior, whether any degree of failure is unacceptable, the cost of designing the additional control infrastructure, whether one can monitor and control the system in real time, and many others. The occurrence of system pathologies is a problem that will certainly plague, perhaps haunt, swarm intelligence researchers and practitioners for many years to come.

7. Discussion

In this chapter, I have considered a few aspects associated with choosing an initial strategy towards designing a particular desired self-organized system. Most problems will involve some element of bottom up simulation (even interactive evolution has to has some generative underlying model); however, I have suggested that there are some situations or features of the problem that may favor a different approach. Ultimately, the decision will be problem dependent, but in Figure 5 I have summarized a few questions that may aid this decision. (They have been organized as a key, similar to those developed by biologists to identify particular species.) This list is not exhaustive but, as mentioned in the introduction, I hope that it might be of some use to those just starting to tackle a particular design problem.

1. Is it possible to define an objective / fitness function mathematically? (In other words, do you know what you are looking for?)

    Yes: go to question 3.

    No: go to question 2.

2. Can system-level behavior be realized in real time?

    Yes: *interactive evolution is a strong possibility*

    No: *either a slow, probably frustrating, process of hand-tuning parameters or systematic search through state space is likely involved, or a technique such as open-ended search*

3. Is there a known analogous system?

    Yes: *emulate known system, tweaking and modifying as necessary, possibly using evolutionary computation to parameterize system*

    No: go to question 4.

4. Does the system involve or require multiple hierarchical levels, or is amenable to their introduction? (In other words, can we chunk?)

    Yes: *some element of top-down engineering may be possible (likely, in conjunction with bottom-up modeling)*

    No: *bottom-up modeling, adopting some of the general principles expounded in the literature* [see text], *may work.*

Figure 5. A key of design strategies. The key, similar to those used by biologists to identify individual species, sets out some of the key criteria, decisions, and constraints that may favor one design strategy over another. It is meant only to hint at some of the general, major decisions that may go into deciding which strategy to use.

There is no doubt that self-organized systems have enormous potential for real-world applications. We are captivated by the lure of "dumb parts, properly connected into a swarm, yield[ing] smart results" [47, p. 13]. If the system complexity resides in the network of interactions, rather than in the individual agents themselves, then modeling agents should be a relatively easy task (Section 1.2) (an extremely appealing notion given that *Boeing* and *Honeywell* estimate that 60–80 % of the development cost of a complex control system is in the software development and not the actual control system design; [40]). We simply program the agents' behavior—just *four* lines of code in Morley & Ekberg's [55] paint booths on a *GM* assembly line (see section 1.6)—and simply let the complexity emerge.

Unfortunately, as we are all aware, "there is no such thing as a free lunch" (attributed to A. Hansen by Hugh-Jones [44]). Yes, the agents may be simple, hopefully cheap, easily designed, and disposable, and there may indeed be a "coexistence of individual simplicity and collective complexity" [16, p. 322]. However, we have to learn not only to design those agents but also, and perhaps more importantly, to connect those agents into the particular network (out of a multitude of possibilities) that works as desired. We do have a list of ingredients of self-organization—agent diversity, positive feedback, and so on [19, 60]—but only a scant knowledge of real design strategies. What is needed is a more rigorous research agenda in this area, a deeper understanding of questions such as

- How much inter-agent diversity should be engineered into these systems?

- What is the relationship between exploration and exploitation? (A question that is strongly related to issues of speed vs. efficiency.)

- What does the role of network structure (e.g., scale-free vs. random) play?

- What is the role of hierarchy and modularity in self-organized systems?

- What potential control strategies are available for self-organized systems?

Since we are dealing with emergent behavior, research in this area may never be easy but there are surely great rewards to be gained. However, if we are to exploit the true utility and power of self-organization in man-made, designed systems we need to rank up the *orders of magnitude* of agent number: millions, perhaps billions, of networked nanorobots, computers, and sensors. Large systems such as these undoubtedly will possess incredible behavioral richness: "Sometimes we expect too much from a technology; at other times we cannot even dream what its impact will be" (6, but see [33, 25]).

Acknowledgments

I thank the former students of my ISyE 8800C and CS 8803 Self-Organization course for discussion of some of the aspects covered in this chapter. Thanks also to Tucker Balch, Paolo Gaudiano and Sergey Malinchik for additional comments on earlier drafts of this manuscript.

References

[1]   Anderson, C., 2002, Self-organization in relation to several similar concepts: are the boundaries to self-organization indistinct?, *Biological Bulletin* 202, 247.

[2]   Anderson, C., 2003, Linking micro- to macro-level behavior in the aggressor-defender-stalker game, in *Proceedings of the Second International Workshop on the Mathematics and Algorithms of Social Insects*, Georgia Institute of Technology, 15th-17th December 2003, edited by C. Anderson and T. Balch. Georgia Institute of Technology (Atlanta), 9.

[3]   Anderson, C., & Bartholdi, J.J., III., 2000, Centralized versus decentralized control in manufacturing: lessons from social insects, in *Complexity and Complex Systems in*

Industry, Proceedings, University of Warwick, 19th-20th September 2000, edited by I.P. McCarthy, I. and T. Rakotobe-Joel. The University of Warwick (Warwick), 92.

[4]   Anderson, C., & McShea, D.W., 2001, Individual versus social complexity, with particular reference to ant colonies. *Biological Reviews* **76**, 211.

[5]   Anderson, C., Franks, N. R., & McShea, D.W., 2001, The complexity and hierarchical structure of tasks in insect societies, *Animal Behaviour* **62**, 643.

[6]   Anonymous, Failure of the Cable 1858, Plaque describing the downfall of the Atlantic Cable Company. National Museum of American History, Behring Center, Washington, D.C.

[7]   Appleby, S., & Steward, S., 1994, Mobile software agents for control in telecommunication networks, *British Telecom Technological Journal* **12**, 104.

[8]   Ball, P., 1999, *The Self-made Tapestry*, Oxford University Press (Oxford).

[9]   Banzhaf, W., 1997, Interactive evolution, *Handbook of Evolutionary Computation*. Oxford University Press (Oxford), **C2.9**, 1.

[10]   Beebe, W., 1921, *Edge of the Jungle*, Henry Holt and Company (New York).

[11]   Beekman, M., Sumpter, D.J.T., & Ratnieks, F.L.W., 2001, Phase transition between disorganized and organized foraging in Pharaoh's ants, *Proceedings of the National Academy of Sciences, U.S.A.* **98**, 9703.

[12]   Bonabeau, E., 1998, Social insect colonies as complex adaptive systems, *Ecosystems* **1**, 437.

[13]   Bonabeau, E., 2002, Predicting the unpredictable, *Harvard Business Review* (March), 109.

[14]   Bonabeau, E., 2003, Don't trust your gut, *Harvard Business Review* (May), 116.

[15]   Bonabeau, E., & Meyer, C., 2001, Swarm intelligence: a whole new way to think about business, *Harvard Business Review* (May), 107.

[16]   Bonabeau, E., & Theraulaz, G., 1995, Why do we need artificial life?, in *Artificial Life: an Introduction*, edited by C.G. Langton, MIT (Cambridge), 303.

[17]   Bonabeau, E., Dorigo, M., & Theraulaz, G., 1999, *Swarm Intelligence: from Natural to Artificial Systems*, Santa Fe Institute on the Sciences of Complexity, Oxford University Press (New York).

[18]   Bonabeau, E., Funes, P., & Orme, B., 2003. Exploratory Design Of Swarms, in *Proceedings of the Second International Workshop on the Mathematics and Algorithms of Social Insects*, , Georgia Institute of Technology, 15th-17th December 2003, edited by C. Anderson and T. Balch, Georgia Institute of Technology (Atlanta), 17.

[19]   Bonabeau, E., Theraulaz, G., Deneubourg, J.L., Aron, S., & Camazine, S., 1997, Self-organization in social insects, *Trends in Ecology and Evolution* **12**, 188.

[20]   Camazine, S., Deneubourg, J.L., Franks, N.R., Sneyd, J., Theraulaz, G., & Bonabeau, E., 2001, *Self-Organization in Biological Systems*, Princeton University Press (Princeton).

[21]   Campbell, D.J., 1988, Task complexity: a review and analysis, *Academy of Management Review* **13**, 40.

[22]   Cao, Y.U., Fukunga, A.S., & Khang, A.B., 1997, Cooperative mobile robotics: antecedents and directions, *Autonomous Robotics* **4**, 1.

[23]   Casti, J. L., 1994, *Complexification*, Harper Collins (New York).

[24]   Couzin, I.D., Krause, J., James, R., Ruxton, G.D., & Franks, N.R., 2002, Collective memory and spatial sorting in animal groups, *Journal of theoretical biology* **218**, 1.

[25]   Crighton, M.J., 2002, *Prey*, Harper Collins (New York).

[26]   Dawkins, R., 1986, *The Blind Watchmaker*, Longman (Harlow).

[27]   Dawkins, R., 1987, The evolution of evolvability, in *Artificial Life*, SFI Studies in the Sciences of Complexity, edited by C.G. Langton, Addison-Wesley (Reading, Mass.), 201.

[28]   Dahlstedt, P., 2001, Creating and exploring huge parameter spaces: interactive evolution as a tool for sound generation, *Proceedings of the International Computer Music Conference, Habana, Cuba.*

[29]   Deneubourg, J.L., Goss, S., Franks, N.R., & Pasteels, J.M., 1989, The blind leading the blind: modelling chemically-mediated army ant raid patterns, *Journal of Insect Behavior* **2**, 719.

[30]   Deneubourg, J. L, Goss, S., Franks, N.R., Sendova-Franks, A., Detrain, C., & Chretien, L., 1991, The dynamics of collective sorting: robot-like ants and ant-like robots in From Animals to Animats, *Proc. of the First International Conference on Simulation of Adaptive Behavior*, edited by J.A. Meyer and S.W. Wilson, MIT Press (Cambridge), 356.

[31]   Detrain, C., Natan, C., & Deneubourg, J.L., 2001, The influence of the physical environment on the self-organising foraging patterns of ants, *Naturwissenschaften* **88**, 171.

[32]   Di Caro, G., & Dorigo, M., 1998, AntNet: distributed stigmergetic control for communications networks. *Journal of Artificial Intelligence Research* **9**, 317.

[33]   Drexler, K.E., 1986, *Engines of Creation*, Anchor Books (Garden City, NY).

[34]   Franks, N.R., Gomez, N., Goss, S., & Deneubourg, J.L., 1991, The blind leading the blind in army ant raid patterns: testing a model of self-organization (Hymenoptera: Formicidae), *Journal of Insect Behavior* **4**, 583.

[35]   Funes, P., Orme, B.. & Bonabeau, E., 2003. Evolving emergent group behaviors for simple humans agents, in *7th European Conference on Articial Life (ECAL 2003): Workshop and Tutorials*, edited by P. Dittrich, and J.T. Kim, 76.

[36]   Gage, D. W., 2000. Minimum-resource distributed navigation and mapping, in *SPIE Mobile Robots XV, Boston MA, November 2000 (SPIE Proceedings Volume 4195).*

[37]   Gautrais, J., Theraulaz, G., Deneubourg, J.L., & Anderson, C., 2002, Emergent polyethism as a consequence of increased colony size in insect societies, *Journal of Theoretical Biology* **215**, 363.

[38]   Goodwin, B., 1994, *How the Leopard Changed Its Spots*, Scribner and Sons (New York).

[39]   Hayashida, N., & Takagi, H. 2000. Visualized IEC: interactive evolutionary computation with multidimensional data visualization, in *Industrial Electronics, Control and Instrumentation (IECON2000)*, Nagoya, Japan, 2738.

[40]   Heck, B.S., Wills, L.M., & Vachtsevanos, G.J., 2003, Software technology for implementing reusable, distributed control systems. *IEEE Control Systems* **23**, 21.

[41]   Hershberger, D., Simmons, R., Singh, S., Ramos, J. & Smith, T., 2002, coordination of heterogeneous robots for large-scale assembly, in *Robot Teams* edited by T. Balch and L.E. Parker, AK Peters (Natick, MA), 369.

[42]   Holland, O., & Melhuish, C., 1999, Stigmergy, self-organization and sorting in collective robotics, *Artificial Life* **5**, 173.

[43]   Hubanks, B., 1998, Self-organizing military logistics, *Embracing Complexity: A Colloquium on the Application of Complex Adaptive Systems to Business*, The Ernst and Young Center for Business Innovation, Cambridge, 51.

[44]   Hugh-Jones, E.M., 1952, Inquest on nationalization, *Ethics* **62**, 169.

[45]   Kauffman, S.A., 1992, *Origins of Order: Self-Organization and Selection in Evolution*, Oxford University Press (Oxford).

[46]   Kelly, K., 1994, *Out of Control*, Addison Wesley (Reading).

[47]   Kelly, K., 1998, *New Rules for the New Economy*, Viking Press (New York).

[48]   Kessler, M.A., & Werner, B.T., 2003. Self-organization of sorted pattern ground, *Science* **299**, 380.

[49]   Kittithrreerapronchai, O., & Anderson, C., 2003, Do ants paint trucks better than chickens? Markets versus response thresholds for distributed dynamic scheduling, in *Proceedings, Swarm Intelligence and its Applications, a special session of the Congress on Evolutionary Computation (CEC2003)*, Canberra, Australia, 8–12th December 2003, 1431.

[50]   Klausmeier, C., 1999, Regular and irregular patterns in semiarid vegetation, *Science* **284**, 1826.

[51]   Lewin, R., 1992, *Complexity: Life at the Edge of Chaos*, Macmillan (New York).

[52]   Lim, I.S., & Thalman, D., 1999, Pro-actively interative evolution for computer animation, *Proceedings of Eurographics Workshop on Animation and Simulation '99 (CAS '99)*, 45.

[53]   Malzberg, B.N., 1973, *Phase IV*, Pocket Books (New York).

[54]   McCormack, J., 1993, Interactive evolution of L-system grammars for computer graphics modeling, in *Complex Systems: from Biology to Computation*, edited by D. Green and T. Bossomaier, ISO Press (Amsterdam).

[55]   Morley, R., & Ekberg, G., 1998, Self-organizing military logistics, *Embracing Complexity: A Colloquium on the Application of Complex Adaptive Systems to Business*, The Ernst and Young Center for Business Innovation, Cambridge, 97.

[56]   Morley, R., & Schelberg, C., 1993, An analysis of a plant-specific dynamic scheduler, *Final Report, Intelligent, Dynamic Scheduling for Manufacturing Systems*, 115.

[57]   Nelson, G.L., 1993, Sonomorphs: an application of genetic algorithms to the growth and development of musical organisms, *Proceedings of the Fourth Biennial Art & Technology Symposium*, 155.

[58]   Parrish, J.A., & Edelstein-Keshet, L., 1999, Complexity, pattern, and evolutionary trade-offs in animal aggregation, *Science* **284**, 99.

[59]   Parrish, J.A., Viscido, S.V., & Grünbaum, D., 2002, Self-organized fish schools: an examination of emergent properties, *Biological Bulletin* **202**, 296.

[60]   Parunak, H.V.D., 1997, "Go to the ant": engineering principles from natural multi-agent systems, *Annals of Operation Research* **75**, 69.

[61]   Reynolds, C., 1987, Flocks, herds, and schools. *Computer Graphics* **21**, 25.

[62]   Schneirla, T.C., 1971, *Army Ants: a Study in Social Organization* (H.R. Topoff, ed.), W.H. Freeman and Company (San Francisco).

[63]   Schoonderwoerd, R., Holland, O., Bruten, J. & Rothkrantz, L., 1996, Ant-based load balancing in telecommunications networks, *Adaptive Behavior* **5**, 169.

[64]   Seeley, T.D., 1997, Honey bee colonies are group-level adaptive units, *American Naturalist* **150**, S22.

[65]   Simon, H.A., 1996, *The Sciences of the Artificial*, MIT Press (Cambridge).

[66]   Sims, K., 1991, Artificial evolution for computer graphics, *Computer Graphics* **25**, 319.

[67]   Sims, K., 1994, Evolving virtual creatures, *Computer Graphics*, Annual Conference Series (SIGGRAPH '94 Proceedings), July 1994, 15.

[68]   Smith, A.R., 1984, Plants, fractals, and formal languages, *Computer Graphics* **18**, 1.

[69]   Smith, J.R., 1991, Designing biomorphs with an interactive genetic algorithm, *Proceedings of the Fourth International Conference on Genetic Algorithms, San Diego*.

[70]   Solé, R., & Goodwin, B., 2000, *Signs of Life*, Basic Books (New York).

[71]   Takagi, H., 1998, Interactive evolutionary computation—cooperation of computational intelligence and KANSEI. *Proceedings of the 5th International Conference on Soft Computing and Information / Intelligent Systems (IIZUKA '98)*, World Scientific, 41.

[72]   Takagi, H., 2000, Active user intervention in an EC search, in *5th Joint conference on Information Sciences (JCIS2000)*, Atlantic City, NJ, 995.

[73]   Ünsal, C., 1993, Self-organization in large populations of mobile robots. Masters thesis, Virginia State University, Petersburg, VA.

[74]   Vaughan, R.T., Støy, K., Sukhatme, G.S., & Matari , M., 2000, Blazing a trail: insect-inspired resource transportation by a robot team, *International. Conference DARS 2000*.

[75]   Werner, B.T., & Hallett, B., 1993, Numerical simulations of self-organized stone stripes, *Nature* **361**, 142.

[76]   Wilson, E. O., 1975. *Sociobiology*, Harvard University Press (Cambridge).

[77]   Wuensche, A., 1994, The ghost in the machine: basins of attraction of random Boolean networks, in *Artificial Life III*, edited by C.G. Langton, SFI Series in the Sciences of Complexity, XVII, Addison-Wesley (Reading), 465.

Chapter 6

Understanding the Complexity of Design

Jonathan R.A. Maier
Georges M. Fadel
Clemson Research in Engineering Design and Optimization Laboratory
Clemson University
gfadel@ces.clemson.edu

1. Introduction

The powerful concept of complexity can be applied to help us understand not only modern engineering systems, but also the design of those systems, and artifacts in general. In this chapter we attempt to establish a two-pronged theoretical framework for understanding the complexity of design. By *design* we mean the activity of designing artifacts in general, not any specific class of artifact.

The first route to understanding the complexity of design is based on a fundamental exploration of what it means for a system to be complex. This avenue is essentially mathematical in character, and for it we rely heavily on the works of Robert Rosen, Nicholas Rashevsky, and Peter Wegner. Having discussed briefly the foundations of this approach, it is then applied to the science of design. In particular, the goal is to show that design in general is a member of the class of systems that are formally described as open and complex, and not a member of the class of systems that are formally described as closed and algorithmic. This amounts to theoretical validation for adopting a paradigm for using an open relational concept, such as affordance, as a basis for design, rather than a closed algorithmic concept such as function. This approach also suggests abstract affordance based descriptive models of design as alternatives to the current function based models of design.

The second route to understanding the complexity of design lies in the study of systems that are in some obvious way complex. This approach is essentially empirical in character. Accordingly, the goal here is to show that design exhibits similar characteristics to other complex systems, in particular, as will be shown, a class of complex systems known as Complex Adaptive Systems (CAS). This constitutes more validation for using a relational as opposed to an algorithmic concept as a basis for design. Also, this suggests that design may be modeled in the same way as other CAS, i.e., in accordance with a cycle in which other CAS are known to operate

In place of algorithms, what is needed for complex systems are structures which are semantically rich and open to interactions. For biology, such a formalism, once he realized such a thing was necessary, was invented by Rashevsky in the form of *relational models* [14], to replace the earlier simplistic *machine metaphor* dating back to Descartes. For computer science, such a formalism was invented by Wegner in the form of *interaction machines* [22, 23], to replace the much more restrictive and simplistic Turing machines [e.g., 21]. For design, we propose that the appropriate formalism is that of *affordance* [7, 9, 11, 12], rather than the much more restrictive and simpler concept of *function*, which is very similar to that of *algorithm*, as explained in this chapter.

The concept of affordance, thus grounded theoretically in the first approach to design complexity (the relational approach), agrees very well with the empirical model for design complexity suggested by the CAS-type approach. The integration of the concept of affordance into the CAS-inspired model for design thus concludes this chapter. However, the terms *complex* and *complexity* have been used before in design and therefore these views need to be distinguished our use of these terms. A review of these approaches is given in the next section.

2. Targeted Approaches to Complexity in Design

2.1. Computational Complexity

Computational complexity attempts to measure the complexity of a given problem in terms of its most compact solution algorithm. This approach is widely used in computer science, and looks primarily at solution algorithms, without considering the many other factors that contribute to the complexity of design. For example, the complexity of design involves relationships and interactions between designers, users, and artifacts. Such two-way relationships cannot be described by a unidirectional algorithm. Therefore, the techniques of computational complexity are not sufficient to describe the broad complexity of design.

2.2. Complexity in Axiomatic Design

Suh defines complexity "as a measure of uncertainty in achieving the specified Functional Requirements". Suh uses the term "Real complexity" to describe the uncertainty associated with the known probability of a solution not completely satisfying the desired objectives. Suh defines "Imaginary complexity" as the uncertainty associated with a designer's lack of knowledge [19, pp. 474-476]. However, there is an important conceptual distinction between complexity as describing the uncertainty of a system—which refers to probability distributions—and complexity as describing the impredicativity of a system—which refers to self-referential causal entailment (see Section 2.3.1). The term "uncertainty" implies that either there is some quantifiable probability involved or that certainty is in fact possible, if only there were more information available.

This situation, as Rosen recognized [16, pp. 86-89], is analogous to the measurement problem in physics. One problem, of uncertainty, involves where a particle is or how fast it is going, which below a certain resolution cannot be described exactly but can be described by a probability distribution. This problem, in and of itself, was enough to wreck the Newtonian paradigm of clockwork-like predictability and determinism of the physical universe. The second and more serious problem is that the location and or velocity of the particle as measured depends on how the measurer measures it. Whereas the former problem is a matter of simple probability, of measurement *uncertainty*, the latter problem is a matter of *complexity*, of causal entanglement between the observer and what is observed. These ideas will be explained more fully in Section 2.3, but suffice it to say that Suh's conception of complexity in design as having to do with uncertainty, while potentially useful, is qualitatively different from the kind of complexity we appear to be confronting in design, which involves interactions and relationships, not just uncertainty.

2.3. Measuring Design Problem Complexity

An approach to measuring the complexity of design problems themselves has been proposed by Dixon and his colleagues [5], based upon the coupling between design targets and design variables. The underlying assumption here is that the more coupled the design problem, the more complex it is. In other words, a situation in which each design variable affects only one design target would be of lowest complexity, and a situation in which each design variable affected several design targets, and thus each target would be affected by several design variables, would be of higher complexity.

This kind of complexity could be modeled by a series of linear equations. How far the coefficient matrix would be away from purely diagonal would define the coupling and thus the complexity in this sense. In contrast, the complexity of design appears to involve more than just mathematical variables and targets; it necessitates the inclusion of relationships between human designers and users who are involved in the determination and mutability of both variables and targets, as well as concepts, testing, and other important aspects of design.

2.4. Measuring Artifact Complexity

As opposed to the previous few methods of measuring problems and solutions, some researchers such as Dixon [5], and Braha and Maimon [2], have tried to measure the complexity of artifacts. Typically such analyses quantify how complicated an artifact is based upon how much effort it takes to describe the artifact using a standardized representation scheme. However, counting parts is not counting complexity. Certainly, the lengthier the description, the more complicated the artifact. But again, only measuring the complexity of the solution algorithm, or the problem, or the artifact, in our opinion misses the larger source of complexity in design, which is the interactions inherent between designers, artifacts, and users, and everything that influences them.

3. Rosen's Approach to the Complexity of Science, Applied to the Complexity of Design

3.1. Impredicativities in Science

Let us return again to the mechanistic paradigm in which design is currently cast, and attempt to understand, with the help of the theoretical biologist Robert Rosen (see [15, 16] for his original arguments), what the power of that paradigm is, but also its limitations. The cornerstones of Newtonian mechanics are Newton's laws of motion and Newton's law of universal gravitation. The mathematical language that describes them both is of course Calculus, which Newton also developed. Newton's laws of motion address the problem of *how a particle responds to an (external) force*, whereas the law of universal gravitation addresses the problem of *how that same particle can exert a force on another particle*. What is most curious about this formulation is that the material parameter that describes a particle's response, its *inertial mass*, is numerically identical to the totally independent (conceptually) parameter that describes the particle's exertion, its *gravitational mass*. This fundamental equivalence was not adequately explained until the time of Einstein, who realized it as a foundation of General Relativity.

Meanwhile, this Newtonian model that equates the behavioral effect of particles, i.e., their reaction (in biological terms, their *phenotype*), with their behavioral causes, i.e., their actions (in biological terms, their *genotype*), experiences severe difficulties in certain common situations, such as the so-called many body problem, the problem of trying to predict the trajectories of several (more than two) particles under mutual gravitational attraction. In such a case, the addition of more than one particle results in a situation where the original particle, although not allowed to push on itself, is allowed to push on other particles which in turn push on the original particle. This is one example of an *impredicativity* (Bertrand Russel's term), meaning that this problem cannot be solved reductionistically, i.e., in closed form, nor in terms of defined probabilities.

The impredicativity, as in all impredicativities, arises at root, from the absolute separation between causes and effects (genotypes and phenotypes), that is, with drawing a line in the sand. Similar impredicativities arise in logic, whenever we try to draw a clear distinction between two classes of related objects, for instances true statements and false statements. For example, the statement,

This statement is false.

is a classic example of an impredicativity. If the statement is true, then it is not false, but if it is not false, then it must be true, but it cannot be true because it says it is false, but if that is true, then it *is* false, but, but, but...*ad infinitum*. The apparent paradox of this statement abides in the fact that, as in all such impredicativities, the statement is *self-referential*.

Mathematically, the many body problem is expressed by networks of differential forms that describe direct and indirect effects of inertia on gravitation, mirroring the network of interacting particles they model. These equations collapse into ordinary rate equations (equations of motion) only when all the differential forms are *exact* (predicative), but not otherwise. The non-exactness of any of these differential forms is one mathematical description of *complexity*, i.e., *impredicativity*, caused by relatively many interacting subsystems, from which the colloquial notion of complexity is derived. To look ahead a bit, this is just another manifestation of Gödel's theorem [8], that the real behavior (truth) of nature is larger than any purely predicative, syntactic, model of it.

More generally, what is happening here is that in any apparently predicative formalism (e.g., what happens with only one or two particles, or whether a statement is true or false), the application of certain larger contexts (e.g., what happens with more than two particles, or whether a statement about falseness is true) generates impredicativities that the original formalism cannot handle. The aforementioned measurement problem in quantum physics is also of this flavor—the original formalism of how the quanta behave (even probabilistically) breaks down in a larger context of when those quanta are actually measured, which depends on how they were measured. Bohr's quip that anyone who is not shocked by quantum mechanics does not understand it refers to the shock over the realization of the reality of this impredicativity, which is anathema to the predicative Newtonian mindset from which most people come until confronted with the evidence of quantum experiments.

Even more generally, what this argument illustrates in brief is that impredicativities, while inconvenient, are essential features of reality. As Gödel proved [8], they are more than just problems with our current formalisms; they are inherent to all formalisms. In other words, if we identify, as Rosen has, *impredicativity* with *complexity*, then reality is complex, and only special models of it are not complex, but these models, while perhaps powerful, will always be to some degree inadequate. For design, therefore, to be exempted from this picture would be a very special case indeed. We should expect to find some impredicativities in design. How frequent, and how serious these impredicativities are, determines whether design can be effectively treated algorithmically versus as a complex science.

The Newtonian paradigm, while powerful, is a formalism, predicative in nature, that can be applied to a certain class of real material systems that exhibit few enough impredicativities for the formalism to be useful. Such real material systems are called *machines*. What is in dispute is not the applicability of the Newtonian paradigm to the *behavior of machines*, which is the traditional focus of the engineering sciences, and is supported by several centuries of positive evidence. What is questionable, however, is the applicability of the Newtonian paradigm to the *design of machines*, which is the focus of design science. Tate and Norlund's finding [20] that design science, even after several decades of research, is still in a pre-paradigm state suggests that the larger mechanistic Newtonian paradigm, while sufficient for many other engineering sciences, is not sufficient for design.

3.2. Implications for Technology and Design

The archetypal example of such a Newtonian machine, especially in recent years, is the Turing machine, a purely predicative formalism that executes an algorithm (as Turing proved, *any* algorithm) with perfectly predictable results. As a historical note, it is worth pointing out that Turing's work was motivated by the recent work of Gödel before him. Whereas Turing set out to contradict Gödel's results, he wound up spectacularly confirming them, by showing that his simple symbol pushing machine could execute any algorithm, but not any mathematics, without loss of truth (e.g., impredicativities). Turing's solution was to invert the problem, to assert that *enough* algorithm would suffice for something larger, something as rich as intelligence. Von Neumann, who helped design the computer architecture to implement Turing's theoretical machine, was of the same persuasion, theorizing some threshold above which "enough" predicativity (such as his cellular automata) would achieve impredicativity (some kinds of "life-like" behavior). This idea, however, is easily debunked by considering the original work of Gödel, as Rosen has done. In a nutshell, Gödel's theorem allows us to discriminate between real number theory, full of impredicativities, with artificial predicative number theory, and proves that the two are indeed distinct. Similarly, the impredicativities of human intelligence, born of the subjective mind, are distinct from any purely predicative algorithmic conception of intelligence.

The essence of the above distinction, between number theory and any formalization of it, between real intelligence and artificial intelligence, between material reality and Newtonian machines, boils down to the simple algorithmic symbol-pushing (syntactic) character of the latter systems as opposed to the complex impredicative (semantic) character of the former systems. An essential feature of the former systems, which can be viewed in one light as the basis for much of their impredicative traits is the interaction inherent in these systems. Along these lines, the computer scientist Wegner recently produced a proof showing that "interaction is an inherently more powerful paradigm for computing than [closed] algorithms" [22]. By interaction, he meant interactions between computer programs with humans, not closed interactions between fixed algorithms and operable data sets, which in general can always be expressed and computed in Turingesque algorithms.

Wegner's proof structure is based explicitly on the proof structure of Gödel. Gödel showed the incompleteness of the set of integers by showing that the set of integers is not recursively enumerable. That the integers are incomplete implies that they are not formalizable. Thus the larger body of number theory, which includes the integers, is not formalizable. Any formalism that attempts to capture number theory (i.e., a complete set), therefore, will leave out essential parts of "real" number theory, which is not formalizable. One consequence of this is that in any formalization of number theory, one can always ask questions about number theory which cannot be answered within that formalism.

Similarly, Wegner showed the completeness of algorithms (expressed on Turing machines) and the incompleteness of interaction (expressed on so-called "interaction machines.") Incompleteness of human interaction follows from the unpredictability of human input, which is modeled by infinite series which are by definition incomplete (because they are infinite). Thus Wegner proved that interaction is a larger, non-formalizable computing paradigm than standard Turingesque closed algorithms. In Rosen's terms, Wegner showed that interaction is complex, and therefore more powerful than closed algorithms. The mechanistic paradigm begins to fail (i.e., is not powerful enough) in design in the same way that it begins to fail in computing, when human interaction is introduced, which is unpredictable by nature.

A similar situation has arisen in the field of economics, where a purely predicative modeling approach has again been shown to be insufficient. The classical assumption in economics is rationality of the human economic agent, in essence, that people rationally make decisions (execute an algorithm) in order to maximize profits. Even when the assumption of perfect rationality, as in the classic Arrow-Debreu equilibrium model [4] is reduced to the "bounded rationality" described by Herbert Simon, this approach fails to describe many empirically evidenced facts of economic life. For example, the Arrow-Debreu rational model mandates *decreasing returns* that drive the economy toward equilibrium., but in real life, often *increasing returns* are observed. The economist W. Brian Arthur has shown that such increasing returns (as well as the common decreasing returns) can be exhibited by models of human economic agents that behave in non-rational, non-algorithmic, i.e., complex ways [1]. His treatment is thus as the economy as a complex adaptive system (see Section 2.4 below), rather than as a simple algorithmic system.

Similar complexities can be found in other scopes of human study. For example, in cognitive psychology, the assumption is again made that human intelligence is basically algorithmic in nature and therefore the attempt is to model human thought, perception, and interaction with the environment with simple predicative Turingesque algorithms [e.g., 17]. Theoretically, such an approach can be repudiated on the same grounds as the Turing test as discussed above, but empirically, these cognitive models have also failed to exhibit semantic qualities of real human minds. Once again, as Gödel showed, no amount of completeness, i.e., predicativity, can amount to incompleteness, i.e., impredicativity.

In design, the semantic, non-rational, non-algorithmic, impredicative, subjective, and unpredictable nature of humanity is inescapable, because artifacts are always designed for human use, usually designed by humans themselves (using computers and other tools), and situated within a larger context of a complex world economy. Some consequences of this are individual differences between consumers and designers, changing mindsets, preferences, needs, and attributes of all these people, non-rational unpredictable behavior, and creativity. Cast in this light, it is quite understandable why approaches to design based on a Newtonian predicative paradigm have failed to rise to the level of the mature design science we desire. The alternative, suggested from these considerations, is a paradigm for design based on the idea that design is complex. Such an approach should further incorporate the complex nature of human interactions (à la Wegner), as opposed to relying solely on closed predicative algorithms. However, there is another route to the same conclusion, from the theory of complex adaptive systems, as discussed next.

4. Understanding Design as a Complex Adaptive System

4.1. History and Overview

The science of complexity has emerged in recent years as a response to the realization that many important phenomena across a wide range of scientific domains possess features that arise from the interaction of many small subsystems, from individuals and corporations in an economy to elementary particles in a large molecule. The key realizations behind this science are three-fold, 1) that many interesting and unsolved problems in science are complex in nature, and not simple, 2) that problems across a wide range of domains have complexity in common, and 3) that complexity itself can be studied.

The domain of interest to the science of complexity is predominately complexity's manifestation in complex systems. A complex system may loosely be defined as a collection of a large group of strongly interacting parts exhibiting non-linear dynamical behavior. Complex systems may be further classified as either non-adaptive or adaptive.

4.2. Complex Adaptive Systems

The concept of a Complex Adaptive System (CAS) has been described in terms of a system "with many different parts which, by a rather mysterious process of self-organization, become more ordered and more informed than systems which operate in approximate thermodynamic equilibrium with their surroundings." [3, pp. 1]. The physicist Murray Gell-Mann identifies the cycle in which all CAS appear to operate [6, pp. 25] as follows:

  I.   Coarse graining of information from the real world
  II. Identification of perceived regularities
  III. Compression into a schema
  IV. Variation of schemata
  V.   Use of the schema
  VI. Consequences in the real world exerting selection pressures that affect the competition among schemata

However, perhaps the most important property of a CAS (that distinguishes it from most of the systems with which engineers are accustomed) is that CAS are *open* systems. CAS are situated; they operate and interact within a larger environment wherein the CAS accepts energy in and exports energy out. Moreover, because the CAS is adaptive, some of the energy in is used to change the internal state of the CAS. Usually this flow of energy in and out is continuous; thus the CAS is continually in a state of flux, constantly adapting to what is usually a changing environment. Another important consequence of CAS being open systems is that the second law of thermodynamics, which is formulated expressly for closed systems, is not applicable. Thus in CAS we often see a *decrease* in entropy (increase in order) over time, sometimes seen as progressive evolution.

Since all of the details of a complex system can in principle never be totally understood, an essential tool for understanding complex systems is to study the system's organization, which is often relatively simple. Understanding the organization of the system can also lead to a better understanding of the system's behavior, since the behavior of a system is strongly affected by the system's internal organization. For example, an important result for business management (a business organization being a complex system) is that the organization of the business strongly affects the behavior of both the business as a whole and the actions of its individual employees [18].

4.3. The Designer-Artifact-User Complex System

One of the strengths of the science of complexity is its ability to handle very large and very complex systems. Thus it frees the investigator from the self-defeating need to over-simplify the problem under study. To study complexity in design, therefore, we need not focus on just artifacts, or just problems, or just solution procedures, as has been done in the past (see Section 2.2). There is in fact total freedom to define the complex system of interest, the "design system," as it were, to encompass all relevant issues and stakeholders. However, in every design there will be different issues and stakeholders at play, but incumbent to every design are at least three major subsystems: 1) the designer(s) of the artifact, 2) the artifact(s) being designed, and 3) the user(s) of the artifact. Thus the complex system of interest in design is the designer-artifact-user (DAU) system.

The properties of the designer-artifact-user system involve many important aspects of design not stated directly in the three component subsystems. First, as in every system, the DAU system is situated in a larger environment. However, the specific environment in which a particular DAU system is situated depends upon the specific design. For example, in the case of the design of a family of consumer appliances, the environment would consist of a corporation, local, regional, national, and global economies, legal regulations, competitor corporations and products, the physical world, etc. Each of these elements of the environment would act upon the DAU system in different ways, in terms of inputs to the DAU system from each entity in the environment, as well as outputs from the DAU system to entities in the environment. A generic situation of this kind is shown in Figure 1.

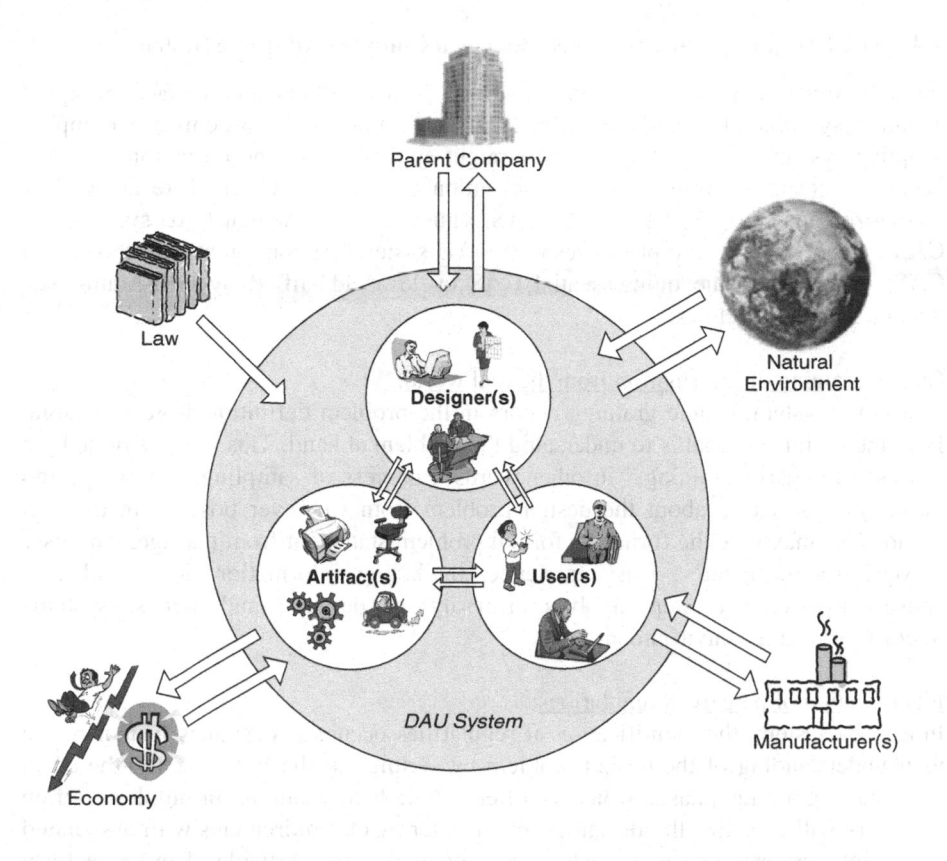

Figure. 1. Generic situated designer-artifact-user (DAU) system

Each of the three basic subsystems within a DAU system need not be singular. In any given design, there may be multiple designers, and/or multiple artifacts being designed, and/or multiple users. By considering the general case of multiple designers, we open the door for important insights into concurrent engineering and collaborative design, two recent and important topics of design research. By considering the general case of multiple artifacts, we open the door for important insights into product family design, another hot topic of design research. And by considering the general case of multiple users, we can consider not just the end user but anyone who might interact with the artifact throughout its life-cycle, thus opening the door for important insights into areas such as human factors, mass customization, and many design-for-x methods. Moreover, by considering designer(s), artifact(s), and user(s) together in the same system, we can also study the interactions between each of these subsystems, and perhaps most importantly, the system behaviors that result from those interactions—for instance the success or failure of the whole design project.

4.4. The Designer-Artifact-User System as a Complex Adaptive System

First, however, it is necessary to show that DAU systems are, in fact, complex adaptive systems. It would be difficult to argue that DAU systems are complex adaptive systems (CAS) merely by a straight application of the definition of CAS, because there is no formal accepted definition of CAS. However, there are widely recognized and studied properties of CAS. Thus we can show that DAU systems are CAS by showing that the properties of a DAU system are consistent with those of a CAS, in particular the quintessential CAS cycle as identified by Gell-Mann (see Section 2.4.2), as follows:

Coarse graining of information from the real world:
In a DAU system, coarse graining occurs in the problem definition stage of design. Here the designer's goal is to understand the problem at hand. This must be done by a process of "coarse graining," in other words a process of sampling, surveying, and gleaning information about the design problem from wherever possible in the real world. This may take the form of a formal problem statement from management, user surveys, the designer's own experience, marketing information, legal and cost constraints, etc. This step involves primarily the designer and user subsystems interacting with the environment.

Identification of perceived regularities:
In a DAU system, the identification of regularities occurs as designers further refine their understanding of the design problem by sorting out the initial data gathered in the coarse graining phase, which is often contradictory and/or incomplete. Often designers will organize the design problem in terms of requirements with associated constraints, criteria, and goals, which may or may not be articulated in some form such as a Requirements List [e.g., 13, pp. 131-135]. Again, this step primarily involves the designer and user.

Compression into schema:
After designers sufficiently understand the problem to continue design work, the broad design space available to the designers must be narrowed in order to arrive at solution concepts. Thus the CAS compression into schema phase is equivalent to the conceptual design phase in a DAU system (where the terms "schema" and "artifact concepts" essentially become interchangeable). This involves exploration of the design space using ideation techniques as well as combination and selection of concepts. The resulting schema is thus a full system solution concept, possibly accompanied by some sort of prototype, physical or otherwise. This step primarily involves the designer and artifact subsystems.

Variation of schemata:
Once an initial system concept is found, the designer must improve, test, and refine that concept in order to arrive at a final production worthy artifact. Often this process requires extensive iteration of earlier phases of the design process. In other cases, the entire design project may be an exercise in variant design, where an artifact already

exists, but the objective is to modify it to suit new circumstances. These design activities within a DAU system are thus equivalent to the variation of schemata phase in the CAS cycle. Again, this step primarily involves the designer and artifact subsystems.

Use of the schema:
In a DAU system, the schema is used when the artifact is released on the market as a finished product. In many systems, this first must be preceded by a manufacturing step. In the discussion above, the manufacturer, since it is not explicitly defined to be a member of the DAU system, exists and interacts with the DAU system from the DAU system's external environment, yet clearly those interactions are important, but perhaps no more or less important than other interactions from the environment, such as from competitors, bodies of law, environmentally conscious issues, etc. And in other cases, such as in one-off products and software, there may be no significant manufacturer at all. At any rate, the schema may be used by a variety of users described by the DAU system's user subsystem, including people involved in any manufacturing process necessary, end users, maintenance and service personnel, etc. This step primarily involves the artifact and user subsystems.

Selection pressures that affect the competition among schemata:
In a DAU system, selection pressure is exerted from the outside environment. This pressure may come from the economy and changing user whims, which would feed back into the DAU system affecting choices and ideas for variant designs. This is especially important for product family design, in particular for what we have termed in other work as "evolving product families" [10]. Different selection pressures may come to bear in an original design exercise, where, for instance, corporate management or marketing people may influence the designer's decisions. It is important in this context not to confuse the competition of competing products in the marketplace—which occurs within a different CAS, the economy [1]—with the subset of competition in the economy that actually feeds back into the DAU system, which is all that is applicable here. This step involves all three subsystems: the designer(s), artifact(s), and user(s).

Implications of DAU systems as CAS:
Having thus established and defined the designer-artifact-user system as the complex design system of interest, we can now proceed to apply insights and ideas from the sciences of complexity in order to further our understanding of design itself, which is our primary objective. First, we may take some comfort from the fact that many troublesome problems in design coincide with equivalent problems in CAS in general. This suggests that our lack of understanding in these areas is not due to any particular lack of knowledge in the design field, but rather the difficult and complex nature of these problems in general. Furthermore, the congruence between unsolved problems in design and unsolved problems in CAS in general is further validation of the appropriateness of studying the complexity of design in general and the use of the DAU complex system in particular.

A list of eight outstanding problems that are common to both design (i.e., the DAU complex system) and CAS in general, organized by phases of the generic CAS cycle is as follows. This list is based upon the list of issues in CAS research deserving further investigation (of which the following is only a subset), as presented by Gell-Mann [6, pp. 26-28]:

Coarse graining of information from the real world:
1. In CAS: An interesting trade-off is "between coarseness for manageability of information and fineness for a better picture of the environment." [6, pp. 26] In design: An analogous trade-off arises between spending a lot time trying to understand the problem, i.e., by extensive user surveys, versus spending less time in this phase although sacrificing understanding of the problem, in order to rush to market for a potential pay-off there.

Identification of perceived regularities:
2. In CAS: A problem is "the tendency of a CAS to err by mistaking regularity for randomness and vice versa." [6, pp. 26] In design: It is difficult to identify true user needs and latent user needs, which is further complicated when users say they want one thing when in reality they actually buy something else. In other words, the difficulty is interpreting user data and other data that describes the problem, which is often incomplete and contradictory.

Compression into schema:
3. In CAS: An issue is "the importance of continual evolution of the observed system with the difficulty inherent in estimating the probability of future histories." [6, pp. 26] In design: Since the marketplace is continually changing, designers must confront the fact that by the time they finish designing the artifact, user preferences and other environmental effects such as market conditions may have changed.
4. In CAS: An important trade-off is "between degree of compression versus time and amount of computation involved." [6, pp. 26] In design: The analogous trade-off is between increasing the number of promising solution concepts and prototypes elaborated, versus time and money spent on them in development.

Variation of schemata:
5. In CAS: "Variation usually proceeds step by step from what is already available, so how can schema change by large jumps?" [6, pp. 27] In design: Most products evolve slowly over time, with modest success. But occasionally a major innovation occurs seemingly out of no-where. How do innovations like this occur, and how can they be engineered intentionally?

Use of the schema:

6. In CAS: A problem is finding a "method of incorporating largely random new data." [6, pp. 27]

   In design: This manifests in the difficulty of designing for the real world. The artifact, once introduced, is subject to all the vagaries of real users and the real marketplace, whereas in order for the designers to design the artifact at all, most of this complexity was lost in the coarse graining phase.

Selection pressures that affect the competition among schemata:

7. In CAS: "Fitness is an elusive concept." [6, pp. 27]

   In design: How can designers "optimize" a design to perform in an environment that is ill defined and approximate (out of the coarse graining phase) and ever changing? What is an appropriate fitness function?

8. In CAS: A problem is "when maladapted schemata occur because of mismatched time scales." [6, pp. 27]

   In design: This occurs when products fail because the market changes faster than products can be designed or redesigned.

Summary of DAU systems as CAS

This description of design as a complex adaptive system thus reinforces the earlier discussion of how the approaches to design based on a Newtonian predicative paradigm have failed to rise to the level of a mature design science. The alternative, again suggested from these considerations, is a paradigm for design based on the idea that design is complex. A high-level conceptual model for the complexity of design is the DAU complex system. However, what has been missing from the discussion of the DAU system thus far is the all important interactions between the various subsystems within the larger DAU system. Recall that from the complexity approach to design following Rosen's (and Wegner's) work, the primacy of impredicative interactions became apparent. Thus we require a deeper understanding of these interactions, which can be generated by applying the powerful concept of affordance.

5. Properties of Affordances within DAU Systems

5.1. Relational Questions and Answers

The preceding discussion suggests that rather than *ad-hoc* or closed form algorithmic approaches to design, in light of the complexity of design, interactions and relational aspects ought to be given preeminence. As a first step, we can refer back to the designer-artifact-user system (Figure 1), and examine the interactions and relationships between the three major subsystems (designers, artifacts, and users). Accordingly, there are three principal categories of interactions to be investigated and to wit, three main associated questions to be answered in this context:

1. What is the nature of the relationship between users and artifacts?
2. What is the nature of the relationship between designers and artifacts?
3. What is the nature of the relationship between designers and users?

The answer to the first question seems obvious. By definition, users use artifacts. Conversely, artifacts are used by users. This answer, however, begs another more subtle question, namely, *What determines how an artifact may be used?* By definition, the *affordances* of an artifact determine how that artifact may be used. Gibson defined affordances as follows:

> The *affordances* of the environment are what it *offers* the animal, what it *provides* or *furnishes*, either for good or ill. The verb *to afford* is found in the dictionary, but the noun *affordance* is not. I have made it up. I mean by it something that refers to both the environment and the animal in a way that no existing term does. It implies the complementarity of the animal and the environment... As an affordance...for a species of animal, however, they have to be measured *relative to the animal*. They are unique for that animal. They are not just abstract physical properties...So an affordance cannot be measured as we measure in physics...An affordance is neither an objective property nor a subjective property; or it is both if you like... Affordances are properties taken with reference to the observer [7].

Interestingly, this idea immediately suggests answers to the remaining two questions. The nature of the relationship between designers and artifacts is that designers create the affordances of artifacts. They specify all the properties (geometries, dynamic behaviors, colors, etc.) that will afford a certain set of uses to a certain set of users. Thus the nature of the relationship between designers and users is that designers must ascertain from users a target set of affordances. Conversely, the users inform the designers of desired uses—what they want the artifact to afford. For an expanded discussion of designing affordances, see [12].

In the context of the designer-artifact-user complex system, artifact-user affordances (AUA) appear as interactions between the artifact and user subsystems, and artifact-artifact affordances (AAA) appear as interactions within the artifact subsystem itself. Interactions between the designer and user subsystems include the information needed to specify which affordances should and should not exist in the artifact under design. Interactions between the designer and artifact subsystems include the specification of the artifact's properties that determine its various affordances internally (i.e., AAA) and externally to the targeted users (i.e., AUA). These interactions within the DAU complex system are shown schematically in Figure 2.

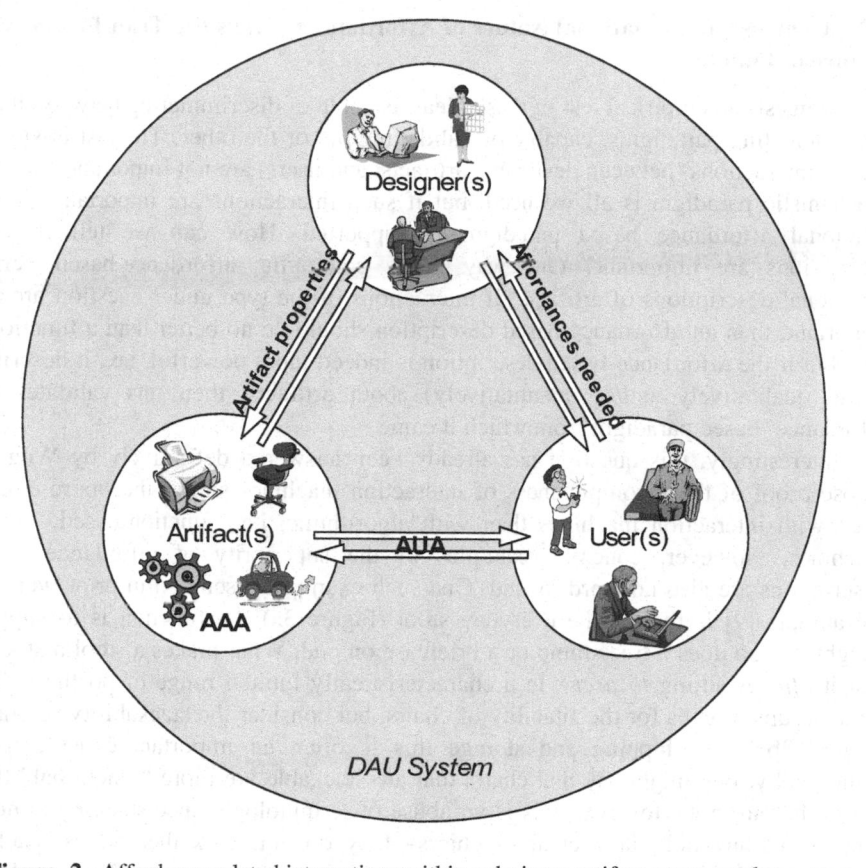

Figure. 2. Affordance related interactions within a designer-artifact-user complex system

To take a step back, the questions and answers outlined above may be contrasted with the kinds of questions and answers that have been addressed by existing approaches to design. One question is, *What is the nature of artifacts?* The answer to that is, they function. According to this view, artifacts are functional in nature. A second question is, *What is the nature of designers?* The answer to that is, they execute solution algorithms (as in Decision Based Design). That a computer can execute such algorithms just as well leads to design automation, and more broadly, classical top-down artificial intelligence. A third question is, *What is the nature of users?* The answer to that is that users are simply economic actors operating under bounded rationality, but rationally, nonetheless. These questions and answers define the tidy algorithmic universe familiar as the standard mechanistic paradigm for design. Under this view, the interactions *between* designers, artifacts, and users discussed previously are not considered explicitly, and design is not complex.

5.2. Contrast: the Relational Nature of Affordances versus the Transformative Nature of Functions

This suggests an empirical test of these ideas, capable of discriminating between these two competing paradigms, capable of validating one or the other. The test is simply this: if interactions (between designers, artifacts, and users) are not important, then the mechanistic paradigm is all we need, but if such interactions are important, then a relational affordance based paradigm is supported. How can we tell if such interactions are important? One way is by comparing affordance-based versus functional descriptions of artifacts. If interactions of the type under question are not important, then an affordance based description should be no better than a functional one, but if the affordance-based description is indeed more powerful, i.e., it describes more (qualitatively and/or quantitatively) about artifacts, then this validates the affordance- based paradigm from which it came.

Interestingly, this question has already been answered definitively by Wegner, whose proof of the incompleteness of interaction machines shows that more can be done with interaction machines than with algorithmic, i.e., function-based, Turing machines. However, concrete examples of the superiority of affordance based descriptions are also not hard to find. One such example, discussed in prior work by the authors [9] is that of the everyday stool (Figure 3a). Its function is to support weight, but so does a tree stump or a briefcase on end. What makes a stool a stool is that it *affords* sitting *to users*, in a characteristically limited range of postures. The same argument goes for the sitability of chairs, but consider the stackability of chairs (Figure 3b). For shipping and storage this is often an important consideration. Colloquially, one might say that chairs that are stackable are more "functional" than chairs that are not. However, this is an abuse of terminology since stacking is not a function of any individual chair or chairs—they do not stack themselves. Rather, users stack chairs, but in order for that to happen the chairs must *afford* stacking. They must possess the *affordance* of stackability.

(a)   (b)   (c)

Figure. 3. Three object lessons in affordance based descriptions versus functional descriptions

Other examples are numerous. Consider a household curling iron (Figure 3c); it does not curl hair by itself—why it is hardly more than a heated shaft and a hinged metal clip. Its function therefore cannot truly be said to curl hair. Yet the curling iron *affords* curling hair because it can be used with the appropriate twisting and pulling motion of the hand, arm, and wrist in order to curl hair. A curling iron thus possesses the affordance of "hair-curlability."

That the artifact affords its purpose, denoted by its very name, is not a trivial observation. The same can be said of an air conditioner, which affords cool (conditioned) air, regardless of whether a convective cooling tower or vapor-compression refrigeration cycle, or thermoelectric effect, or some other process is used to achieve it [9]. Suffice it to say that these few examples, together with their ultimate inductive conclusion embodied in Wegner's theorem, validate on descriptive theoretical grounds the appropriateness of a relational affordance based paradigm for design, and more generally a view of design as being fundamentally complex.

6. Summary Remarks

This chapter has laid forth a theoretical discussion of the complex nature of design and has argued that the concept of affordance, because of its relational character, is more appropriate for design than the transformative concept of function. The concept of affordance is a powerful tool to help understand specific processes in design, within a larger context of design as a complex adaptive system. The issue of putting this theoretical understanding into practical use has been another focus of our work [12] and is on-going.

References

[1]   Arthur, W. B., S. N. Durlauf, and D. A. Lane, (eds.), 1997, *The Economy as an Evolving Complex System II*, Perseus Press (New York).

[2]   Braha, D. and O. Maimon, 1988, *The Measurement of Design Structural and Functional Complexity*, IEEE Transactions on Systems, Man, and Cybernetics - Part A: Systems and Humans, **28**(4):527-535.

[3]   Cowan, G. A., D. Pines, and D. Meltzer, (eds.), 1994, *Complexity: Metaphors, Models, and Reality*, SFI Studies in the Sciences of Complexity, Addison-Wesley (New York).

[4]   Debreu, G., 1957, *Theory of Value: An Axiomatic Analysis of Economic Equilibrium*, Wiley (New York).

[5]   Dixon, J., M. Duffey, et al., 1988, "A Proposed Taxonomy of Mechanical Design Problems," Proceedings of ASME DETC '88, Computers in Engineering Conference, 41-46.

[6]   Gell-Mann, M., 1994, Complex Adaptive Systems, in in *Complexity: Metaphors, Models, and Reality*, edited by G. A. Cowan, D. Pines and D. Meltzer, SFI Studies in the Sciences of Complexity, Addison-Wesley (New York).

[7]   Gibson, J. J., 1979, *The Theory of Affordances*, in The Ecological Approach to Visual Perception, Houghton Mifflin (Hopewell, NJ).

[8]   Gödel, K., 1931, *Metamathematics*, van Norstrand (New York).

[9]   Maier, J. R. A. and G. M. Fadel, 2001a, "Affordance: The Fundamental Concept in Engineering Design," Proceedings of ASME DETC-01, Design Theory and Methodology Conference, Pittsburgh, PA, September 9-12, 2001.

[10]   Maier, J. R. A. and G. M. Fadel, 2001b, "Strategic Decisions in the Early Stages of Product Family Design," Proceedings of ASME DETC '01, Design for Manufacturing Conference, Pittsburgh, PA, September 9-12, 2001.

[11]   Maier, J. R. A. and G. M. Fadel, 2002, "Comparing Function and Affordance as Bases for Design," Proceedings of ASME DETC '02, Design Theory and Methodology Conference, Montreal, Canada, September 29-October 2, 2002.

[12]   Maier, J. R. A. and G. M. Fadel, 2003, "Affordance Based Methods for Design," Proceedings of ASME DETC '03, Design Theory and Methodology Conference, Chicago, IL, September 2-6, 2003.

[13]   Pahl, G. and W. Beitz, 1996, *Engineering Design: A Systematic Approach*, 2nd edition, Springer-Verlag (New York).

[14]   Rashevsky, 1960, *Mathematical Biophysics: Physico-Mathematical Foundations of Biology*, Volume 2, 3rd edition, Dover (New York).

[15]   Rosen, R., 1991, *Life Itself*, Columbia University Press (New York).

[16]   Rosen, R., 2000, *Essays on Life Itself*, Columbia University Press (New York).

[17]   Rosenbloom, P. S., J.E. Laird, and A. Newell (eds.), 1993, *The SOAR Papers: Research on Integrated Intelligence*, MIT Press (Cambridge, MA).

[18]   Senge, P. M., 1990, *The Fifth Discipline: The Art and Practice of the Learning Organization*, Currency Doubleday (New York).

[19]   Suh, N. P., 2001, *Axiomatic Design: Advances and Applications*, Oxford University Press (New York).

[20]   Tate, D. and M. Nordlund, 2001, "Research Methods for Design Theory," Proceedings of ASME DETC '01, Design Theory and Methodology Conference, Pittsburgh, PA, September 9-12, 2001.

[21]   Turing, A. M., 1959, *Can a Machine Think?*, Mind, **59**:236.

[22]   Wegner, P., 1997, *Why Interaction is More Powerful than Algorithms*, Comm. ACM, **40**(5):80-91.

[23]   Wegner, P., 1998, *Interactive Foundations of Computing*, Theoretical Computer Science, **192**(2):315-351.

Chapter 7

Spiraling out of Control:
Problem-Solving Dynamics in Complex Distributed Engineering Projects

Jürgen Mihm
WHU, Burgplatz 2, 56179 Vallendar, Germany
jmihm@whu.edu

Christoph H. Loch
INSEAD, Boulevard de Constance, 77305 Fontainebleau, France
christoph.loch@insead.edu

1. Introduction

Over the last two decades, the New Product Development (NPD) process has become an important focus of interest for industrial companies as well as academic researchers. The success of NPD process analyses [e.g., 6, 8] reflects this increased importance.

Under the pressure of foreign competition, many companies have realized the significance of NPD, not only for their economic well-being, but also for their very survival. In many industries, not only are 75-85% of life cycle production costs determined during the design process [7], but a delay of six months in NPD cycle time can reduce product profitability by a third over its life cycle [24], or, in fast clockspeed industries, such as notebook computers, by over a half [22]. Moreover, it has long been known that R&D investments boost business growth [e.g., 28].

The identification of NPD as a key success factor has given rise to a multitude of NPD management techniques, among which multi-functional teams have become the most prominent. However, in spite of advances in practice and theory, the NPD process remains plagued by failures and missed performance targets [e.g., 30, 43, 44]. Large engineering projects, in particular, seem to be prone to failure. Specifically, the pathology of "problem-solving oscillations" [25] seems to be recurring in large projects across many industries, as the following examples suggest.

We interviewed the project manager for the design of a new engine in a large automotive OEM. Such a project involves several hundred engineers and incorporates many component design choices concerning, for example, the mechanical configuration, materials, electronics and software. The project took off with one set of basic choices, only to run into trouble three months later because of some component performance hitch. So, another approach was tried, which also ran into trouble. The cycle of new design starts and failures repeated itself for another three times until the project manager took the decision to take the first set of parameters and work it all through, despite the obvious shortcomings. He commented, "In the end, we were where we had started out." (Figure 1).

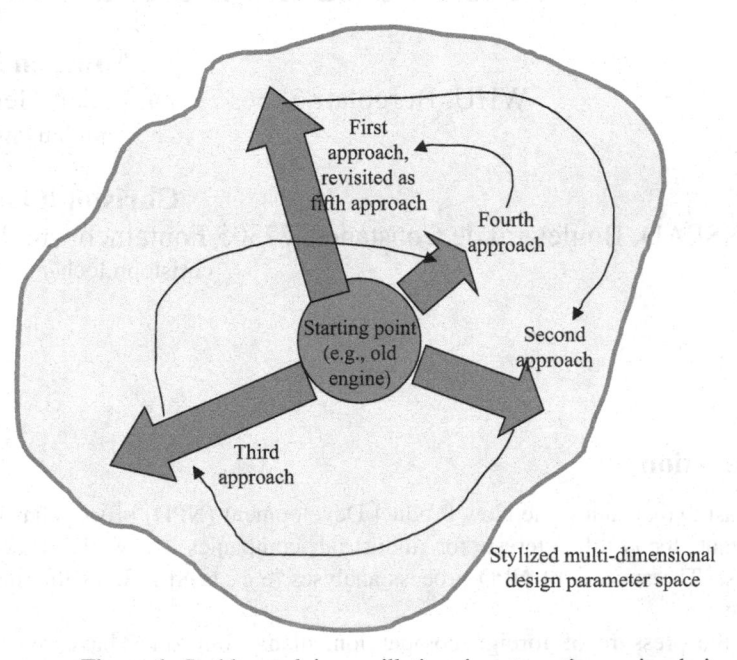

Figure 1. Problem-solving oscillations in automotive engine design.

In another automotive company, Yassine *et al.* [54] observed similar behavior. Progress oscillated between being ahead and behind schedule, and one engineer commented, "We just churn and chase our tails until someone says that they won't be able to make the launch date." Similarly, Terwiesch and Loch [44] and Terwiesch *et al.* [45] examined the effect of engineering change orders (ECOs) in the final phase of yet another car development project. They found that ECOs snowballed from one component to another, sometimes in cycles, causing budget and schedule overruns.

These effects are not confined to the automotive industry. Iansiti [16] documented the development progress of Microsoft Office, measured as the number of outstanding coding errors (bugs). The number of reported bugs did not simply converge to a marketable level, but exhibited large ups and downs. When developers worked to eliminate the bugs, new ones were introduced into the code, or old bugs became exposed, as performance problems required the introduction of new code. Bugs waxed and waned for over two years until the product finally converged. The same happened repeatedly for later products, not only at Microsoft, but also at many software companies [8].

In the aeronautics industry, the same phenomenon of problem-solving oscillations exists [17]: in the design of the Boeing 767-F, half of the engineering labor cost budget was spent on redoing work that the engineers had done before, because the original work did not lead to a satisfying systems result. About 25-30% of design decisions required rework. In some instances, up to 15 iterations were required for some of the decisions to reach a stable state, causing cost overruns and loss of quality. Klein *et al.* come to the conclusion that "the dynamics of current collaborative design processes are daunting, and have led to reduced design quality, long design cycles and needlessly high costs." Also in other industries, it is reported that rework can consume up to 50% of the engineering capacity and up to one third of the development budget [6, 35].

Consider one more example from the semiconductor industry. Hamilton [11] reports how the design of Intel's Itanium chip design went around in a circle. The team found itself in a "nightmarish world where a change to one module would ripple through the work of several hundred other people, leaving more problems in its wake." The design only finally converged after a manager introduced improved methods for quickly discovering the ripple effects of component design decisions.

What do these examples have in common? They all show that problem-solving oscillations in design are a major problem in large NPD projects. They suggest that instability is inherent to many design processes. Thus, the lack of control over the final product performance, experienced by many development managers, may be due to *systemic* reasons more than individual shortcomings.

Unfortunately, the technology management community (managers and academic researchers) does not have a clear understanding of which (combinations of) parameters have the ability of driving spirals of mutually amplifying changes that result in oscillations and divergence. In the dominant research paradigm of the social sciences over the last decades, the world of interactions is viewed as linear: if we do twice as much of A, we get roughly x times B. Thus, conclusions about the right choices can easily be drawn. Unfortunately, complex NPD projects represent systems of *interacting* design decisions in a *distributed* way: although there is a global architecture, complex NPD projects always have a substantial number of *local* component decisions because no one knows how to centrally optimize problems of the relevant complexity. Distributed interactive systems fundamentally behave in a *nonlinear* way – doing twice as much of A may give us 0.1B in one instance and 5 times B in another. In nonlinear systems, small local changes may have large global effects.

In the next section, we explain how the problem-solving oscillations described in this section systematically arise in complex systems, such as NPD projects. In Section 3, we will develop a number of actions that engineering management can take to avoid, or at least mitigate, system oscillations. In the final section, we draw conclusions from these insights.

2. Causes of Problem-Solving Oscillations

Large NPD projects represent examples of complex systems of many parts with many interactions, or, more formally, collections of components and activities that are "made up of a large number of parts that interact in non-simple ways, ... [such that] given the properties of the parts and ... their interactions, it is not a trivial matter to infer the properties of the whole" [34, pp. 195]. The characteristics of complex systems are fundamentally prone to problem-solving oscillations.

2.1 Why problem-solving iterations arise

A typical feature of complex systems is that the overall problem has to be partitioned into pieces in order to be manageable – economists call this "bounded rationality", or the inability to consider all system parts and their interactions in making a decision [e.g., 50]. This is also well-known in Operations Research [see, e.g., 37]. Thus, individuals or departments are assigned pieces of the problem, coordinated by a system architecture with defined interfaces. These individuals act locally to do the best they can with ("optimize") the pieces of the problem for which they are responsible. Ideally, in a perfectly modular system, the individuals can work in true separation from one another, but many interdependencies remain in most large NPD projects. In other words, the individuals influence one another, and while they may be aware of the influences, they cannot fully take them into account in their local decisions (that would amount to a perfect system optimization). As the component designs evolve over time, ongoing problem choices in other groups make the requirements for a particular group inherently unstable [e.g., 46, 50].

Consider the example in Figure 2, the climate control system (CCS) of a car. The CCS supplies the passenger cabin with outside air and controls the temperature and humidity of the air. This particular, relatively simple, system consists of 13 components (such as the radiator; the entire engine counts as one "component" for the purpose of the CCS) and 20 interactions (represented by arrows), not counting any interactions with the rest of the car. Interactions represent exchanges of material flows, mechanical force, energy (e.g., electricity), information, and the competition for space (the latter is not shown in the Figure).

The interactions can be represented in the form of a design structure matrix (DSM, see [9, 40]). The DSM lists all activities (which we can group by component) along the top (information providing) and along the side (information receiving). For example, an entry in column A (engine) and row B (cooling circuit) of the matrix represents a dependence (of the cooling circuit); entries both in column B and row C (radiator) and in column C and row B an interdependence (the components depend on each other). In large projects, we typically find *dependence cycles* – for example, the auxiliary heater influences the heating circuit, which influences the main unit (with motors and valves), which, in turn, affects the auxiliary heater through signals and space constraints.

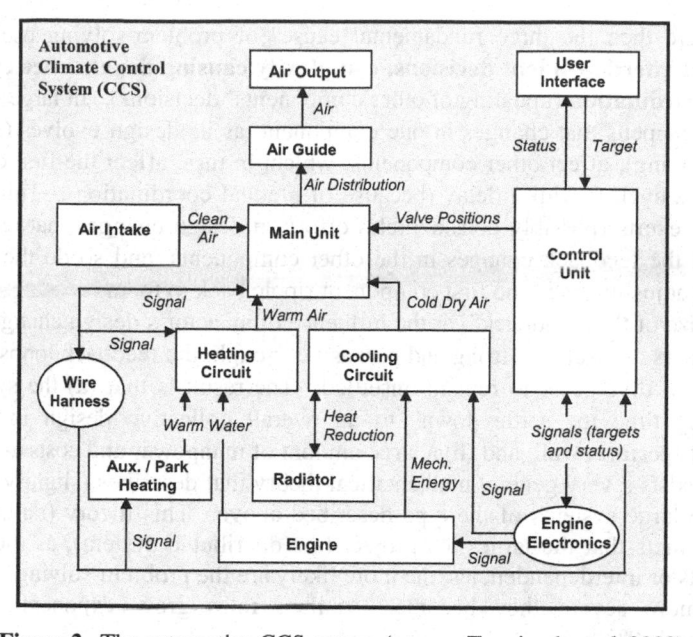

Figure 2. The automotive CCS system (*source*: Terwiesch *et al.* 2002).

The interdependencies require *coordination mechanisms* among the players: as one individual makes changes to his component, or sub-problem, the other players must be informed in order to take the changes into account in their own decisions. Coordination mechanisms range from static *a priori* rules (such as design handbooks) over formal design meetings and engineering changes to frequent communication in cross-functional teams [e.g., 1]. One specific way to exchange design information is through prototypes [6, 8]. Often, problems are discovered when a prototype is assembled from individual components. In addition to official mechanisms, coordination frequently happens when the individual component owners meet informally (at random encounters, at lunch, etc.), or even accidentally, and exchange information about the respective state of their designs.

The different coordination mechanisms vary in their "channel capacity" or their ability to convey ill-defined and preliminary information; static channels have the lowest capacity, and informal face-to-face communication the highest [1]. Thus, conventional wisdom has long advocated using high-bandwidth communication channels in complex projects. Unfortunately, the high bandwidth channels also consume a lot of time. No communication channel provides perfect coordination because the organization quickly suffocates under information overload. In all large and complex NPD projects, therefore, information travels gradually, and thus, component engineers receive updates of their peers' decisions with a delay. Modern communication media (such as e-mail, CAD files, electronic meetings, videoconferencing) do *not* ensure that all project members are immediately informed of all important events. They merely force the individual engineers to choose what they pay attention to.

These are then the three fundamental causes of problem-solving oscillations – **distributed interdependent decisions, complexity causing dependence cycles, and gradual coordination** (updating of other components' decisions). In large projects, it inevitably happens that changes in one component, as its design evolves (distributed problem solving), affect other components, which, in turn, affect the first component (dependence cycles) with a delay (because of gradual coordination). Thus, the first component exerts (possibly considerable) effort on design changes that are obsolete because of the feedback changes in the other components, and so do the others, as continuing adjustments in the first component ripple back to them.

If the size of the "feedback", or the influence of an actor's design change reflected through others on itself, is strong and pervasive enough, the feedback loops can cause the system to diverge or to remain unsettled. The result is that (a) the system may take a long time to "settle down" to an overall collective design in which all components perform well, and (b) a large amount of manpower and costs are wasted.

There exists a very general mathematical theory that describes (slightly simplified models of) large systems of the type described above. This theory (random matrix theory) predicts that the larger the project (the distributed system), as measured by components or interdependencies, the more likely are the problem-solving oscillations and the more severe they become – failure rates grow exponentially before approaching 100%.[1]

Random matrix theory, thus, predicts that a complex NPD project (featuring either many components, or many interactions, or both) *inherently* has a high probability of problem-solving oscillations, of instability, and of a long design conversion time. The larger the system, the worse the oscillations become. In other words, large complex projects are hard to manage, and often fail, even when none does anything wrong. This prediction is largely borne out by empirical evidence – large projects often *do* fail, and companies undertaking them must either develop special expertise or abandon them (for example, there is a trend in the software industry to keep projects smaller, see [31]).

2.2 The insufficiency of modularity and rich communication channels

One often hears the assertion, "If we could just communicate effectively, we would avoid all this churning." This is also often proposed in NPD literature [e.g., 1, 13]. However, we have already alluded to the fact that modern, rich communication channels will never eliminate problem-solving oscillations because the availability of ubiquitous information does not remove the processing bottleneck of individual component designers or groups. Even if *all* design changes were posted *every day* on

---

[1] Take the DSM and write in the element of the i-th row and the j-th column by how much component i would change its design [in a suitable metric], in order to maintain its best component solution, when component j changes its design by a small amount. This is called the Jacobian matrix of the linearized system, and it describes the system's behavior in the sense that its largest [real] eigenvalue represents the size of the feedback of a component change through other components for itself. That is, if the largest eigenvalue is positive, the system tends to oscillate and diverge, while a negative largest eigenvalue characterizes a dampened system in which oscillations quickly die out. Random matrix theory formally shows that the largest eigenvalue grows as the system size grows (everything else being equal), see [25, 26].

a central electronic bulletin, the component designers would *still* not have the time to process them. As one manager remarked to us, "We simply do not have the time and luxury to have everyone fully informed all the time of everything that is going on."

A subtle and devious effect of oscillations, especially in the presence of rich communication channels, lies in *equivoque*, or the inability of the participants to understand and interpret what is going on. As the reasons for problem-solving oscillations are not widely understood, it is not uncommon for project teams to be caught up in endless cycles of rippling changes, regardless of intensive communication, to ask themselves why they are facing such difficulties, and to doubt their competence. Seemingly unexplainable problems can cause severe stress and dysfunction [e.g., 45, 52].

A second widely discussed tool to reduce complexity is *modularity*. A modular system is one with few, and well defined, interfaces cutting across the modules (component groups) and functions of the product [e.g., 47]. For example, software modules are classes or subroutines that have a clear interface for evoking them. In car development, modularization comes in the form of mechanical "chunks" with clear interfaces to the rest of the car. For example, a car engine is developed largely independently of the body.

Modularity, in effect, reduces complexity itself by dividing the complex system into several smaller subsystems, which do not (or barely) interact. Making the subsystems smaller already strongly reduces the oscillations effect (as random matrix theory predicts), without any additional improvements. Why does not everyone build modular systems, if they are so helpful? The answer is that the design restrictions imposed by modularity reduce system performance and compactness, especially for products incorporating new technologies that are not yet fully understood. In particular, modularization limits the search space of the design team, which may result in a sub-optimal solution [10]. Suffering these performance disadvantages may well make a product uncompetitive [e.g., 48]. Thus, modularity is not always an option.

A more subtle form of modularity is a "feed-forward" structure of dependencies among the components.[2] This means that the dependencies "order" the components into a natural sequence that will never produce any feedback dependencies to previous components. Such a structure prevents any dependence cycles and, thus, any oscillations. But again, this puts heavy restrictions on the permissible product design (only strictly sequential dependencies), and is rarely feasible for a competitive product [e.g., 49].

---

[2] Such a feed-forward dependence structure means that the DSM is lower-triangular, that is, all entries are in the lower left half under the diagonal, reflecting the fact that the components can be re-numbered in such a way that no dependence points back to a lower-numbered component. Traditional project planning methods, such a PERT or critical path, also assume a feed-forward dependence structure.

3. What Can We Do? – Managerial Actions to Mitigate Oscillations

In the previous section, we argued that one way to reduce problem-solving oscillations is to attack their root, complexity in itself, by limiting the system size or the number of interdependencies (modularity). While such complexity reduction is not always feasible, there is a range of less radical levers that engineering managers have at their disposal to at least mitigate oscillations. Some of these are well-known to managers, but are usually discussed in different contexts, and have not been clearly linked to the roots of problem-solving oscillations.

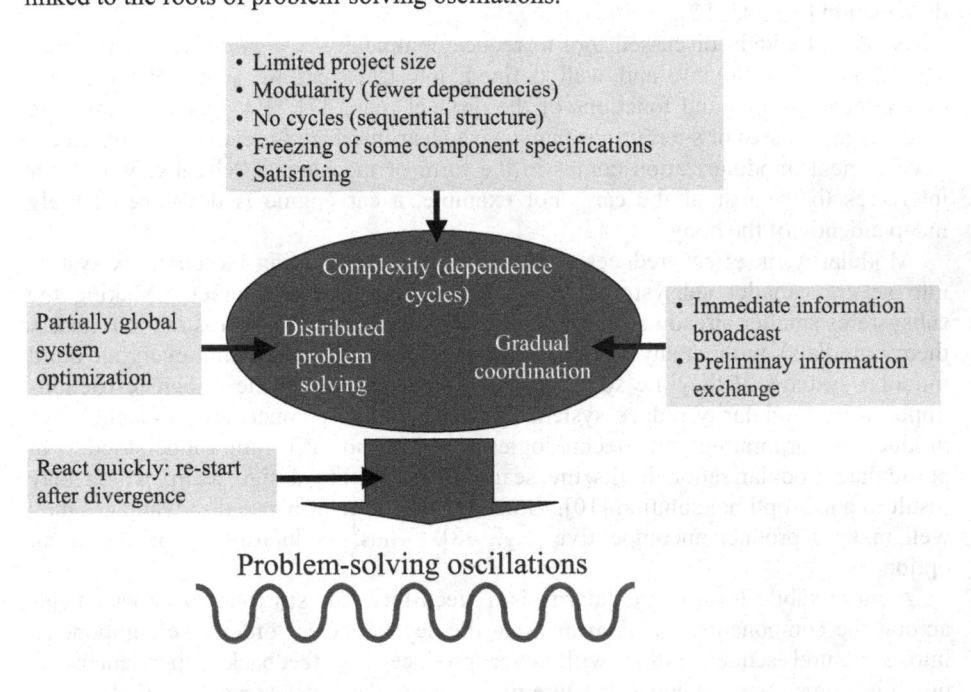

Figure 3. Actions mitigating problem-solving oscillations.

Figure 3 summarizes, together with the complexity reduction measures from Section 2, seven types of actions that managers can take; we discuss them in detail in the remainder of this section. The first two (re-start and freeze) are the standard results that happen when problem-solving oscillations are not consciously managed. The third and fourth (global optimization and immediate communication) represent attempts to eliminate the two conditions for oscillations (which are required in addition to complexity), namely, distributed problem solving and gradual coordination. The last three (satisficing, communicating preliminary information and introducing a coordinating hierarchy) have not previously been discussed in the context of problem-solving oscillations and are somewhat counterintuitive.

3.1 Re-start after system has diverged

An oscillating system may not converge to a satisfactory design at all, but rather "spiral out of control", that is, the system may exhibit components that are forced into less and less adequate solutions by the restrictions from other components. A "diverged" design does not, of course, reach the market – the design is stopped, and either the project is scrapped, or design starts anew from a known starting point. The trick is to stop a project quickly if divergence becomes apparent. We have seen many projects that continued to linger on, although it was clear that they would not lead to reasonable results. The waste of time and resources of indecision can be much worse than the pain caused by decisively re-starting.

What makes this difficult to do is, again, equivoque, as in Section 2. It is not clear exactly how to define the point when divergence of a project becomes apparent. For a while, it is not apparent what the outcome of the project will be, whether it will oscillate a bit longer and then converge, or fail. During this phase, project organizations experience enormous stress. The tendency to hang on for too long is called "over-commitment" – hope, social pressure, or an unwillingness to give up what one has achieved, may prevent the decision to stop [see 39]. Sometimes, a new decision maker is needed to look at progress with an unbiased eye, and make the call [4]. Installing a regular review process and introducing quality gates for NPD projects can help the organization to utilize and translate into action previous experience with oscillating projects.

3.2 Freeze the specifications of some components

It is an important architectural choice as to what should be optimized for the system and what components and interfaces are less important. Such "secondary" system elements may be fixed at some point during the development process. Freezing specifications stops short of segmenting the design (or reducing complexity itself, as modularization does). It defines which optimizations across interfaces have precedence over other optimizations. Holding some components and interfaces fixed reduces the size of the part of the design system that contributes to oscillations. This is a way of bringing a spiraling project under control, albeit at the cost of performance, as changing the fixed components may be beneficial.

Take the example of developing an integrated entertainment system (CD, radio, cassette, TV, GPS, and links to the phone) as part of a new car model. After a long "back and forth" about the best operating system (OS) for the software, the team had to settle on one (freeze the decision) although the choice was not the one with the highest performance (and hotly contested). However, without the freezing, the many other components of the project had no chance of converging to a design.

More generally, some of the luxury car manufacturers have traditionally given design and feature decisions more emphasis than they have to production issues. As a result, feedback loops from production back to product elements of the design process are limited. This eliminates some problem-solving oscillations, at the expense of production cost.

Experience plays an important role in freezing decisions. An organization developing a next generation product, based on a well-known architecture and well-understood technologies, can predict many aspects of the system's performance. The organization can choose in advance ranges of design parameters that are likely to yield high performance. Thus, many parameters can be frozen (i.e., ranges do not need to be considered) without trading off performance. In contrast, when freezing is used in novel projects, it is often not understood what performance is sacrificed; rather, the freezing is defensive in order to get the project's progress under control.

3.3 Satisfice

Like freezing specifications, satisficing substitutes for truly reducing complexity. It achieves this by requiring the individual component designer to forgo the last bit of local component performance improvement.

In many design projects, significant initial progress is made quickly, while a large fraction of iterations are linked to the last few percent of performance (fine-tuning and perfecting). By not insisting on the last few percent, it often turns out that the number of iterations, and thus problem-solving oscillations, significantly decrease [23]. Because of the non-linear deterioration in system convergence, a small compromise in component performance may buy a disproportionally large improvement in design iterations and design conversion time. "Enough is enough" turns out to be a good motto for fast design convergence.

Of course, satisficing can be dangerous to overall system performance if not used consciously and in a controlled fashion. Designers have to be aware that, for some components, even small reductions in overall performance are not acceptable. For others, there is large leeway. Therefore, the overall architect (or the project leadership) needs to specify in advance which elements must be fully optimized and which components are allowed to compromise. This explicit reasoning for satisficing is important also psychologically, to prevent a creeping spirit of mediocrity ("I didn't insist on the best solution not because I wasn't ambitious enough but for the good of everybody.").

The well-known method of "design to cost" can be used to target satisficing. Design to cost efforts select those product components and features that result in the largest benefit for the customer. Thus, the method can help in selecting those components that can be subject to satisficing (either no benefit, or already sufficient benefit).

3.4 Design components for partial system optimization

Large, complex projects are characterized by distributed problem solving. In other words, the only way we know how to accomplish the task is to cut the overall problem into pieces (components), with as few and well defined interfaces as possible, and work on them separately. For example, there is an engineer who designs door handles, and he designs the best door handles in the world! The components have local performance measures. Unfortunately, these local performance measures take on a life of their own (e.g., over-designing the door handle with no customer benefit but a cost penalty), and they lead to problem-solving oscillations, as we have seen.

Now, imagine that the component engineers could take into account the effects of their decisions on the other components, trading off their own local performance for system performance. Of course, this is not possible because it would amount to global system optimization again (and it is a feature of large, complex projects that we do not know how to accomplish system optimization). However, it is often possible partially, for larger subsystems. For example, an entire interface (a large module) of a software product or the engine of a car (a complex system in itself) may be optimized as a whole. Optimizing larger system chunks amounts to reducing the effective number of "components" of the system (i.e., fewer distributed optimizations happen), which simultaneously improves system performance and reduces oscillations. As for specification freezing, the ability to optimize larger system chunks is greatly aided by experience and the use of well-known architectures and technologies.

Having system optimization in mind when designing components is sometimes surprisingly difficult to achieve. Power struggles, pride in one's own work, social identity and emotional relationships, in addition to a "rational" assessment of efficiency, influence behavior. First, it is well established [e.g., 18, 42] that group identity and perceived conflicts over resources or power have a strong impact on cooperation. In other words, if a design trade-off must cut across perceived group boundaries (such as different engineering disciplines, or the functional boundary engineering versus marketing), the actors may refuse to collaborate. If a common group identity exists for the information exchanging parties, information will be exchanged more freely. In one case, an electronics engineer in a car development project commented, "We work within a few micro inches, but these mechanical metal bashers, they work with tolerances of a quarter of an inch. Isn't that ridiculous!"

Second, personal relationships are required in order to overcome a general reluctance to share information. For example, Uzzi [49] investigated buyer-supplier relationships in the New York fashion industry, and found that ongoing relationships tended to make economic exchanges reliable, and to eliminate opportunistic behavior and free-riding (however, if the relationships persisted too long, their purpose took on an identity of its own, independent of, and sometimes to the detriment of, the economic rationale). Similarly, Bensaou and Anderson [3] found, in a study of buyer-supplier relationships, that those with personal ties performed better.

3.5 *Immediate communication broadcasts*

In addition to attacking complexity itself and distributed problem solving, managers can also attack the third reason for oscillations, namely communication delays (and thus gradual coordination). We stated in Section 1.2.2 that it is obviously impossible to inform every one of all relevant events in a project – information inundation and overload would result. However, as for system optimization (Section 1.3.4), it is often possible to do this partially, by quickly communicating a component design change to *a few critical* other components with the strongest dependence.

The DSM of interdependencies in a design project is usually only thinly populated, with typical densities of about 10% [e.g., 38]. That is, of all possible dependencies in the matrix,[3] only 10% actually exist. It is, therefore, often possible to quickly communicate along a few important dependencies, thus avoiding work that is based on obsolete assumptions about other components.

---

[3] In a system of N components, the fully populated DSM matrix contains $N(N-1)$ dependencies.

Executing this focused quick communication requires process discipline and awareness. Awareness, or the knowledge of the parties in the project of the nature of their interdependencies, may seem trivial and self-evident. It may be trivial, perhaps, if all personnel working in the project are experienced and have worked in similar projects before, but in most projects, a large fraction of personnel is not experienced. We have, in our work, encountered numerous examples of project workers who negatively impact fellow workers simply because they were not even aware of the fact that an interaction existed. Thus, it is an important first investment in a project to educate the participants about the overall project architecture and about interdependencies, and to give them an opportunity to explore the precise content of the interdependencies. The DSM is one widely known tool for this education. Such initial education is an important investment.

There are different methods for ensuring quick communication. Co-location is still the most effective, but not everyone in a project of several hundred people can be co-located. CAD technologies help by making certain dependencies (such as spatial conflicts) quickly visible. Fast compilers and "daily builds" help to update information quickly in software development [e.g., 8].

3.6 Exchange preliminary information

In addition to quickly communicating *final* design choices, oscillations can be dramatically dampened by also communicating *preliminary* information, that is, an intermediate status of an evolving design, which is still likely to change [23]. Exchanging preliminary information helps because it, again, reduces the amount of work other components do, based on obsolete information that they regard as final and reliable.

Preliminary information makes dependencies and uncertainty explicit by labeling a design status as unfinished. Preliminary information is either *unstable* (I estimate that the engine will deliver 4 gallons of hot water per hour to the heating unit, but don't rely on this estimate because it may change, so the heating unit may have to find additional heating energy, or not) or *imprecise* (I don't know yet how much hot water the engine will deliver; it may vary between 2 and 6 gallons per hour), or both [see 45]. Communicating the type of uncertainty in the design status can help the dependent components to adjust their own component decisions and, thus, avoid wasted effort. Signaling the preliminary nature of the information also helps to reduce equivoque, as it makes the component designers aware of the overall status of the project.

Similar to global system optimization (Section 1.3.4), preliminary information often fails because of biases in human behavior. Engineers have a tendency not to communicate preliminary information to their peers, sometimes because of perfectionism, and often because having to reverse a decision may be viewed as a lack of technical understanding and expertise. Therefore, engineers often hold back information until they can be sure that their choice is justified.

Regular meetings (formal and informal) can foster the exchange of preliminary information. However, as we discussed in Section 1.2.2, additional communication comes at a cost. First, meetings cost time, and second, the information exchanges become a subject of reciprocity considerations (see Section 1.3.4): "I will not move before you give me your final specifications, and you will not move before I give you mine." This is sometimes called coordination fever. Reciprocal relationships can help to boost cooperation, but may also cause spirals of retaliation, thus reducing information exchanges.

3.7 Introduce a coordinating hierarchy

Problem-solving oscillations represent an area of unexpected benefits from introducing a hierarchy. We have already argued that forcing modularity on a design, or ignoring interdependencies that cut across groups, may substantially compromise overall system performance. A hierarchy offers the following compromise: Form groups of highly connected components and communicate frequently, and without delay, within those groups, and communicate less frequently at defined points in time, through a coordinating manager, across groups. The grouping produces some of the benefits of modularity (because the groups correspond to smaller subsystems), and the periodic coordination across groups avoids the downside of too much treating the groups in isolation [26].

This works well if it is naturally possible to form the overall design team into groups in which most interactions are concentrated. NPD organizations often do this intuitively – they structure the project organization around subsystems that are highly connected internally and less connected to other subsystems [29, 38].

This is, however, subtle, and may have unexpected negative effects. First, the structure of the groups and the hierarchy itself influence communication (e.g., communication across groups becomes much harder as group identities are formed) and coordination patterns and, thus, the outcome of the design effort. Therefore, the project leadership must ensure that important interdependencies are not overlooked.

Second, the effort to concentrate design interdependencies within the groups may go too far – the groups (sub-problems) themselves may get so intertwined and complex in the process, that they themselves start oscillating. For example, manufacturing engineers, applying the method of design-for-assembly, have noticed "design origami": in its drive for reducing the number of parts, design-for-assembly sometimes produces parts that are so complex that they drive total cost up instead of down [e.g., 41].

Third, a highly split structure limits the search space of the organization: as the sub-problems of the groups become smaller, each performs a more and more local solution search. As a whole, this may substantially reduce the overall amount of search, and result in an incremental and low-performing system [32]. In short, the organization needs to seek a balance between decentralization and centralization. While no one knows where the "optimal balance" lies, knowing that neither extreme is desirable is already helpful.

4. Discussion and Conclusion

In this chapter, we started with the observation that large scale distributed activities in organizations tend to lead to problem-solving oscillations, or repeated "cycling" of the problem-solving evolution. Such oscillations imply design process instability and unpredictable design conversion times. The examples in the Introduction, as well as theoretical work, suggest that these problems are pervasive in large, complex design projects.

Problem-solving oscillations are driven by a combination of distributed interdependent decisions, complexity causing dependence cycles, and gradual coordination (delayed status updates across components). Theoretical considerations suggest that oscillations are, to some degree, inevitable in large, complex projects, and cannot be entirely avoided.

However, there are management techniques (summarized in Figure 3) that can *mitigate* them. The radical conclusion would be simply not to undertake large projects. However, less extreme measures are available. First, the degree of distributed interdependent decisions can be reduced by collaboration and, at least partial, system optimization (for example, by optimizing larger, aggregated chunks rather than only components). Second, complexity itself can sometimes be reduced (system modularity or a feed-forward structure of dependencies) or managed (by freezing components or satisficing with respect to component performance, in order to avoid iterations from perfectionism). Third, modern information technologies make coordinating information more freely available to project workers, and can contribute to faster information exchange *if* they are used in a focused way, and if people understand the advantage of communicating. Specifically, project workers often have to change their attitudes (of not wanting to be seen as changing their minds) in order to be motivated to exchange preliminary information. Thus, product and organizational architecture and collaboration *behavior* have a great impact on the problem-solving convergence of complex engineering projects.

In some industries, a trend is emerging to restructure design processes toward becoming more robust in a sense that is consistent with our discussion. For example, "functional design" is being discussed in automotive engineering. The idea is that functionally defined, rather than physical component, units have fewer interfaces – to give one example, the functional unit "driver interface" characterizes a set of increasingly interacting components (driven by electronic integration), such as steering, acceleration and deceleration, gear shifting, control of car movement and status, communication and entertainment. Approaching the design from this angle is hoped to allow better optimization of this set of functions (incorporating many components), and, simultaneously, allow clearer definition of the interfaces between this functionally defined chunk and the rest of the car. The effectiveness of this approach will have to emerge (the benefit is currently not proven, especially in the light of the fact that automobiles are becoming more complex).

Many companies believe that new technologies (such as 3D-CAD and distributed databases) will drive productivity in NPD projects. This chapter argues that new technologies improve the potential design performance, but also make project execution harder by increasing interdependencies and complexity. Breakthrough improvements will come from a *combination* of new technologies and creative thinking about the drivers of problem-solving oscillations.

References

[1]   Adler, P. S., 1995, Interdepartmental Interdependence and Coordination: The Case of the Design/Manufacturing Interface, *Organization Science,* 6 (2) 147 – 167.

[2]   Allen, T. J., 1966, Studies of the problem-solving process in engineering design, *IEEE Transactions on Engineering Management,* **13 (2)**, 72.

[3]   Bensaou, M., E. Anderson, 1999, Buyer-Supplier Relations in Industrial Markets: When Do Buyers Risk Making Idiosyncratic Investments? *Organization Science* 10 (4), 460-481.

[4]   Boulding, W., Morgan, R., Staelin, R., 1997, Pulling the plug to stop the new product drain, *Journal of Marketing Research* 34 (1), 164 – 177.

[5]   Brooks, R., Maes, P. (ed.), 1994, *Artificial Life IV,* MIT Press (Cambridge).

[6]   Clark, K. B., Fujimoto, T., 1991, *Product Development Performance: Strategy, Organization and Management in the World Auto Industry,* Harvard Business School Press (Boston, MA).

[7]   Creese, R. C., Moore, T., 1990, Cost Modeling for concurrent engineering, Cost Engineering **32(6)**, 23.

[8]   Cusumano, M. A., Selby, R. W., 1995, *Microsoft Secrets.* New York: the Free Press.

[9]   Eppinger, S. D., Whitney, D. E., Smith, R. P., Gebala, D. A., 1994, A Model-Based Method for Organizing Tasks in Product Development, *Research in Engineering Design* 6, 1-13.

[10]   Ethiraj, S.K., Levinthal, D., 2004, Modularity and Innovation in Complex Systems. *Management Science,* forthcoming.

[11]   Hamilton, D.P., 2001, Circuit Break: Gambling It Can Move Beyond the PC, Intel Offers a New Microprocessor. *Wall Street Journal,* May 29, A1.

[12]   Hartman, J., Wernecke, J., 1996, *The VRML 2.0 Handbook - Building Moving Worlds on the Web,* Addison-Wesley Developers Press (Reading).

[13]   Hauptman, O., Hirji, K. K., 1996, The Influence of Process Concurrency on Project Outcomes in Product Development: an Empirical Study with Cross-Functional Teams, *IEEE Transactions in Engineering Management* 43, 153 - 164.

[14]   Heudin, J.C. (ed.), 1998, *Virtual Worlds - Proceedings of the First Int. Conf. on Virtual Worlds,* Springer-Verlag Lecture Notes in Computer Science (Berlin), **1434**, 5.

[15]   Husbands, P., Harvey, I. (ed.), 1997, *Fourth European Conference on Artificial Life,* MIT Press (Cambridge).

[16]   Iansiti, M., 1990, Microsoft Corporation: Office Business Unit, HBS Case 9-691-033.

[17]   Klein, M., Braha, D., Syama, H., Bar-Yam, Y., 2003, Editorial: Special Issue on a Complex System Perspective on Concurrent Engineering, *Concurrent Engineering Research and Applications,* **11(3)**, 163.

[18]   Kramer, R. M., 1991, Intergroup relations and organizational dilemmas: the role of categorization processes. *Research in Organizational Behavior* **13**, 191 – 228.

[19]   Krueger, M.W., 1991, *Artificial Reality II,* Addison-Wesley (Reading).

[20]   Langton, C.G., 1988, Artificial Life, in *Artificial Life,* edited by C.G. Langton, SFI Studies in the Sciences of Complexity, Addison-Wesley (Reading), **6**, 1.

[21]   Langton, C.G. (ed.), 1994, *Artificial Life III,* SFI Studies in the Sciences of Complexity, Addison-Wesley (Reading), **17**.

[22]   Loch, C. H., 1999, Acer Mobile Systems Unit (A) and (B), INSEAD case 03-99-4825.

[23]   Loch, C. H., Mihm, J., Huchzermeier, A., 2003, Concurrent Engineering and Design Oscillations in Complex Engineering Projects, *Concurrent Engineering Research and Applications,* **11(3)**, 187.

[24] McDonough III, E. F., 1993, Faster New Product Development: Investigating the Effects for Technology and Characteristics of the Project Leader and Team, *Journal of Product and Innovation Management*, **10(3)**, 241.

[25] Mihm, J., Loch, C.H., Huchzermeier, A., 2003a, Problem Solving Oscillations in Complex Engineering Projects, *Management Science*, **49(6)**, 733.

[26] Mihm, J., Loch, C.H., Huberman, B., Wilkinson, D., 2003b, Hierarchies and Problem Solving Oscillations in Complex Organizations, INSEAD-*WHU Working Paper*.

[27] Mihm, J., 2003, Complexity in New Product Development, DUV (Wiesbaden).

[28] Morbey, G. K., 1988, R&D: Its relationship to company performance, *Journal of Product Innovation Management* 5(3), 191 – 200.

[29] Morelli, M. D., Eppinger, S. D., Gulati, R. K., 1995, Predicting Technical Communication in Product Development Organizations, *IEEE Transactions on Engineering Management* 42, 215 - 222.

[30] Morris, P. W. G., Hugh, G. H., 1987, *The Anatomy of Major Projects*, Wiley (Chichester).

[31] Reel, J.S., 1999, Critical success factors in software projects, *IEEE Software* 16(3), 18 – 23.

[32] Rivkin, J.W., Siggelkow, N., 2003, Balancing search and stability: interdependencies among elements of organizational design, *Management Science* 49 (3), 290 - 311.

[33] Sanchez, E., Tomassini, M. (ed.), 1996, *Towards Evolvable Hardware - The Evolutionary Engineering Approach*, Springer-Verlag Lecture Notes in Computer Science (Berlin), **1062**.

[34] Simon, H. A., 1969, *The Sciences of the Artificial*. Cambridge, MA: MIT Press.

[35] Soderberg, L. G., 1989, Facing up to the Engineering Gap, *The McKinsey Quarterly*, Spring, 3 – 23.

[36] Sutherland, I., 1965, The Ultimate Display, *Proceedings IFIP Congress*, 506.

[37] Sobieszczanski-Sobieski, J., Agte, J. S., Sandusky, R. R., 1998, Bi-level integrated system synthesis (BLISS). American Institute of Aeronautics and Astronautics (AIAA) Report NASA/TM-1998-208715.

[38] Sosa, M. E., Eppinger, S.D., Rowles, C.M. 2003, Identifying modular and integrative systems and their impact on design team interactions, *Transactions of the ASME*, **125**, 240.

[39] Staw, B.M., Ross, J., 1987, Behavior in escalation situations: antecedents, prototypes and solutions, *Research in Organizational Behavior* 9, 39 – 78.

[40] Steward, D. V., 1981, *Systems Analysis and Management: Structure, Strategy and Design*. Petrocelli Books, New York.

[41] Stoll, H. W., 1986, Design for Manufacture: an Overview, *Applied Mechanics Review* 39 (9), 1356 – 1364.

[42] Tajfel, H., 1982, Social psychology of intergroup relations, *Annual Review of Psychology* 33, 1 - 39.

[43] Tatikonda, M. V., Rosenthal, S. R., 2000, Technology Novelty, Project Complexity and Product Development Project Execution Success, *IEEE Transaction on Engineering Management*, **47(1)**, 74.

[44] Terwiesch, C., Loch, C. H., 1999, Measuring the Effectiveness of Overlapping Development Activities, *Management Science*, **45(4)**, 455.

[45] Terwiesch, C., Loch, C. H., De Meyer, A., 2002, Exchanging Preliminary Information in Concurrent Engineering: Alternative Coordination Strategies, *Organization Science* 13 (4), 402 - 419.

[46] Thomke, S. H., 1997. The role of flexibility in the development of new products, *Research Policy*, **26**, 105 - 119.

[47] Ulrich, K.T., 1995, The role of product architecture in the manufacturing firm, *Research Policy* 24, 419 – 440.

[48] Ulrich, K.T., Ellison, D.J., 1999, Holistic customer requirements and the design-select decision, *Management Science* 45, 641 – 658.

[49] Uzzi, B., 1997, Social Structure and Competition in Interfirm Networks: The Paradox of Embeddedness, *Administrative Science Quarterly* 42, 35 - 67.

[50] Van Zandt, T, 1999, Decentralized information processing in the theory of organizations, in: Sertel, M. (ed.), *Contemporary Economic Issues* Vol. 4, London: MacMillan, 125 - 160.

[51] Verna, D., Grumbach, A., 1998, Can we define Virtual Reality? The MRIC Model, in *Virtual Worlds*, edited by J.C. Heudin, Springer-Verlag Lecture Notes in Computer Science (Berlin), **1434**, 41.

[52] Weick, K. E., 1993, The Collapse of Sensemaking in Organizations: the Mann Gulch Disaster, *Administrative Science Quarterly* 38 (4), 628 – 652.

[53] Womack, J. P., Jones, D. T., Roos, D., 1990, *The machine that changed the world: The story of lean production*, Rawson Associates (New York).

[54] Yassine, A., Joglekar, N., Braha, D., Eppinger, S., Whitney, D., 2003, Information Hiding in Product Development: The Design Churn Effect, *Research in Engineering Design*, 14 (3), 131-144.

Chapter 8

The Dynamics of Collaborative Design: Insights From Complex Systems and Negotiation Research

Mark Klein
Center for Coordination Science
Massachusetts Institute of Technology
Cambridge, MA 02139
m_klein@mit.edu

Hiroki Sayama
University of Electro-Communications
Tokyo, Japan
sayama@cx.hc.uec.ac.jp

Peyman Faratin
Laboratory for Computer Science
Massachusetts Institute of Technology
Cambridge MA 02139
peyman@mit,edu

Yaneer Bar-Yam
New England Complex Systems Institute
Cambridge MA 02138
yaneer@necsi.org

1. Introduction

Almost all complex artifacts nowadays, including physical artifacts such as airplanes, as well as informational artifacts such as software, organizations, business processes and so on, are defined via the interaction of many, sometimes thousands of participants, working on different elements of the design. This *collaborative design* process is challenging because strong interdependencies between design decisions make it difficult to converge on a single design that satisfies these dependencies and is acceptable to all participants. Current collaborative design approaches are as a result typically characterized by heavy reliance on expensive and time-consuming processes, poor incorporation of some important design concerns (typically later life-cycle issues such as environmental impact), as well as reduced creativity due to the tendency to incrementally modify known successful designs rather than explore radically different and potentially superior ones.

Research on negotiation focuses on understanding what local behaviors are to be expected from (relatively small numbers of) self-interested agents attempting to come to agreements in the face of interdependencies. Complex systems research compliments this perspective by attempting to understand the global dynamics that emerge as the collective effect of many such local decisions. These two perspectives, when brought together, have we believe much to offer to a understanding of the dynamics of collaborative design. The remainder of this paper is dedicated to exploring some of these insights.

2. A Model of Collaborative Design

Let us first establish a working definition of collaborative design. A design (of physical artifacts such as cars and planes as well as behavioral ones such as plans, schedules, production processes or software) can be represented as a set of *issues* (sometimes also known as *parameters*) each with a unique value. A complete design for an artifact includes issues that capture the *requirements* for the artifact, the *specification* of the artifact itself (e.g. the geometry and materials), the *process* for creating the artifact (e.g. the manufacturing process) and so on through the artifacts' entire life cycle. If we imagine that the possible values for every issue are each laid along their own orthogonal axis, then the resulting multi-dimensional space can be called the *design space*, wherein every point represents a distinct (though not necessarily good or even physically possible) design. The choices for each design issue are typically highly *interdependent*. Typical sources of inter-dependency include shared resource (e.g. weight, cost) limits, geometric fit, spatial separation requirements, I/O interface conventions, timing constraints etc.

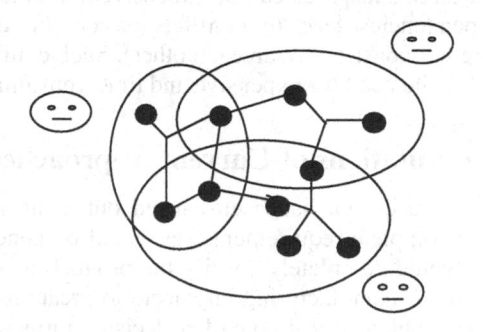

Figure 1. A Model for Collaborative Design

Collaborative design is performed by multiple participants (representing individuals, teams or even entire organizations), each potentially capable of proposing values for design issues and/or evaluating these choices from their own particular perspective (e.g. manufacturability). Figure 1 below illustrates this model: the small black circles represent design issues, the links between the issues represent design issue inter-dependencies, and the large ovals represent the design *subspace* (i.e. subset of design issues) associated with each design participant. In a large artifact like a commercial jet there may be millions of components and design issues, hundreds to thousands of participants, working on hundreds of distinct design subspaces, all collaborating to produce a complete design.

Some designs are better than others. We can in principle assign a *utility* value to each design and thereby define a *utility function* that represents the utility for every point in the design space (though in practice we may only be able to assess *comparative* as opposed to *absolute* utility values). A simple utility function might look like the following:

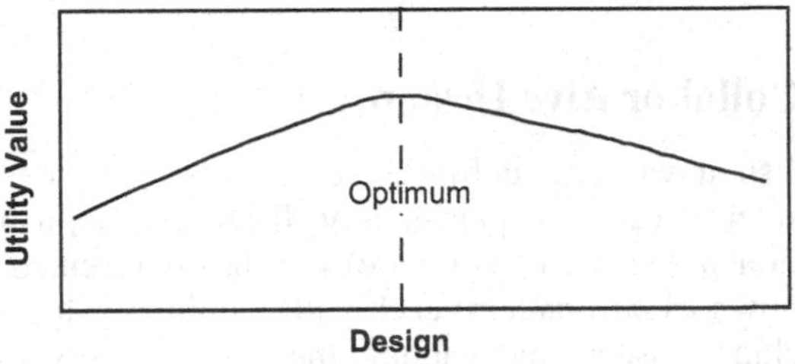

Figure 2. A simple utility function, with a single optimum

The *goal* of the design process can thus be viewed as trying to find the design with the optimal (maximal) utility value, though often optimality is abandoned in favor of 'good enough'.

The key challenge raised by the collaborative design of complex artifacts is that the design spaces are typically huge, and concurrent search by the many participants through the different design subspaces can be expensive and time-consuming because design issue interdependencies lead to conflicts (when the design solutions for different subspaces are not consistent with each other). Such conflicts severely impact design utility and lead to the need for expensive and time-consuming design rework.

3. Strengths and Limitations of Current Approaches

Traditionally, collaborative design has been carried out using a serialized process, wherein for example a complete requirements set would be generated, then given to design engineers who would completely specify the product geometry, which in turn would then be given to the manufacturing engineers to create a manufacturing plan, and so on. This has the problem that if an earlier decision turns out to be sub-optimal from the perspective of someone making dependent decisions later on in the design process (e.g. if a requirement is impossible to achieve, or a particular design geometry is very expensive to manufacture): the process of revising the design is slow and expensive, and often only the highest priority changes are made. The result is designs that tend to be poor from the standpoint of later life-cycle perspectives, including for example environmental concerns such as recyclability that are becoming increasingly important.

More recently, several strategies have emerged for better accounting for the interdependencies among collaborative design participants. These include concurrent engineering and least-commitment design:

Concurrent engineering involves the creation of multi-functional design teams, including representatives of all important design perspectives, for each distinct design subspace. Design decisions can be reviewed by all affected design perspectives when they are initially being considered, so bad decisions can be caught and revised relatively quickly and cheaply. While this approach has proven superior in some ways to traditional serial design, it does often incur an overwhelming burden on engineers as they have to attend many hours of design meetings and review hundreds of proposed changes per week [6].

Least-commitment design is a complimentary approach that attempts to address the same challenges by allowing engineers to specify a design incompletely, for example as a rough sketch or set of alternatives, and then gradually make the design more specific, for example by pruning some alternatives [9, 13]. This has the advantage that bad design decisions can be eliminated before a lot of effort has been invested in making them fully specific, and engineers are not forced to make arbitrary commitments that lead to needless conflicts.

While the adoption of these approaches has been helpful, major challenges remain. Consider for example the Boeing 767-F redesign program [6]. Some conflicts were not detected until long (days to months) after they had occurred, resulting in wasted design time, design rework, and often even scrapped tools and parts. It was estimated that roughly half of the labor budget was consumed dealing with changes and rework, and that roughly 25-30% of design decisions had to be changed. Since maintaining scheduled commitments was a priority, design rework often had to be done on a short flow-time basis that typically cost much more (estimates ranged as high as 50 times more) and sometimes resulted in reduced product quality. Conflict cascades that required as many as 15 iterations to finally produce a consistent design were not uncommon for some kinds of design changes. All this in the context of Boeing's industry-leading concurrent engineering practices. The dynamics of current collaborative design processes are thus daunting, and have led to reduced design creativity, a tendency to incrementally modify known successful designs rather than explore radically different potentially superior ones.

Improving the efficiency, quality and creativity of the collaborative innovative design process requires, we believe, a much better understanding of the dynamics of such processes and how they can be managed. In the next section we will review of the some key insights that negotiation and complex systems research offers for this purpose.

4. Insights from Complex Systems and Negotiation Research

A central focus of complex systems research is the dynamics of distributed networks, i.e. networks in which there is no centralized controller, so global behavior emerges solely as a result of concurrent local actions. Such networks are typically modeled as multiple nodes, each node representing a state variable with a given value. Each node in a network tries to select the value that maximizes its consistency with the influences from the other nodes. The dynamics of such networks emerge as follows: since all nodes update their local state based on their current context (at time T), the choices they make may no longer be the best ones in the new context of node states (at time T+1), leading to the need for further changes.

The negotiation literature adds the following refinement to this model. Each one of the nodes is *self-interested*, i.e. attempts to maximize its own local utility, at the same time it is seeking a satisfactory level of consistency with the nodes it is inter-dependent with. A central concern of negotiation research is designing the rules of encounter between inter-dependent nodes such that each node is individually incented to make decisions that maximize *social welfare*, i.e. the global utility of the collected set of local decisions. In this case, we can define global utility simply as the sum of node utilities plus the degree to which the inter-node influences are satisfied.

Is this a useful model for understanding the dynamics of collaborative design? We believe that it is. It is straightforward to map the model of collaborative design presented above onto a network. We can map design participants onto nodes, where each participant tries to maximize the utility of the subsystem it is responsible for, while ensuring its decisions satisfy its dependencies (represented as the links between nodes) with other subsystems. As a first approximation, it is reasonable to model the utility of a design as the local utility achieved by each participant plus a measure of how well all the decisions fit together. Even though real-world collaborative design clearly has top-down elements early in the process, the sheer complexity of many design artifacts means that eventually no one person is capable of keeping the whole design in his/her head and assessing/refining its global utility. Centralized control of the design decisions becomes impractical, so the design process is dominated perforce by concurrent subsystem design activities (performed within the nodes) done in parallel with subsystem design consistency checks (assessed by seeing to what extent inter-node influences are satisfied). We will assume, for the purposes of this paper, that individual designers are reasonably effective at optimizing their individual subsystems.

The key factor determining network dynamics is the nature of the influences between nodes. There are two important distinctions: whether the influences are *linear* or not, and whether they are *symmetric* or not. We will consider each one of these distinctions in turn, with an important side trip into the negotiation literature to understand the dilemmas raised by the presence of self-interested agents. This will be followed by a discussion of subdivided network topologies, and the role of learning. Unless indicated otherwise, the material on complex systems presented below is drawn from [2].

4.1. Linear vs. Non-Linear Networks

Non-Linearity Produces Multi-Optimum Utility Functions: If the value of nodes is a linear function of the influences from the nodes linked to it, then the system is linear, otherwise it is non-linear. Linear networks have a single *attractor*, i.e. a single configuration of node states that the network converges towards no matter what the starting point, corresponding to the global optimum. Their utility function thus looks like that shown in Figure 2 above. This means we can use a 'hill-climbing' approach (where each node always moves directly towards increased local utility) because local utility increases always move the network towards the global optimum.

Non-linear networks, by contrast, are characterized by having utility functions with multiple peaks (i.e. local optima) and multiple attractors, as in Figure 3:

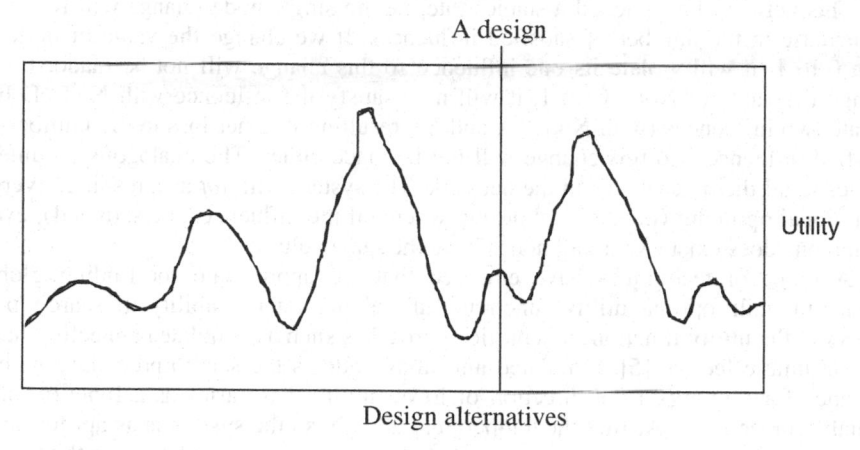

Figure 3. A multiple optima utility function, characteristic of non-linear networks

A key property of non-linear networks is that search for the global optima can *not* be performed successfully by pure hill-climbing algorithms, because they can get stuck in local optima that are globally sub-optimal. Consider, for example, what happened in Figure 3 above. Hill-climbing took the design to the top of a local optimum, which has substantially lower utility than some other designs.

To make this concrete, let us examine the following simple example: a network consisting of binary-valued nodes where each node is influenced to have the same value as the nodes it is linked to, and all influences are equally strong (Figure 4):

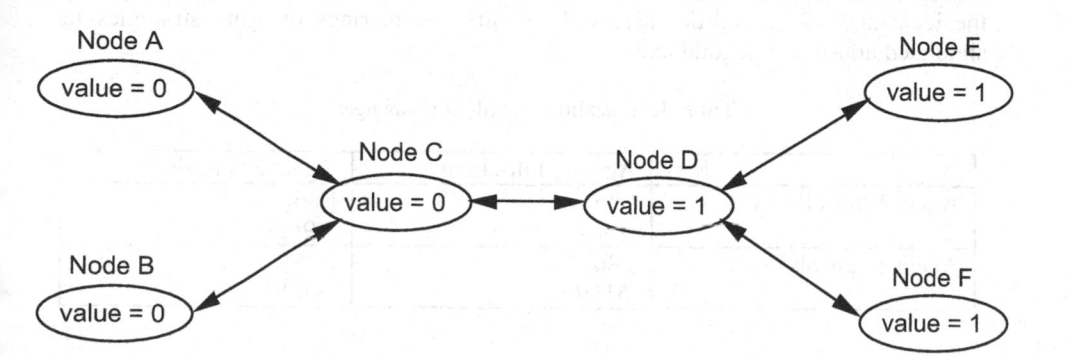

Figure 4. A simple network illustrating how networks can get stuck in local optima

Node A, for example, is influenced to have the same value as Node C, while Node C is influenced to have the same value as Nodes A, B and D. For simplicity's sake, we assume that the global utility is determined solely by the degree to which the inter-node influences are satisfied. We can imagine using this network to model a real-world situation wherein there are six subsystems being designed, with two equally optimal design options for each, and we want them to use matching interfaces.

This network has reached a stable state, i.e. no single node change will result in an increase in the number of satisfied influences. If we change the value of node A from 0 to 1, it will violate its one influence so this change will not be made. If we change the value of Node C to 1, it will now satisfy the influence with Node D but violate two influences (with Nodes A and B), resulting in a net loss in the number of satisfied influences, so this change will not be made either. The analogous argument applies to all the other nodes in the network. The system will *not* as a result converge on a global optimum (i.e. an ideal design where all the influences are satisfied), even though one does exist (where all nodes have the same value).

A range of techniques have emerged that are appropriate for finding global optima in multi-optima utility functions, all relying on the ability to search past valleys in the utility function. Stochastic approaches such as simulated annealing have proven quite effective [5]. Simulated annealing endows the search procedure with a tolerance for moving in the direction of lower utility that varies as a function of a virtual 'temperature'. At first the temperature is high, so the system is as apt to move towards lower utilities as higher ones. This allows it to range widely over the utility function and possibly find new higher peaks. Since higher peaks generally tend to also be wider ones, the system will spend most of its time in the region of high peaks. Over time the temperature decreases, so the algorithm increasingly tends towards pure hill-climbing. While this technique is not provably optimal, it has been shown to get close to optimal results in most cases.

A Social Dilemma with Self-Interested Agents: Annealing runs into a dilemma, however, when applied to systems with self-interested agents. Let us assume that at least some actors are 'hill-climbers', concerned only with maximizing their local utilities, while others are 'annealers', willing to accept, at least temporarily, lower local utilities as part of the exploratory process. We can use a simulation approach to explore what happens. Table 1 summarizes the results for such experiments, giving the local and global utilities achieved for different pairings of agent strategies in simulated non-linear negotiations:

Table 1. Annealing vs. hill-climbing agents

	Agent 2 hill-climbs	Agent 2 anneals
Agent 1 hill-climbs	[.86] .73/.74	[.86] .99/.51
Agent 1 anneals	[.86] .51/.99	[.98] .84/.84

In this table, the cell values are laid out as follows:

[<global optimality>]
<agent 1 optimality >/<agent 2 optimality>

Details on the negotiation results described in this paper are available, unless otherwise specified, in [7, 8].

These results show that, while annealers increase *global* utility, and are therefore highly desirable, annealers always fare *individually* worse than hill-climbers when both are present. Hill-climbing is thus a 'dominant' strategy: no matter what strategy the other agent uses, it is individually more rationale to be a hill-climber. If all agents do this, however, then they forego the higher individual utilities they would get if they both annealed. Individual strategic considerations thus drive the system towards the strategy pairing with the *lowest utility values*.

What can be done about this? This pattern of utility values is an instance of a well-known phenomenon in game theory known as the "prisoner's dilemma" [10]. It has been shown that this dilemma can be avoided if there are repeated interactions between agents [1]. The idea is simple. Each agent uses an annealing strategy at first, but if it determines that the agent it is negotiating with is using hill-climbing, it itself then switches to hill-climbing for its future negotiations with that agent, thereby forcing them both into the 'lose-lose' quadrant of Table 1. It turns out that this 'tit for tat' approach incents annealing behavior in all agents, *assuming* that they negotiate with each other multiple times. This idea can be refined with the addition of a 'reputation mechanism', wherein agents consult a database of previous negotiations (in addition to their individual experience) in order to determine whether the agent they currently face tends to be an annealer or hill-climber. Ideally, however, we would prefer to find a way to incent annealing behavior within the context of a single negotiation, without the requirement of multiple interactions. Can this be done?

Some apparently reasonable approaches are, it turns out, quite ineffective. One approach, for example, is what we can call 'adaptive' annealing. A negotiation typically consists of a relatively large number of offers and counter-offers, resulting in increasingly better interim agreements that eventually are accepted as final by both parties. An agent could therefore in principle switch in mid-stream from being an annealer to being a hill-climber if it determines that the other agent is being a hill-climber. Determining the strategy type of the agent you are negotiating with is in fact relatively easy: an annealer tends to accept a much higher percentage of interim proposals than a hill-climber. The problem with this approach is that determining the type of an agent in this way takes *time*. Our simulations have shown that the divergence in acceptance rates between annealers and hill-climbers only becomes clear *after most of the utility has been committed*, so it is too late to fully recover from the consequences of having started as an annealer if you negotiated with a hill-climber. Hill-climbing therefore remains the dominant strategy. Another possibility is for annealers to simply be less concessionary, i.e. less willing to accept utility-decreasing interim agreements. This in fact allows us to eliminate the poor annealer payoffs that underlie the prisoner's dilemma, but only at the cost of radically reduced global utility. In both cases, we are unable to incent agent strategies that optimize the global utility of the outcome.

Resolving the prisoners' dilemma within the scope of a single negotiation can be achieved, however, through the use of what we call a 'parity-enforcing annealing mediator'. Rather than requiring that the agents anneal, we move the annealing into a third party we call a mediator. In this approach, possible agreements are generated (in our experiments they were generated by the mediator, but this is a not a critical part of the scheme) and then voted on by the negotiating agents. The mediator is a kind of annealer: it is endowed with a time-decreasing willingness to at least temporarily follow up on design proposals that one or both agents voted against. Agents are free to remain hill-climbers in their voting behavior, and thus avoid making harmful concessions. The mediator, by virtue of being willing to provisionally pursue utility-decreasing agreements, can traverse valleys in the agents' utility functions and thereby lead the agents to win-win solutions. Paradoxically, using a mediator that occasionally *ignores* agent preferences leads to outcomes that are better for both agents.

Achieving maximal global utilities in this scheme requires that agents be able to annotate their votes with strength information. A binary scheme is sufficient, wherein agents annotate their accept votes as being either *strong* or *weak*. This allows the possibility of 'over-rides', wherein the mediator pursues an interim agreement that was strongly preferred by one agent and weakly rejected by another. Over-rides are important because such agreements are likely to increase global utility. Agents might of course be tempted to exaggerate in such contexts, marking every vote as being a strong one. But this possibility can be foiled by enforcing running parity on the number of times each agent over-rides the other. This works for the following reason. One can think of this procedure as giving agents 'tokens' that they can use to gain over-rides. A truthful agent spends its tokens exclusively on over-rides that truly offer it a strong local utility increase. An exaggerator, on the other hand, will spend tokens even when the utility increment it derives is relatively small. At the end of the day, the truthful agent has spend its tokens more wisely and to better effect.

Lessons: How do these insights apply to collaborative design? Generally speaking, linear networks represent a special case (only a tiny fraction of all possible influence relationships are linear), but they have proven adequate for modeling what has been called *routine* design. Routine design involves highly familiar requirements and design options, as for example in automobile brake or transmission design [3]. In these contexts, designers can usually start the design process near enough to the final optimum that the process acts as if it has a single attractor. Previous research on design dynamics has focused on this class of design model, generating such useful results as approaches for identifying design process bottlenecks [12] and for fine-tuning the lead times for design subtasks [4].

Rapid technological and other changes have made it increasingly clear, however, that many of the most important collaborative design problems (e.g. concerning software, biotechnology, or electronic commerce) involve *innovative* design, radically new requirements, and unfamiliar design spaces. It is often unclear how to achieve a given set of requirements. There may be multiple very different good solutions, and the best solution may be radically different than any that have been tried before. For such cases non-linear networks seem to represent a more accurate model of the collaborative design process.

This has important consequences. One is a tendency to stay with well-known designs. When a utility function has widely separated optima, once a satisfactory optimum is found the temptation is to stick to it. This design conservatism is exacerbated by the fact that it is often difficult to compare the utilities for radically different designs. We can expect this effect to be especially prevalent in industries, such as commercial airlines and power plants, which are capital-intensive and risk-averse, since in such contexts the cost of exploring new designs, and the impact of getting it wrong, can be prohibitive.

Another consequence is that collaborative design as currently practiced is probably quite prone to getting stuck in local optima that may be significantly worse than radically different alternatives. Annealing-like processes potentially applicable to addressing this problem are widely used in human collaborative design settings. 'Brainstorming', for example, with its emphasis on not pruning candidate solutions too quickly, can be viewed as a kind of annealing. Designers are, however, generally much more strongly encouraged to create a good design for their own subsystems, than to concede to make someone else's job easier. This incentive structure leads to the "prisoner's dilemma" described above.

The prisoner's dilemma can, as we have seen, be avoided if we assume that agents have multiple negotiation encounters and use a 'tit for tat' scheme for deciding when to be concessionary or not. Such schemes are probably used, in fact, by many designers in collaborative settings. The relative infrequency of major negotiations, the absence of reputation databases, and high turnover in personnel may, however, sabotage the efficacy of such strategies. It seems likely, in addition, that many engineers make some use of the other approaches we described above, being adaptive or simply highly sparing in how much they concede. These are, after all, apparently reasonable strategies. They do not, however, have the desired result of fostering the discovery of more optimal overall designs. Mediation, as we have seen, has the potential of resolving the prisoner's dilemma, and it in fact has an important place in current collaborative design practice. Senior engineers, and in some cases teams of such engineers (sometimes called "change boards") are often called upon to mediate situations where the achievement of satisfactory global utility appears to be threatened. Engineers with that level of experience are, however, a scarce resource, so this tactic is typically reserved for only the most serious problems.

In brief, it appears likely that current collaborative design practice, particularly for highly innovative design, is prone to getting stuck in unnecessarily suboptimal solutions. We will discuss possible solutions to these problems in the section "How We Can Help" below.

4.2. Symmetric vs. Asymmetric Networks

Asymmetry Allows Non-Convergence: Symmetric networks are ones in which influences between nodes are mutual (i.e. if node A influences node B by amount X then the reverse is also true), while asymmetric networks do not have this property. Asymmetric networks (if they have cycles in them; see below) add the complication of having *dynamic* attractors, which means that the network does not converge on a *single* configuration of node states but rather cycles indefinitely around a relatively small *set* of configurations. Let us consider the simplest possible cyclic asymmetric network: the 'odd loop' (Figure 5):

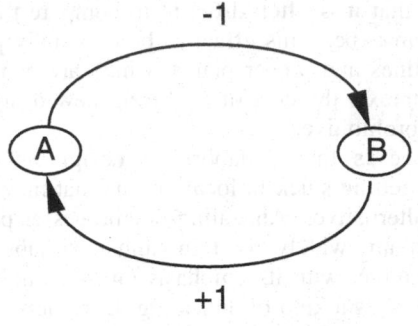

Figure 5. The simplest possible cyclic asymmetric network – an 'odd loop'

This network has two links: one where node B influences node A to have the same value, and another where node A influences node B to have the opposite value. Imagine both nodes have the initial value 1, and update each other in parallel. The states of the two nodes will proceed as follows:

State	Value of Node A	Value of Node B
Initial state	1	1
State 1	1	-1
State 2	-1	-1
State 3	-1	1
State 4	1	1

After one time step (state 1) node A will cause node B to 'flip' to –1, and node B will leave node A unchanged. After a second iteration (state 2) node A leaves node B unchanged, but node B causes the value of node A to flip. If we trace this far enough we find that the system returns to its initial state (State 4) and thus will repeat *ad infinitum*. If we plot the state space that results we get the following simple dynamic attractor:

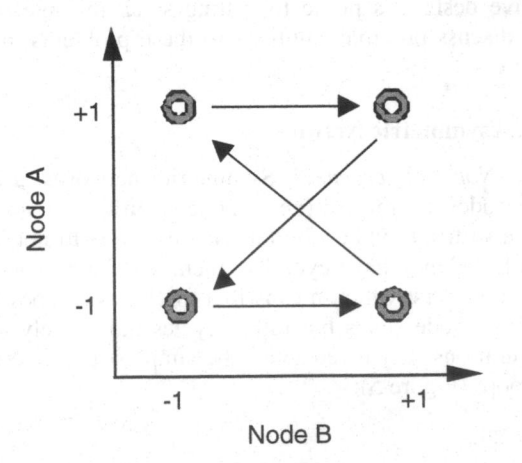

Figure 6. The dynamic attractor for the odd loop

More complicated asymmetric networks will produce dynamic attractors with more complicated shapes, including ones where states are *never exactly repeated*, but the upshot is the same: the system will not converge. One can always of course stop the system at some arbitrary point along its trajectory, but there is no guarantee that the design utility at that point will be better than that at any other point because the system, unlike the symmetric case, does not necessarily progress monotonically towards higher utility values. This can be understood in the following way. Every utility function can, in principle, be 'compiled' into a (symmetric) network that will progress monotonically towards higher utility values as long as the individual nodes perform local optimization. The opposite, however, is not true. There are many networks (including most asymmetric ones) that do *not* correspond to any well-formed utility function, so their sequences of states clearly can not be viewed as progressing towards a utility optimum [2].

If a network is *acyclic* however (also known as a *feed-forward* network, wherein a node is never able to directly or indirectly influence its own value), it has a well-defined utility function and thus will not have a dynamic attractor.

Lessons: How does this apply in collaborative design settings? Traditional serialized collaborative design is an example of an asymmetric feed-forward network, since the influences all flow uni-directionally from the earlier product life cycle stages (e.g. design) to later ones (e.g. manufacturing) with only weak feedback loops if any. In such contexts the attractors should be static and convergence should always occur, given sufficient time. In such settings we may not, however, expect particularly optimal designs. It is typically very difficult, given the bounded rationality of human beings, for designers earlier in the design life cycle to ensure that the designers later on in the life cycle will be able to produce near-optimal solutions for their very different but highly dependent problems. This is in fact the rational underlying the adoption of concurrent engineering approaches. 'Pure' concurrent engineering, where all design disciplines are represented on multi-functional design teams, encourage roughly symmetric influences between the participants and thus can also be expected to have convergent dynamics with static attractors. Current collaborative design practice, however, is a hybrid of these two approaches, and thus is likely to have the combination of asymmetric influences and influence loops that produces dynamic attractors and therefore non-convergent dynamics.

This, moreover, is a fundamental problem. As noted above, it is in principle straightforward to compute the proper inter-node influences given a global utility function. In design practice, however, we do *not* know the global utility function, especially once we have reached the realm of detailed design. The space of possible designs, and the cost of calculating their individual utility values, is simply too large. At best the global utility function is revealed to us incrementally as we generate and compare different candidate designs. The influence relationships between designers are, as a result, invariably defined directly based on experience and our knowledge of design decision dependencies. But such a heuristic approach can easily lead to the creation of influence networks that do not instantiate a well-formed utility function, and thus display dynamic attractors.

Dynamic attractors were found to not to have a significant effect on the dynamics of at least some routine (linear) collaborative design contexts [4], but may prove more significant in innovative (non-linear) collaborative design. It may help explain, for example, why it sometimes takes so many iterations to fully propagate changes in complex designs [6].

5. Subdivided Networks

Subdivision Can Speed Convergence: Another important property of networks is whether or not they are sub-divided, i.e. whether they consist of sparsely interconnected 'clumps' of highly interconnected nodes, as for example in Figure 7:

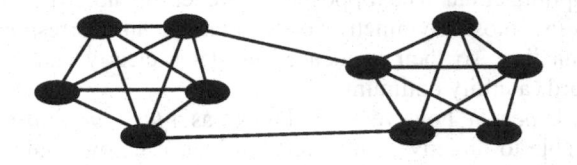

Figure 7. An example of a subdivided network

When a network is subdivided, node state changes can occur within a given clump with only minor effects on the other clumps. This has the effect of allowing the network to explore more states more rapidly. Rather than having to wait for an entire large network to converge, we can rely instead on the much quicker convergence of a number of smaller networks, each one exploring possibilities that can be placed in differing combinations with the possibilities explored by the other sub-networks [11].

Lessons: This effect is in fact widely exploited in design communities, where it is often known as *modularization.* This involves intentionally creating subdivided networks by dividing the design into subsystems with pre-defined standardized interfaces, so subsystem changes can be made with few or any consequences for the design of the other subsystems. The key to using this approach successfully is defining the design decomposition such that the utility impact of the subsystem interdependencies on the global utility is relatively low, because standardized interfaces rarely represent an optimal way of satisfying these dependencies. In most commercial airplanes, for example, the engine and wing subsystems are designed separately, taking advantage of standardized engine mounts to allow the airplanes to use a range of different engines. This is almost certainly *not* the optimal way of relating engines and wings, but it is good enough and simplifies the design process considerably. If the engine-wing interdependencies were crucial, for example if standard engine mounts had a drastically negative effect on the airplane's aerodynamics, then the design of these two subsystems would have to be coupled much more closely in order to produce a satisfactory design.

6. Imprinting

Imprinting Captures Successful Influence Patterns: One common technique used to speed network convergence is *imprinting,* wherein the network influences are modified when a successful solution is found in order to facilitate quickly finding (similar) good solutions next time. A common imprinting technique is reinforcement learning, wherein the links representing influences that are satisfied in a successful final configuration of the network are strengthened, and those representing violated influences weakened. The effect of this is to create fewer but higher optima in the utility function, thereby increasing the likelihood of hitting such optima next time.

Lessons: Imprinting is a crucial part of collaborative design. The configuration of influences between design participants represents a kind of 'social' knowledge that is generally maintained in an implicit and distributed way within design organizations, in the form of individual designer's heuristics about who should talk to whom when about what. When this knowledge is lost, for example due to high personnel turnover in an engineering organization, the ability of that organization to do complex design projects is compromised. It should be noted, however, that imprinting reinforces the tendency we have already noted for organizations in non-linear design regimes to stick to tried-and-true designs, by virtue of making the previously-found optima more prominent in the design utility function, and thus may be counter-indicated for challenges requiring highly innovative designs.

7. How Can We Help?

What can we do to improve our ability to do innovative collaborative design? We will briefly consider several possibilities suggested by the discussion above.

Information systems are increasingly becoming the medium by which design participants interact, and this fact can be exploited to help monitor the influence relationships between them. One could track the volume of design-related exchanges or (a more direct measure of actual influence) the frequency with which design changes proposed by one participant are accepted as is by other participants. This can be helpful in many ways. Highly asymmetric influences could represent an early warning sign of non-convergent dynamics. Detecting a low degree of influence by an important design concern, especially one such as environmental impact that has traditionally been less valued, can help avoid utility problems down the road. A record of the influence relationships in previous failed and successful design projects can be used to help better manage future projects. This will require being able to determine which influences were critical in these previous efforts. If a late high-impact problem occurred in a subsystem that had a low influence in the design process, for example, this would suggest that the relevant influence relationships should be modified in the future. Incentive mechanisms can be put in place that reward engineers not just for producing good subsystem designs, but also for participating in what are believed to be productive patterns of mutual influence with other designers. Note that this has the effect of making a critical class of normally implicit and distributed knowledge more explicit, and therefore more amenable to being preserved over time, as well as transferred between projects and even organizations.

Information systems can also potentially be used to help assess the degree to which the design participants are engaged in routine (i.e. optimization-driven) vs innovative (i.e. highly exploratory) design strategies. We could use such systems to estimate for example the number and variance of design alternatives being considered by a given design participant. This is important because, as we have seen, a premature commitment to a routine design strategy that optimizes a given design alternative can cause the design process to miss other alternatives with higher global optima. Tracking the degree of innovative exploration can be used to fine-tune the use of innovation-enhancing interventions such as incentives, competing design teams, introducing new design participants, and so on. As with simulated annealing, it will probably make sense to encourage more conceding and exploration early on in the design process, and gradually transition to hill-climbing as time goes on.

The prisoner's dilemma incentive structure that leads to suboptimal designs can be addressed in at least two ways that are probably under-utilized in current practice. One is by the introduction of reputation mechanisms. If we simply make information available on which designers have a history of conceding sparingly, we are likely to find an increase in concessionary behavior, and therefore improved design outcomes, even in the absence of explicit (e.g. salary) incentives. Another possibility is the wider use of mediators. Mediators in collaborative design contexts have traditionally been senior engineers capable of dictating the *content* of a design outcome. Our work on negotiation algorithms suggests, however, that mediators can be effective by guiding the design *process*, for example as we suggested above by occasionally having the agents follow up on design options that one or both rejected, and by enforcing rough parity in the number of mixed wins. Process-oriented mediation does not require the same depth of domain expertise as content-oriented mediation, and it is therefore likely that designers can be trained to provide that for each other, and that such mediation can become much more widely available as a result.

Finally, information systems can be used to track the history of design alternatives explored and thereby detect the design loops that indicate a non-convergent design process.

8. Conclusions

Existing collaborative design approaches have yielded solid but incremental design improvements, which has been acceptable because of the relatively slow pace of change in requirements and technologies. Consider for example the last 30 years of development in Boeing's commercial aircraft. While many important advances have certainly been made in such areas as engines, materials and avionics, the basic design concept has changed relatively little (Figure 8):

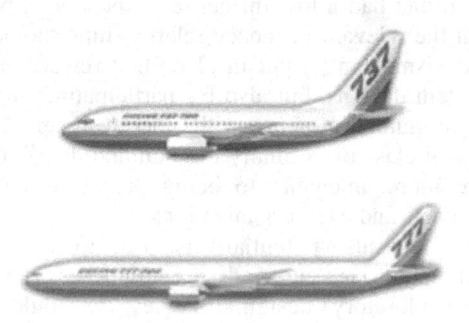

Figure 8. The Boeing 737 (inaugurated 1965) and the Boeing 777 (1995)

Future radically innovative design challenges, such as high-performance commercial transport, will probably require, however, substantial changes in design processes:

Figure 9. The Boeing Sonic Cruiser (under development)

This paper has begun to identify what recent research on negotiation and complex systems can offer in this regard. The key insights are that important properties of collaborative design dynamics can be understood as reflecting two basic facts: (1) collaborative design is a kind of distributed network, and (2) the agents in this network are self-interested and respond to local incentives. This is powerful because this means that our growing general understanding of networks and negotiation can be applied to help us better understand and eventually better manage collaborative design regardless of the domain (e.g. physical vs informational artifacts) and type of participants (e.g. human vs software-based).

This insight leads to several others. Most prominent is the suggestion that we need to fully embrace an influences- and incentives-centric perspective on how to manage complex collaborative design processes. It is certainly possible for design managers to have a very direct effect on the content of design decisions during preliminary design, when a relatively small number of high-level global utility driven decisions are made top-down by a small number of players. But once the detailed design of a complex artifact has been distributed to many players, the global utility impact of local design changes is too difficult to assess, and design decisions are too voluminous and complex to be made top-down, so the dominant drivers become local utility maximization plus fit between these local design decisions. In this regime encouraging the proper influence relationships and concession strategies becomes the primary tool available to design managers. If these are defined inappropriately, we can end up with designs that take too long to create, do not meet important requirements, and/or miss opportunities for significant utility gains through more creative (far-ranging) exploration of the design space.

Acknowledgements

This work was supported by the National Science Foundation and the Defense Advanced Research Projects Agency.

References

[1]   Axelrod, R. (1984). The Evolution Of Cooperation, Basic Books.
[2]   Bar-Yam, Y. (1997). Dynamics of complex systems. Reading, Mass., Addison-Wesley.
[3]   Brown, D. C. (1989). Making design routine. Proceedings of IFIP TC/WG on Intelligent CAD.
[4]   Eppinger, S. D., M. V. Nukala, et al. (1997). "Generalized Models of Design Iteration Using Signal Flow Graphs." Research in Engineering Design 9(2): 112-123.
[5]   Kirkpatrick, S., C. D. Gelatt, et al. (1983). "Optimization by simulated annealing." Science 220: 671-680.
[6]   Klein, M. (1994). "Computer-Supported Conflict Management in Concurrent Engineering: Introduction to Special Issue." Concurrent Engineering Research and Applications 2(3).
[7]   Klein, M., P. Faratin, et al. (2002a). Using an Annealing Mediator to Solve the Prisoners' Dilemma in the Negotiation of Complex Contracts. Agent-Mediated Electronic Commerce (AMEC-IV) Workshop, Bologna Italy, Springer.
[8]   Klein, M., P. Faratin, et al. (2002b). Negotiating Complex Contracts. Autonomous Agents and Multi-Agent Systems, Bologna Italy, AAAI Press.
[9]   Mitchell, T. M., L. I. Steinberg, et al. (1985). "A Knowledge-Based Approach To Design." IEEE Transactions on Pattern Analysis and Machine Intelligence PAMI(7): 502-510.
[10]   Osborne, M. J. and A. Rubinstein (1994). A course in game theory. Cambridge, Mass., MIT Press.
[11]   Simon, H. A. (1996). The Sciences of the Artificial. Cambridge MA USA, MIT Press.
[12]   Smith, R. P. and S. D. Eppinger (1997). "Identifying controlling features of engineering design iteration." Management Science 43(3): 276-93.
[13]   Sobek, D. K., A. C. Ward, et al. (1999). "Toyota's Principles of Set-Based Concurrent Engineering." Sloan Management Review 40(2): 67-83.

Chapter 9

Modularity in the Design of Complex Engineering Systems

Carliss Y. Baldwin
Kim B. Clark
Harvard Business School
Boston, MA 02163

1. Introduction

In the last decade, the concept of modularity has caught the attention of engineers, management researchers and corporate strategists in a number of industries. When a product or process is "modularized," the elements of its design are split up and assigned to modules according to a formal architecture or plan. From an engineering perspective, a modularization generally has three purposes:

- To make complexity manageable;
- To enable parallel work; and
- To accommodate future uncertainty.

Modularity accommodates uncertainty because the particular elements of a modular design may be changed after the fact and in unforeseen ways as long as the design rules are obeyed. Thus, within a modular architecture, new module designs may be substituted for older ones easily and at low cost.

This chapter will make three basic points. First, we will show that *modularity is a financial force* that can change the structure of an industry. Then, we will explore the *value* and *costs* that are associated with constructing and exploiting a modular design. Finally we will examine the ways in which modularity shapes organizations and the risks that it poses for particular firms.[1]

---

[1] Some of the arguments and figures in this paper are taken from Baldwin and Clark, 2000. The figures are reprinted by permission.

2. The Financial Power of Modularity

To demonstrate the financial power of modularity, let us begin by looking at some data from the computer industry. Figure 1 is a graph of the market values (in 2002 constant US dollars) of substantially all the U.S. based public corporations in the computer industry from 1950 to 2002. The firms are aggregated into sixteen subsectors by primary SIC code. The SIC codes included in the database and their definitions are listed in Table 1. IBM, Intel and Microsoft are shown separately.

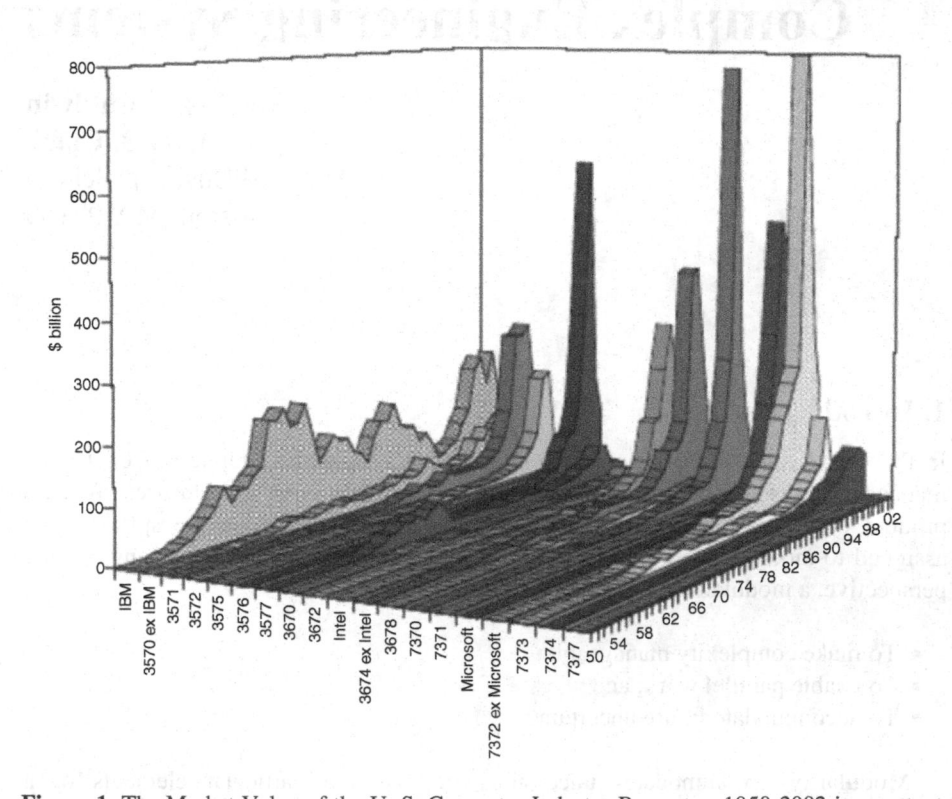

Figure 1. The Market Value of the U. S. Computer Industry By sector, 1950-2002 in constant 2002 US dollars

Table 1. SIC Codes Included in the Database

SIC Code	Category Definition	Start Date (1)
3570	Computer and Office Equipment	1960
3670	Electronic Components and Accessories	1960
3674	Semiconductors and Related Devices	1960
3577	Computer Peripheral Devices, n.e.c.	1962
3678	Electronic Connectors	1965
7374	Computer Processing, Data Preparation and Processing	1968
3571	Electronic Computers	1970
3575	Computer Terminals	1970
7373	Computer Integrated Systems Design	1970
3572	Computer Storage Devices	1971
7372	Prepackaged Software (2)	1973
3576	Computer Communication Equipment	1974
3672	Printed Circuit Boards	1974
7370	Computer Programming, Data Processing, and Othe Services	1974
7371	Computer Programming Services	1974
7377	Computer Leasing	1974

(1)   Start date is the first year in which six or more are present in the categor
(2)   This category had six firms in 1971, dipped to five in 1972, and back to six in 1973.

Figure 1 tells a story of industry evolution that runs counter to conventional wisdom. In economics the dominant theories of industry evolution describe a process of pre-emptive investment by large, well-capitalized firms, leading to stable market structures and high levels of concentration over long periods of time.[2] These theories

---

[2] The original theory of pre-emptive investment leading to industry concentration, with supporting historical evidence, was put forward by Alfred Chandler [11, 12]. A complementary theory of concentration following the emergence of a "dominant design" was put forward by William Abernathy and James Utterback [1]. Modern formulations of these theories and some large-scale empirical tests have been developed by John Sutton [46] and Steven Klepper [27]. Oliver Williamson (1985, Ch. 11) has interpreted the structures of modern corporations (unified and multi-divisional) as responses to potential opportunism (the hazards of market contracting). It is our position that the basic "task structures" and the economic incentives of modular design (and production) systems are different from the task structures and incentives of classic large-volume, high-flow-through production and distribution systems. Therefore the organizational forms that arise to coordinate modular design (and production) may not ressemble the classic structures of the modern corporation.

are backed up by a great deal of supporting empirical evidence going back to the late 19th Century. The data in Figure 1, by contrast, show that while IBM dominated the industry in the 1950s and 1960s, in the 1970s and 1980s the computer industry "got away" from IBM. (IBM's market value is the blue "mountain range" at the far left of the chart.) In 1969, 71% of the market value of the computer industry was tied up in IBM stock. By 2002, IBM was no longer dominant, and the largest firm (Microsoft) accounted for less than 20% of the total value of the industry.

Underlying Figure 1 is a pattern of extreme turbulence at the level of firms. The entire database spanning the years 1950 to 2002 contains about 2,700 firms. Of these, only about 1,100 survived in 2002. Thus around 1,600 or 60% of the firms that entered the computer industry over five decades no longer exist: they went bankrupt, were acquired, or moved out of the industry. Not surprisingly (for those who lived through it), much of this turnover occurred between 1997 and 2002, the years of the Internet Bubble and Crash. Around 1,200 firms entered during these six years, while 1,100 failed or were acquired.

The figure also shows that market values were initially concentrated in a few firms, but are now spread out over across sixteen industrial categories. Whole industries have come and gone. For example, the original computer category, SIC 3570, "Office and Computer Equipment", once included firms like Digital Equipment Corporation, Sperry Corporation, Data General and NCR, as well as IBM. This category has virtually disappeared: IBM has been reclassified into SIC 7370, "Computer Programming and Data Processing," and the other firms mentioned have failed or been acquired. By 2002, Hewlett Packard was the only firm of any size remaining in this once-pre-eminent category. Conversely, in 1970, SIC 7372, "Packaged Software," included only 7 firms with a combined market capitalization of just over $1 billion. In 2002, this category had grown to 408 firms with a combined market cap of almost half a trillion dollars ($490 billion).

Volatility and turbulence can be observed at the level of the whole industry as well. Figure 2 shows the total market value of all firms in the industry for the sample period 1950 – 2002. The chart is dominated by the Internet Bubble and Crash, which created and then destroyed $2.5 trillion in the space of five years (1997 – 2002). Apart from the Bubble, the industry as a whole has experienced significant value increases over time. From 1960 to 1996, even as value was being dispersed and redistributed over many, many firms, aggregate value kept pace. Then around 1997, the aggregate value of this group of firms seemed to spin out of control. More value was created and destroyed in a few years than the whole industry had managed to create over its entire history. The causes of this remarkable pattern are the subject of ongoing research, but we have no explanations for it as yet.

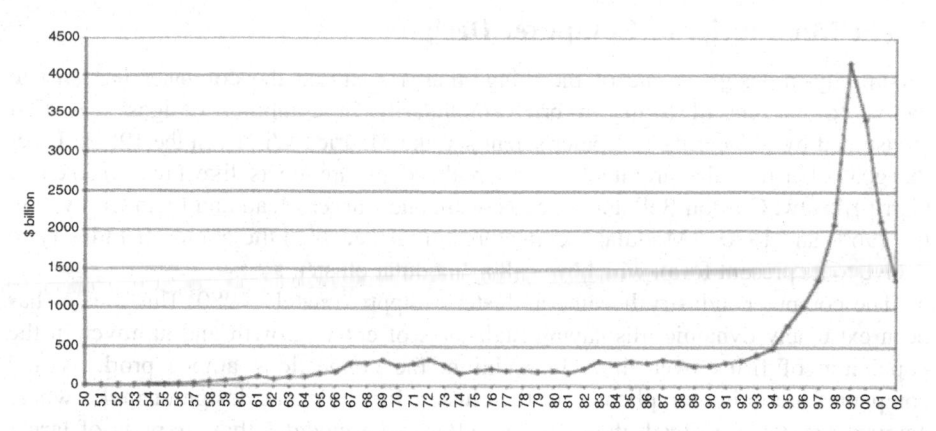

Figure 2. The Market Value of the Computer Industry Aggregated, 1950-2002, in constant 2002 US dollars

In summary, the computer industry presents us with a pattern of industry evolution involving more firms, more products, and (mostly) increasing value created over time. As a result of this pattern, the computer industry today consists of a large *cluster* of over 1,000 firms, no one of which is very large relative to the whole. In addition, the total market value of the industry is now spread widely but very unevenly across the sixteen sub-industries. (See Figure 3.) We contend that modularity in the design of complex computer systems is what allowed this *creation* of value, the *dispersion* of value across so many firms, and finally new *concentrations* of value to take place. We will expand on this argument in the sections that follow.

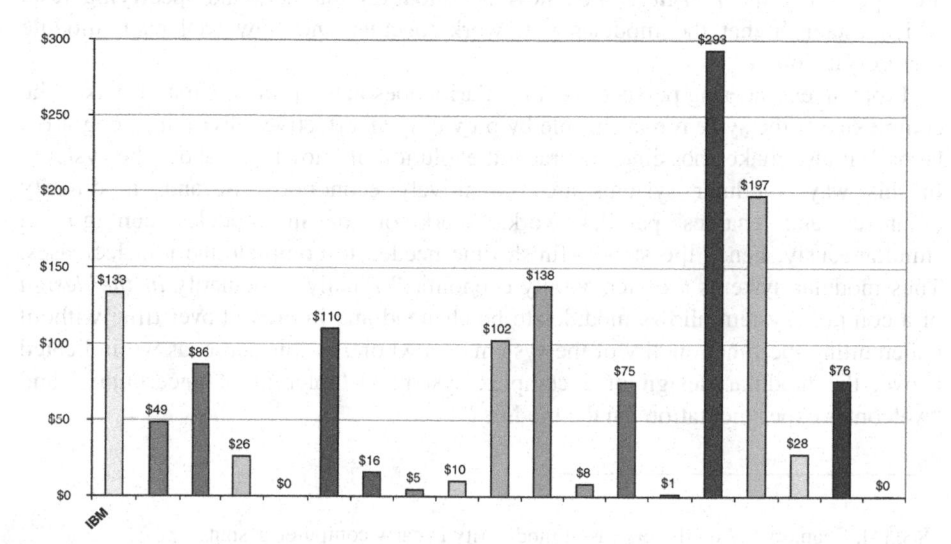

Figure 3. The Distribution of Market Value in the U. S. Computer Industry as of 2002 By sector, in constant 2002 US dollars

3. The Modularity of Computer Designs

Modularity-in-design is one of the things that has caused the computer industry to evolve to its present form. In brief, modularity in computer designs was first envisioned by pioneering computer scientists like Maurice Wilkes in the 1950s. Later the goal of a modular architecture was realized by architects like Fred Brook and Gerrit Blaauw, Gordon Bell and Allen Newell, and Carver Mead and Lynn Conway in the 1960s and 1970s.[3] Modular architectures in turn enabled the computer industry to evolve to its present form, which we call a "modular cluster".

The computer industry became a cluster in approximately 1980. This cluster has been extremely dynamic, displaying high rates of entry, growth and turnover in the population of firms over time. In addition, the connections among products and companies are quite complicated in the cluster. Firms do not design or make whole computer systems; instead, they design and/or make *modules* that are parts of larger systems. These modules include hardware components like computers, microprocessors and disk drives; software components like operating systems and application programs; as well as process components like fabrication, assembly, systems integration, and testing.

Modules, in fact, are always distinct parts of a larger system. They are designed and produced independently of one another, but must function together as a whole. Modularity allows tasks—both design tasks and production tasks—to be divided among groups, which can work independently and do not have to be part of the same *firm*. Compatibility among modules is ensured by "design rules" that govern the architecture, the interfaces, and the standardized tests of the system. Thus "modularizing" a system involves specifying its *architecture*, that is, what its modules are; specifying its *interfaces*, i.e., how the modules interact; and specifying *tests* which establish that the modules will work together and how well each module performs its job.

From an engineering perspective, modularity does many things. First, it makes the complexity of the system manageable by providing an effective "division of cognitive labor."[4] It also makes possible the graceful evolution of knowledge about the system.[5] In this way, modular systems are "cognitively economic." Second, modularity organizes and enables parallel work. Work on or in modules can go on simultaneously, hence the start-to-finish time needed to complete the job decreases. Thus modular systems are "temporally economic." Finally, modularity *in the design* of a complex system allows modules to be changed and improved over time without undercutting the functionality of the system as a whole. In this sense, as we indicated above, the modular design of a complex system is "tolerant of uncertainty" and "welcomes experimentation" in the modules.

---

[3] See [4], Chapters 6-7 on the origins of modularity in early computer designs.
[4] O.E. Williamson, 1999, "Human Action and Economic Organization," mimeo, University of California, Berkeley; quoted in M. Aoki, *Towards a Comparative Institutional Analysis*, 2001, MIT Press, Chapter 4.
[5] See [21].

4. Modularity in Design, Production and Use

Humans interact with artifacts in three basic ways: they design them; produce them; and use them. There are, as a result, three basic types of modularity: modularity-in-design, modularity-in-production, and modularity-in-use. We will discuss these "modularities" in reverse order.

A system of goods is *modular-in-use* if consumers can mix and match elements to come up with a final product that suits their taste and needs. For example, consumers often buy bed frames, mattresses, pillows, linens, and covers made by different manufacturers and distributed through different retailers. The parts all fit together because different manufacturers make the goods in standard sizes. These standard dimensions constitute design rules that are binding on manufacturers, wholesalers, retailers, and users. Modularity-in-use thus supports customization of the system to suit the needs and tastes of the end-user.

Manufacturers have used *modularity-in-production* for a century or more. Carmakers, for example, routinely arrange to manufacture the components of an automobile at different sites and bring them together for final assembly. They can do so because they have completely and precisely specified how the parts will interact with the vehicle. The engineering specifications of a component (its dimensions, tolerances, functionality, etc.) constitute a set of design rules for the factories that supply the parts. Such process modularity is fundamental to mass production.

However, the fact that, in a complex system, the elements of use or the tasks of production have been split up and assigned to separate modules does not mean that the *design* of the system is modular. Indeed systems that are modular-in-use or modular-in-production may rest on designs that are tightly coupled and centrally controlled. For example, Intel Corporation famously imposes a "copy exactly" rule on its fabrication plants. The production of chips can go on independently at separate sites because the layout of the plants and the work processes within the plants are the same. Thus Intel's "copy exactly" plants are modular-in-production but not modular-in-design. In a similar vein, a sectional sofa is a suite of furniture that is modular-in-use. Purchasers can combine and recombine the elements of the suite at will. But those elements must be designed as one interdependent whole, or the patterns and shapes will not form a pleasing ensemble. Thus the sectional sofa suite is modular-in-use, but not modular-in-design.

A complex engineering system is modular-in-design if (and only if) the *process of designing it can be split up and distributed across separate modules*, that are coordinated by design rules, not by ongoing consultations amongst the designers. Of all the "modularities", modularity-in-design is the least well understood and has the most interesting economic consequences. This is because new designs are fundamentally *options* with associated economic *option value*. Modularity-in-design multiplies the options inherent in a complex system. This in turn both increases the total economic value of the system and changes the ways in which the system can evolve. In the rest of this chapter, we will explain how to map and measure the option value of modularity-in-design.

5. Designs as Options

A fundamental property of designs is that at the start of any design process, the final outcome is uncertain. Once the full design has been specified and is certain, *then the development process for that design is over.*

Uncertainty about the final design translates into uncertainty about the design's eventual value. How well will the end-product of the design process perform its intended functions? And what will it be worth to users? These questions can never be answered with certainty at the beginning of any substantive development process. Thus the ultimate value of a design is unknown when the development process begins.

Uncertainty about final value in turn causes new designs to have "option-like" properties. In finance, an option is "the right but not the obligation" to choose a course of action and obtain an associated payoff. In engineering, a new design creates the ability but not the necessity—the right but not the obligation—to do something in a different way. In general (if the designers are rational), the new design will be adopted only if it is better that its alternatives. Thus the economic value of a new design is properly modeled as an option using the methods of modern finance theory.

The option-like structure of designs has three important but counterintuitive consequences. In the first place, when payoffs take the form of options, taking more risk creates more value.[6] Risk here is defined as the *ex ante* dispersion of potential outcomes. Intuitively, a risky design is one with high technical potential but no guarantee of success. "Taking more risk" means accepting the prospect of a greater *ex ante* dispersion. Thus a risky design process is one that has a very high potential value conditional on success but, symmetrically, a very low, perhaps negative, value conditional on failure.

What makes the design an option, however, is that the low-valued outcomes do not have to be passively accepted. As we said, the new design does not have to be adopted; rationally, it will be adopted only if it is better than the alternatives, including the *status quo* alternative. In effect, then, the downside potential of a risky design is limited by the option to reject it after the fact. This means that "risk" creates only upside potential. More risk, in turn, means more upside potential, hence more value.[7]

The second counterintuitive result is that when payoffs take the form of options, seemingly redundant efforts may be value-increasing. Two attempts to create a new design may arrive at different endpoints. In that case, the designers will have the option to take the better of the two. The option to take the better of two or best of several outcomes is valuable.[8] Thus when faced with a risky design process, which

---

[6] This is a basic property of options (See [34]).

[7] It follows, of course, that if a risky design is "hardwired" into a system so that it must be implemented regardless of its value, then the design process loses its option-like properties. In such cases, "taking more risk" in the sense defined above, will not increase, and may decrease value.

[8] Stulz, 1982, first analyzed the option to take the higher-valued of two risky assets. Sanchez [40] worked out the real option value of parallel design effort in product development.

has a wide range of potential outcomes, it is often desirable to run multiple "design experiments" with the same functional goal. These experiments may take place in parallel or in sequence, or in a combination of both modes.[9] But whatever the mode, more risk calls for more experimentation.

The third result is that *options interact with modularity in a powerful way*. By definition, a modular architecture allows module designs to be changed and improved over time without undercutting the functionality of the system as a whole. This is what it means to be "tolerant of uncertainty" and to "welcome experiments" in the design of modules. As a result, modules and design experiments are economic complements: an increase in one makes the other more valuable.[10] (Below we will derive this result in the context of a formal model.)

The effect of modularity-in-design on options and option value is depicted in Figure 4. Here we envision a system that is making the transition from being one interdependent whole to being a set of modules governed by design rules. The system goes from having one large design option (i.e., to take the whole design or leave it) to having many smaller options—one per module. Thus the act of *splitting* a complex engineering system into modules multiplies the valuable design options in the system. At the same time, this *modularization* moves decisions from a central point of control to the individual modules. The newly decentralized system can then evolve in new ways.

Notice, however, that by modularizing, one barrier to entry by competitors, the high costs of developing an entire complex engineering system (like an automobile, a computer, or a large software package) are reduced to the costs of developing individual modules. Thus the modularization of a large, complex system, even as it creates options and option value, also sows the seeds of increased competition focused on the modules. We shall revisit this issue at the end of the chapter.

---

[9] See [30].

[10] This is the definition of economic complementarity used by Milgrom and Roberts [35] and Topkis [47]. The complementarity of modularity and experimentation was first demonstrated by Baldwin and Clark [2; 4, Chapter 10].

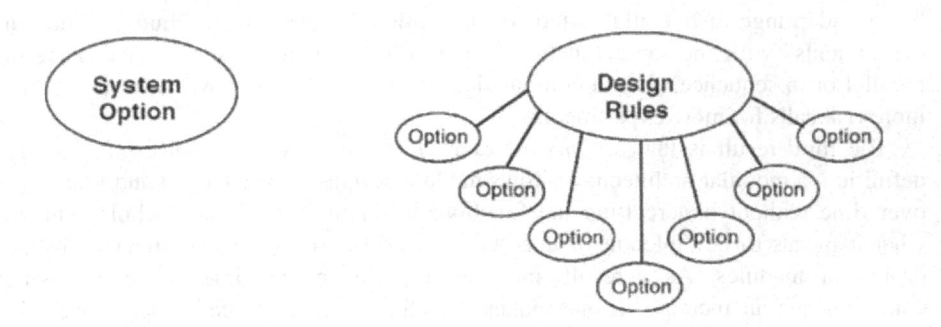

Figure 4. Modularity Creates Design Options

6. Mapping the Design of a Complex Engineering System

We will now look at modularity-in-design more carefully. To help us do so, we will represent the design of a complex system using the methods of Design Structure Matrix (DSM) Mapping. In this mapping technique, the system is first characterized by listing a set of design parameters for the system. The design parameters are then arrayed along the rows and columns of a square matrix. The matrix is filled in by checking—*for each parameter*—which other parameters affect it and which are affected by it. For example, if Parameter A affects the choice of Parameter B, then we will put a mark "**x**" in the cell where the *column of A* and the *row of B* intersect. We repeat this process until we have recorded all parameter interactions. The result is a map of the dependencies that affect the detailed structure of the artifact. For example, Figure 5 is a DSM map of the dependencies in the design for a laptop computer system circa 1993.[11]

DSM maps are well known in the engineering professions. They can be constructed for any artifact or complex system, whether it is tangible or intangible. Thus there are DSM maps of products, like computers and automobiles, and DSM maps of both production processes and design processes. Many such maps have been constructed by Steven Eppinger and his colleagues at MIT.

The DSM map in Figure 5 indicates that the laptop computer design has four blocks of very tightly interrelated design parameters corresponding to the (Disk) Drive System, the Main Board, the LCD Screen, and the Packaging of the machine. There is also a scattering of dependencies ("**x**'s") outside the blocks. The dependencies arise both above and below the main diagonal blocks, thus the blocks are *interdependent*.

---

[11] The DSM methodology was invented by Donald Steward. The DSM map shown in Figure 5 was prepared by Kent McCord and published in McCord and Eppinger [32]. Reprinted by permission.

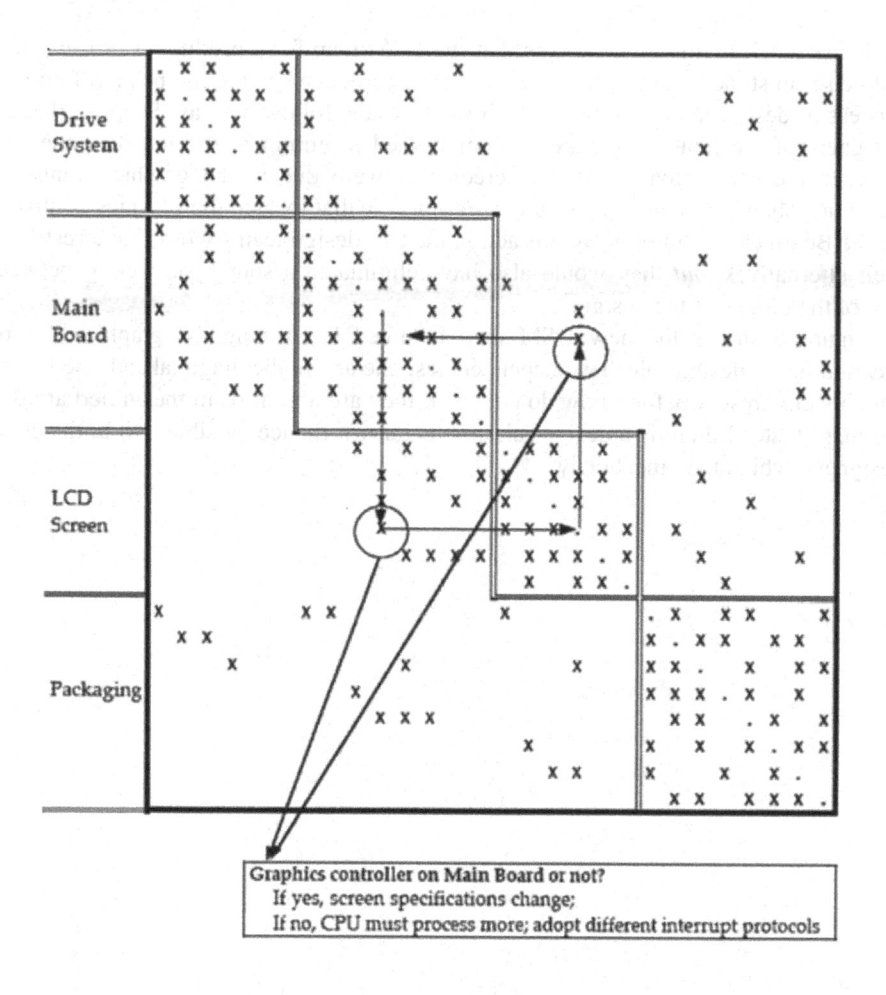

Figure 5. Design Structure Matrix Map of a Laptop Computer

Cycling and *iteration* are needed to resolve design interdependencies. For example, as shown in the figure, the location of the computer's graphics controller creates dependencies between the Main Board and the LCD Screen and vice versa.

Because of these dependencies, there will be ramifications of any choice made at this particular point: these are indicated by the arrows in the diagram. If two teams were working on the different components, they would have to confer about the location of the graphics controller in order to coordinate their design choices. But unforeseen consequences might arise later, causing the initial choice to be revisited. There would then be further consequences: new arrows would arise, which, through the chain of dependencies, might wander all over the map. Such cycling is the inevitable consequence of an interdependent design structure.

However, it is important to note that the DSM map for a product or process need not be set in stone forever. Dependencies and interdependencies can be modified by a process of design rationalization, which works in the following way. Suppose that the designers of the laptop computer system wished to eliminate the interdependencies between the Main Board and the Screen that were due to the graphics controller location. They could do so by setting a *design rule* that located the graphics controller on the Board (for example). By this action the two design teams would have restricted their alternatives, *but* they would also have eliminated a source of cycling between two of the blocks of the design.

Figure 6 shows the new DSM map obtained by turning the graphics control location into a design rule. Two dependencies, one above the diagonal and one below, which were present before, now do not exist: they are absent from the circled areas in the map. Instead there is a design rule that is known (hence "visible") to both sets of designers, which they must obey.

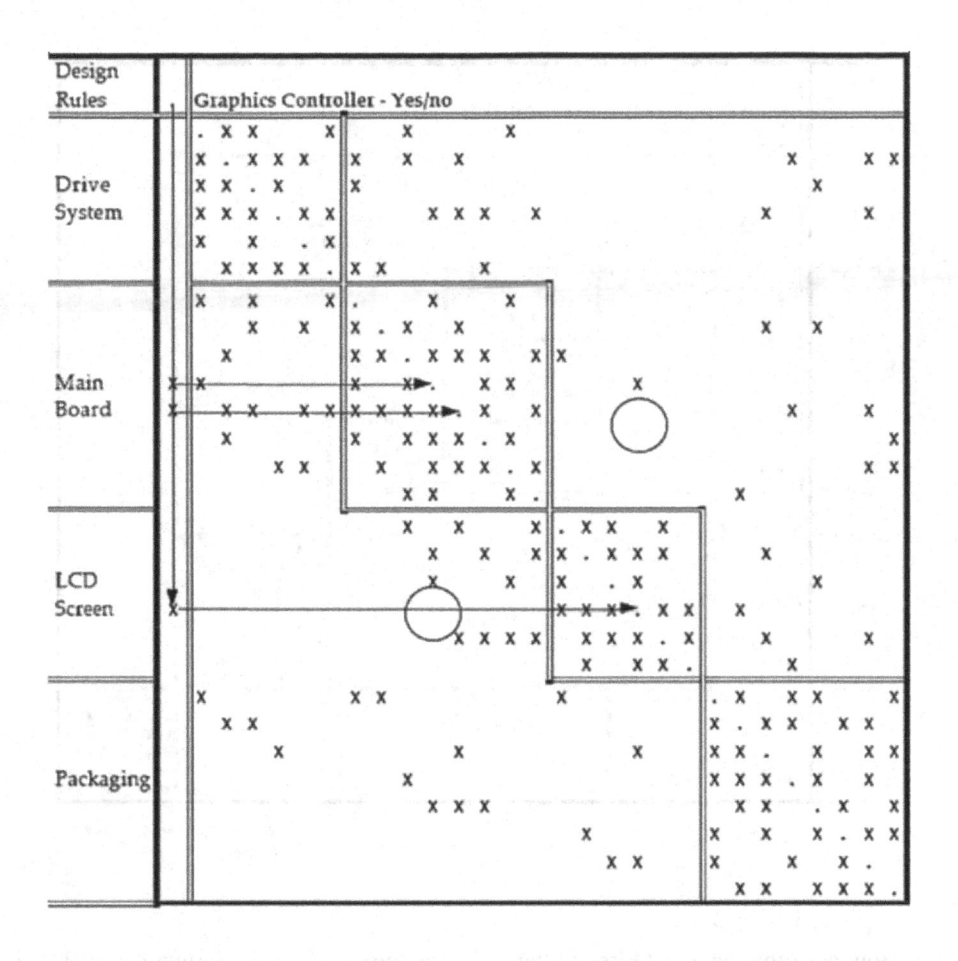

Figure 6. Eliminating Interdependencies by Creating a Design Rule

Carrying this process through to its logical conclusion results in a radically different structure: a *modular structure* as shown in Figure 7. Here we have the same highly interdependent blocks as before: the Drive System, the Main Board, the LCD Screen, and Packaging. And within those blocks essentially nothing has changed, the pattern of interdependency is the same. *But the out-of-block dependencies both above and below the main diagonal have all disappeared.*

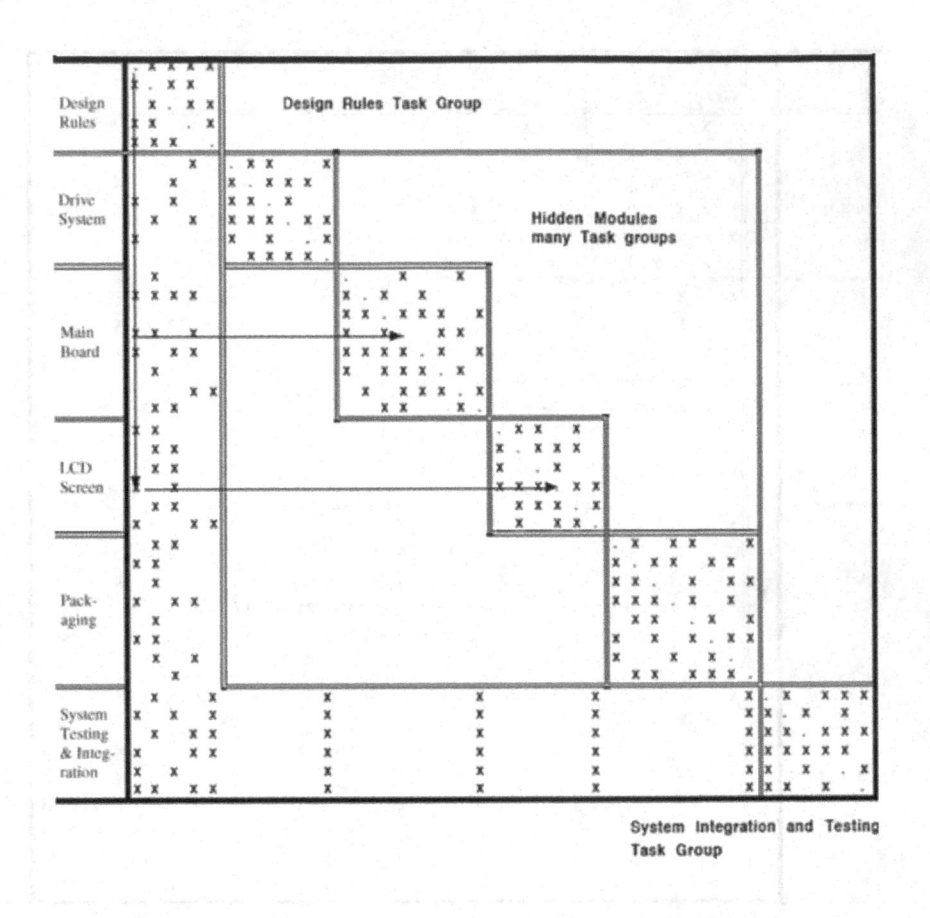

How does that happen? First, in the new structure, each of the former out-of-block dependencies has been addressed by a design rule. Thus, there is now a new "Design Rules" block (not drawn to scale), whose parameters affect many of the parameters in the component blocks. Those dependencies are indicated by the "x"s in the vertical column below the Design Rules block. (Design rule parameters are often called "standards.")

By obeying the design rules, teams working on the designs of each of the component blocks—which are now modules—can maintain conformity with the other parts of the system. But note that there has been another, earlier stage in the process in which the design rules were established.

Furthermore, the new process, as shown in Figure 7, delivers *four separate items*, which must still be integrated into a functioning whole system. No set of design rules is perfect, and unforeseen compatibility problems are often revealed in the latter stages of a modular design process. For these reasons, a "System Integration and Testing" (SIT) block appears in the lower right corner of the modular DSM. This block is affected by the design rules and by some parameters of the hidden modules. But decisions taken in the SIT block, by definition, will not affect choices in the prior blocks. (If they do, then the structure is no longer modular).

Therefore, a modular design structure has three characteristic parts:

- *design rules*, which are known and obeyed by teams responsible for individual modules;

- so-called *hidden modules* that "look to" the design rules, but are independent of one another as work is proceeding; and

- a *systems integration and testing module* in which the hidden modules are assembled into a system, and any remaining, minor problems of incompatibility are resolved.

A complex system design may go from being *interdependent* to being *modular* in the following way. The "architects" of the system must first identify the dependencies between the distinct components and address them via a set of design rules. Second, they must create encapsulated or "hidden" modules corresponding to the components of the system. And third, they must establish a separate system integration and testing activity that will assemble the modular components and resolve unforeseen incompatibilities.

7. The Design Hierarchy Representation

A DSM map is one way to represent a modular system: a design hierarchy is another.[12] A design hierarchy shows which modules are affected by which other modules. (See Figure 8.) At the very top of the design hierarchy are the system-wide design rules: these must be obeyed (hence "are visible to") all modules in the system.

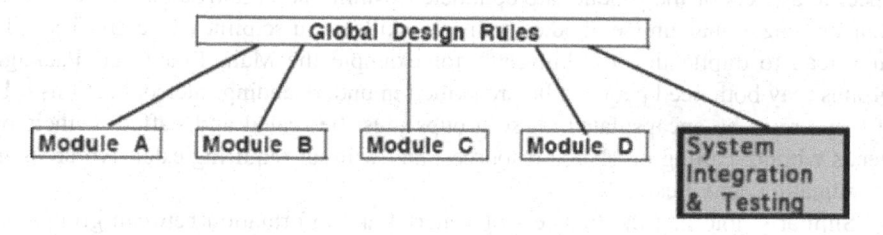

Figure 8. A Two-level Modular Design Hierarchy

Below the system-wide design rules, there may be "architectural modules," which establish design rules for certain subsystems. There are no architectural modules in Figure 8, but in most complex engineering systems there are one or two layers of

---

[12] See [13, 31].

architectural modules. For example, operating systems like Microsoft Windows and Unix are architectural modules in a computer system.

Finally, at the bottom of the design hierarchy are the hidden modules of the system: these must obey the design rules, hence "look to" them. But the hidden modules' own parameters are "encapsulated": they do not affect, and hence do not need to be known to those working on other modules. Hidden modules are thus the primary source of option value in a modular system. (Even so, depending on the rules governing intellectual property, much of the value created in the hidden modules may be captured by the companies that control architectural modules and/or design rules.)

8. Modular Organizations

The "modularized" architectures depicted in Figures 7 and 8 lead naturally to a "modularized" organizational structures.[13] In a recent paper which addressed the nature of transactions we postulated that the activities represented by the x's in Figures 5 and 6 naturally map onto organizations since each interaction captured on the DSM represents a transfer of material or information or both.[14] In this way it is natural to look at Figure 6 and see a 'traditional' organization structured around specific technologies or disciplines. The within-block interactions represent interactions that are internal to each organizational unit while the out-of-block interactions can be viewed as interactions that require coordination across units. In this system, each organizational unit would have liaison personnel whose function was to assure that activities in each unit of the overall endeavor remained synchronized and coordinated.

In the modularized design of Figures 7 and 8 however, many of the cross-unit liaison and coordination functions have been eliminated. This is done in two ways: through design rules and through encapsulation. Design rules (standards) ensure that decisions that affect multiple units are fixed and communicated ahead of time, and not changed along the way. Encapsulation means that all activities associated with specific aspects of the product are conducted within one organizational unit – even if that organizational unit embodies multiple skillsets, disciplines or activities. This may lead to duplication of skillsets – for example the Main Board and Packaging groups may both need people who are skilled in understanding interconnections – but, if they are to be encapsulated, these groups must be scaled and staffed to their own needs without calling on shared resources, and without requiring extensive cross-unit coordination activities.

Similarly note that the transfers of material and information between groups have been simplified to a simple handoff from each module task group to a final systems integration and testing group. If these handoffs as well as the design rules can be standardized and codified, then there is no need for the various groups to reside in the

---

[13] See [41].
[14] See [5].

same company: armslength transactions betweens several firms can cost-effectively replace complex coordinating flows of information within a single firm.

An educational analogy might be useful here. Instead of a laptop computer, Figure 6 might as easily represent the traditional departmental or discipline-based organizational structure of a university (like Harvard). A discipline-based organizational structure is well-suited to teaching courses, but it is ill-suited to carrying out broadly-based research initiatives that cut across many disciplines. Indeed, in order to conduct interdisciplinary research, the traditional departmental structure requires many cross-unit interactions as shown in Figure 6. These interactions are generally time-consuming and prone to cycling. In addition, many task-relevant interactions may get lost in the shuffle and not take place at all.

By contrast, a center-based or project-based collaborative structure gathers participants from multiple disciplines and organizes them into self-contained teams as in Figure 7. Even though some interactions are lost, and there may be duplication of resources across centers, the new organizational structure imposes a much smaller coordination burden on the overall research endeavor. The reduced coordination costs in turn can offset the opportunity losses and costs of implementing the more modular, team-based structure. If the coordination cost savings are large, then a "more modular" organization is a good idea. But there is always a tradeoff—some things are lost while other things are gained.

9. Modular Operators

A key benefit of systems with modular designs is that, especially at the lower levels of the design hierarchy, such systems can evolve. The information that is encapsulated in the hidden modules can change, as long as the design rules are obeyed. Therefore, as we said earlier, modular systems are "tolerant of uncertainty" and "welcome experimentation" *in their modules.* In contrast, the design rules of a modular system, once established, tend to be both rigid and long lasting.

There are certain generic design actions one can apply to a modular system. Following the lead of John Holland of the University of Michigan,[15] we have labeled such actions "operators." In our prior work, we identified six modular operators, and analyzed the sources of their economic value. In particular, given a modular structure, one can:

- *split* any module;

- *substitute* a newer module design for an older one;

- *exclude* a module;

- *augment* the system by adding a module that was not there before;

---

[15] See [23].

- collect common elements across several modules and organize them as a new level in the hierarchy (modular *inversion*); and

- create a "shell" around a module so that it works in systems other than one for which it was initially designed (modular *porting*).

Figure 9 shows how each of these operators affects the structure of a modular system.

We must emphasize that we regard our list of six operators as the beginning of a useful taxonomy. The list is by no means exhaustive. Indeed three other operators have been identified in empirical investigations of design evolution. These are:

- the *linking* of two pre-existing modules;[16]

- the *recombining* of two previously separate modules (this is the opposite of splitting); and

- *embracing and extending* a pre-existing module (this operator was famously used by Microsoft on Sun's version of Java).

The important thing to understand is that *operators correspond to search paths in the design space of a complex engineering system.* These search paths in turn are *options* in the so-called "value landscape" of the complex system. As options, the operator/search paths can be valued using fairly standard analytic techniques from finance. Thus, for example, the decision to *split* a complex system (or subsystem) into several modules can be valued. The decision to *augment* the system by designing several variants of a module customized for different users or purchasers can also be valued. In the next section, we will describe the economic structure of option values for the modules of a complex engineering system.

---

[16] Bala Iyer, 2003, private communication.

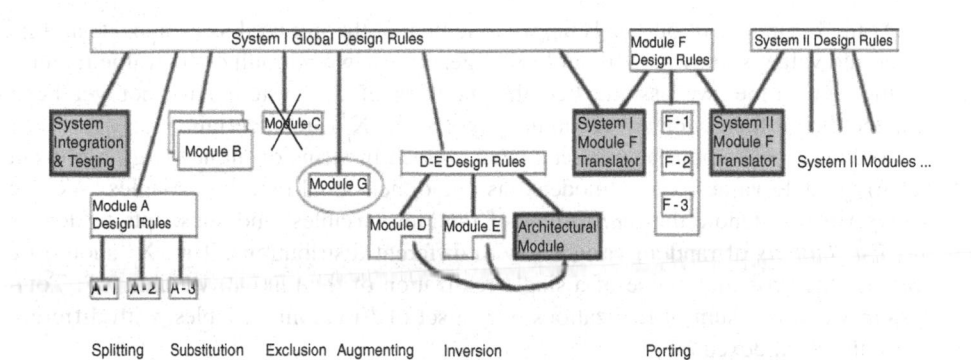

Splitting   Substitution   Exclusion   Augmenting   Inversion   Porting

We started with a generic two-level modular design structure, as shown in Figure 8, but with six modules (A, B, C, D, E, F) instead of four. (To display the porting operator, we moved the "System Integration & Testing Module" to the left-hand side of the figure.) We then applied each operator to a different set of modules.

- Module A was Split into three sub-modules.
- Three different Substitutes were developed for module B.
- Module C was Excluded.
- A new Module G was created to Augment the system.
- Common elements of Modules D and E were Inverted. Subsystem design rules and an architectural module were developed to allow the inversion.
- Module F was Ported. First it was split; then its "interior" modules were grouped within a shell; then translator modules were developed.

The ending system is a three-level system, with two modular subsystems performing the functions of Modules A, D, and E in the old system. In addition to the standard hidden modules, there are three kinds of special modules, which are indicated by heavy black borders and shaded interiors:

- System Integration & Testing Module
- Architectural Module
- Translator(s) Module(s)

Figure 9. The Effect of the Six Operators on a Modular System

10. Option Values in a Complex Engineering System

We begin this section by introducing some notation. We assume that the total economic value of a complex engineering system can be expressed as the sum of a minimal system value, S_0, plus the incremental value added by the performance of each of J modules. Equation 1 thus denotes the *ex post* value that will be realized once the system's design is complete:

$$\text{Economic Value of the system} = \quad S_0 + \sum_{j=1}^{J} X_j^1; \tag{1}$$

At the beginning of our analysis, we assume that the minimal system exists and its economic value is known. Without loss of generality, we normalize that value to zero. At the same time, we assume that the modules of the system have not yet been realized; hence their eventual economic payoffs (the X_j^1) are uncertain.

As long as economic payoffs can be expressed in terms of money (e.g., a present value), module values can be modeled as one-dimensional random variables. We use superscripts to denote the *realizations* of random variables, and subscripts to denote the *distributions* of random variables with different distributions. Thus X_j^1 should be read as "the economic value of a single realization of the random variable X_j". Total system value is a sum of realizations over a set of J random variables, with different distributions, indexed by j.

The realization of a module design is the outcome of a development effort targeted at that module. The realization in turn can have positive or negative value. A design with negative value is not worth incorporating into the system: it subtracts more functionality than it adds. At the end of a design interval, the developers can observe the realization for each module and compare that value to zero. If the new module design has positive value, it will be added to the system, and the system's value will increase by that amount. If the new module design has zero or negative value, it can be discarded, and the developers can try again. In this fashion, the developers can mix and match old and new module designs. The ability to accept or reject a particular realization is the developers' basic option and the focus of our analysis.

For simplicity, we assume that the firm or firms developing the complex system are risk-neutral expected-value maximizers, and that design intervals are short enough that we can ignore the time value of money. In that case, the *ex ante* economic value of the entire system (whose *ex post* value is given by equation 1) can be expressed as follows:

$$V_J = S_0 + \text{Emax}(X_1^1, 0) + \text{Emax}(X_2^1, 0) + ... + \text{Emax}(X_J^1, 0) ; \qquad (2)$$

Equation 2 indicates that each module's realized value will be compared to a benchmark equal to zero. If the new module design has value greater than the benchmark, the new design will be incorporated into the system, otherwise it will be rejected. Thus the expectation of the value of the new design is the *maximum* of its realization and zero. The expectation of the maximum of a random variable and a scalar is larger than the expectation of the random variable alone, thus the option to reject module designs adds to the economic value of the system.

Equation 2 is very general. We can gain further insight by specializing the assumptions. For example, as a thought experiment, consider a system with a total of N design parameters, and think of allocating the parameters into J distinct modules of different sizes. Let X_α denote the economic value of a module of size αN where α is less than (or equal to) one and the set of αs sums to one:

$$\sum_{j=1}^{J} \alpha_j = 1$$

For purposes of illustration, assume that X_α is a normally distributed random variable with mean zero and variance $\sigma^2 \alpha N$: $X_\alpha \sim N(0, \sigma^2 \alpha N)$. In this case, the *variance* of a module's value will be proportional to the number of design parameters in the module. Roughly speaking, the dispersion of outcomes increases as a module's "complexity" measured by αN goes up.

Define z_α as:

$$z_\alpha = \frac{X_\alpha}{\sigma(\alpha N)^{1/2}}$$

z_α is a standard normal variant with mean zero and variance one: $z_\alpha \sim N(0, 1)$.

Substituting for the Xs in terms of z in equation 2, suppressing S_0, and collecting terms, we have:

$$V_{\underline{\alpha}} = V(\alpha_1, ..., \alpha_J; \sigma, N) = \sigma N^{1/2}(\alpha_1^{1/2} + \alpha_2^{1/2} + ... + \alpha_J^{1/2})\text{Emax}(z, 0)$$

$$(3)$$

Here Emax(z, 0) is the expectation of the right half of a truncated standard normal distribution and equals .3989. Note that the system value depends on the *elements* of the vector of $(\alpha_1, ..., \alpha_J)$ as well as the system parameters σ and N. This underscores the fact that impact of modularity cannot be captured by a single summary measure or statistic (e.g., the average degree of modularity). The *details* of the modular structure (i.e., the elements of the vector• $\underline{\alpha}$) affect the system's option value in important and nonlinear ways.

Several results follow directly from equation 3. For example, we can compare the value of a modularized system to the value of the corresponding unmodularized system:

Proposition 1. Under the assumptions given above, let an engineering system of complexity N be partitioned into J independent modules of complexity $(\alpha_1 N, \alpha_2 N, ..., \alpha_J N)$ respectively. The modularized system has value:

$$V_{\underline{\alpha}} = (\alpha_1^{1/2} + \alpha_2^{1/2} + ... + \alpha_J^{1/2})V_1$$

$$(4)$$

relative to V_1, the value of the corresponding unmodularized system.

Proof.
By definition, a one-module design has both J and α_j equal to one. Thus $V_1 = \sigma\, N^{1/2}$ Emax(z, 0). Collecting terms and substituting in equation (3) yields the result. QED.

From the fact that $< \alpha_1 N,\ \alpha_2 N,...,\ \alpha_j N >$ are fractions that sum to one, it follows that the sum of their square roots is greater than one. Thus, as expected, under these very specialized assumptions, a modular design is "always" more valuable than the corresponding non-modular design. Moreover, additional modularization (the splitting) increases value: if a module of size α is split into sub-modules of size β and γ, such that $\beta + \gamma = \alpha$, then the two modules' contribution to overall value will rise because $\beta^{1/2} + \gamma^{1/2} > \alpha^{1/2}$.

In this fashion, higher degrees of modularity can increase the value of a complex design *through option value*. This result is a special case of a well-known theorem, first stated by Robert Merton in 1973.[17] For general probability distributions, assuming aggregate value is conserved, Merton showed that a "portfolio of options" is more valuable than an "option on a portfolio."

Up to this point in our thought experiment, we have assumed that designers will create only *one* new design per module. However, as we indicated above, an important fact about options is that "duplication of effort," in the sense of mounting several design experiments aimed at the same target, may be desirable. Pursuing several experimental designs gives developers the opportunity to select the best outcome after the fact. How much economic value does this does this option create? Is it worth the cost? The answer to this question, it turns out, depends on both the modular structure of the overall system and the technical potential inherent in each module.

To quantify the value of parallel experimentation, let us suppose that in each of J modules, the designers initiate k_j independent design efforts. When the designs are complete, the designers then have the *option* in each module to select the best of the k_j outcomes for the final design. Thus let $Q(k_j)$ denote the expected value of the highest realization of k independent draws from a standard normal distribution as long as the realization is greater than zero. Formally:

$$Q(k) = k \int_0^\infty z[N(z)]^{k-1} n(z) dz$$

where N(z) and n(z) are respectively the standard normal distribution and density functions.[18]

---

[17] See [34].
[18] This distribution of the best of k realizations is well known in statistics: it is the distribution of the "maximum order statistic of a sample of size k." Our expectation differs from the standard one, however, because it is taken only over the range of values above zero. See Lindgren, 1968, on order statistics in general.

By similar reasoning as above, the total value of a system with J modules and k_j experiments on the jth module is:

$$V(\underline{a}, \underline{k}; \sigma, N) = \sigma N^{1/2} \sum_{j=1}^{J} \alpha_j^{1/2} Q(k_j) = \sum_{j=1}^{J} \sigma_j Q(k_j)$$

(5)

The rightmost expression simply notes that the option value of each module is the product of the module's dispersion parameter, σ_j, times a "highest draw expectation" for a standard normal variate. We have tabulated Q(k) for values of k up to 50: the results are shown in Table 2. Using the tabulated values it is straightforward to operationalize this valuation methodology.

Table 2. Tabulated Values of Q(k) for k = 1, ..., 50

k	Q(k)	k	Q(k)	k	Q(k)	k	Q(k)	k	Q(k)
1	0.3989	11	1.5865	21	1.8892	31	2.0565	41	2.1707
2	0.681	12	1.6293	22	1.9097	32	2.0697	42	2.1803
3	0.8881	13	1.668	23	1.9292	33	2.0824	43	2.1897
4	1.0458	14	1.7034	24	1.9477	34	2.0947	44	2.1988
5	1.1697	15	1.7359	25	1.9653	35	2.1066	45	2.2077
6	1.2701	16	1.766	26	1.9822	36	2.1181	46	2.2164
7	1.3534	17	1.7939	27	1.9983	37	2.1293	47	2.2249
8	1.4242	18	1.82	28	2.0137	38	2.1401	48	2.2331
9	1.4853	19	1.8445	29	2.0285	39	2.1506	49	2.2412
10	1.5389	20	1.8675	30	2.0428	40	2.1608	50	2.2491

Equation 5 applies to systems wherein the modules are asymmetric. If modules are symmetric, then it will be optimal to run the same number of experiments on each module. The 2J arguments in equation 5 then collapse to two, and the value of the system as a whole, denoted V(j,k), becomes:

$$V(j,k) = \sigma(Nj)^{1/2} Q(k)$$

(6)

Figure 10 graphs this function for different values of j and k. The way to read this chart is as follows. The vertical axis shows system value as a function of two variables. The first variable, on the right-hand axis, is the number of modules in the system. In the figure, this variable ranges from 1 to 25. The second variable, on the left axis, is the number of design experiments, i.e. R&D projects, per module. This variable also ranges from 1 to 25. The surface shown on the vertical axis indicates the value of different combinations of modules and experiments. As we go out along the middle of this surface, we see the value of running one experiment on one module, two experiments on each of two modules, and so on, until at the far corner, we have 25 experiments on each of 25 individual modules in the system.

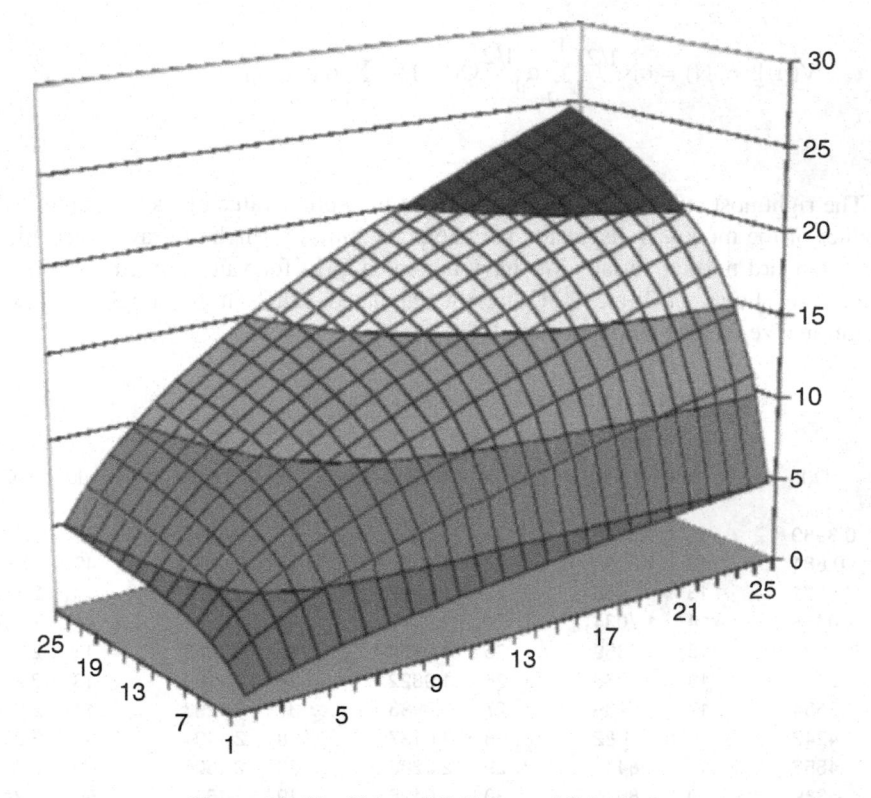

Figure 10. The Value of Splitting and Substitution

The figure shows that there is strong complementarity between design modularity and experimentation. More modules make more experiments more valuable, and vice versa. The two things go together.

Although this function is the result of a thought experiment (and no real system is symmetrical to this degree), the result is nevertheless compelling. The amount of economic value being created by the combination of modules and experiments is really very large. The values are calculated relative to the value of a single experiment in a non-modular system; we see, in effect, that a complex project organized as 25 modules with 25 experiments per module can obtain approximately 25 times the value of the same project organized as a single, interdependent whole. These values are hypothetical, and do not recognize the costs of creating a modular architecture, running experiments, or integrating and testing the modular system. But a value multiple of 25 will pay for a lot of engineering costs! In other words, the incentives afforded by the combination of modularity and experimentation are so great that in a free-market economy, if it is possible to modularize, someone will do so in order to capture this value.

The graph is also surprising in the way the factors interact. Modules and experiments work together, and therefore, as one increases modularity, and one should experimentation, that is, R&D, as well, in order to get the best outcome. This is costly, as we shall see. But there is another kind of "cost" involved: the cost of potential innovation ignored. Often engineers with the encouragement of managers will modularize a complex engineering system in order to reduce its cognitive complexity or shorten its development time. Neither group may understand what the modularization implies for the value of innovation and experimentation within the system. Yet if the firm that initially modularizes the system ignores this value, *there will be strong incentives for other firms to enter the market offering their own new module designs.*

Something like that happened to IBM after the company introduced the first modular computer, System/360, in the mid-1960s. System/360 was a powerful and popular modular system, a *tour de force* in terms of product design, marketing, and manufacturing. However, IBM's top managers did not understand the value of the options that had been created by its own new modular design. They did not increase inhouse design efforts, and as a result, left profitable opportunities "on the table." Before long, new firms moved in and seized these opportunities. This was the start of the pattern of industry evolution and value migration depicted at the beginning of this chapter.

Many of the new firms that entered the computer industry in the wake of System/360 were staffed by former IBM engineers. Engineers who had worked on System/360 and its successors could see the module options very well, and knew the design rules of the system. Thus when IBM's top managers did not fund their projects, they took those projects elsewhere. Beginning in the early 1970s, scores and then hundreds of engineers left IBM and joined others in founding companies that supplied "plug-compatible" modules for IBM's System/360 and 370. As it happens, one of IBM's main R&D labs was located in San Jose, California, and the exodus of engineers from the San Jose labs was one of the key factors that contributed to the emergence of what we now call Silicon Valley.

11. The Costs of Modularity

We have yet to address the costs of modularity. In fact, the costs of creating and exploiting a modular system can be a significant offset to the value that is created.

There are, first of all, the costs of making an interdependent system modular: the cost of creating and disseminating design rules. The DSM mapping techniques, discussed above, show how painstaking the process of modularization must be, if it is to succeed in creating truly independent modules. *Every important cross-module dependency must be understood and addressed via a design rule.* Obviously the density of the dependencies matters here. Modular breakpoints (interfaces) need to go at the "thin crossing points" of the interdependency graph. Some systems are naturally more "loosely-coupled" than others: they have more thin crossing points hence relatively more potential modules.

For example, circuits, the physical system on which computers are based, are one- and two-dimensional; whereas mechanical solids are three-dimensional. Clearly it is harder to split up complex, curved, 3-dimensional designs, and to create flexible interfaces for them: there are more dependencies to manage, and the tolerances are much tighter. Thus modularizing an automobile's design is a tougher engineering problem than modularizing a circuit design: the cost of creating a modular architecture and related interfaces will be higher. This has led some scholars, like Daniel Whitney at MIT, to predict that autos and airplanes will achieve only limited modularity in practice.[19] The option values inherent in these tightly coupled systems will be low relative to systems that can more easily be modularized.

It is also costly to run the experiments needed to realize the potential value of a modular system and to design the tests needed to determine if particular modules are compatible with the system, and which one performs best. Indeed figuring out how many and what kind of experiments to run and how to test the results are important sub-problems within the overall option valuation problem. The interaction of option value and the costs of experimentation and testing modules causes each module in a large system to have a unique value profile. For example, Figure 11 shows how value profiles may differ across modules as a function of the number of experiments. Some modules (those that are hidden and have high technical potential) can support a lot of experiments; others (those that are visible and/or have low technical potential) can support few or none.

Size	Visibility	Examples
large	hidden	disk drive; large application program
large	visible	microprocessor; operating system
small	hidden	cache memory; small application program
small	visible	instruction set; internal bus

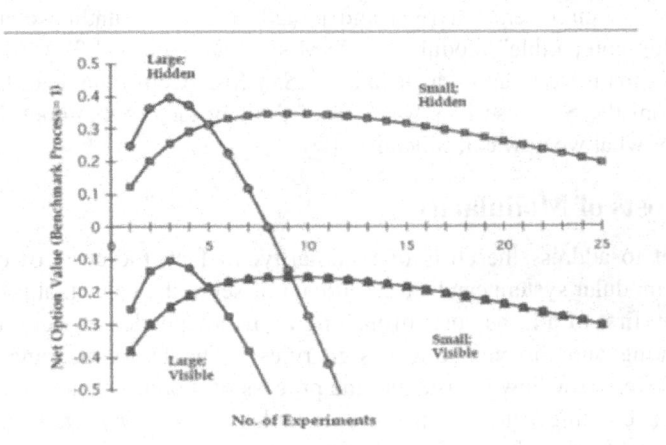

Figure 11. The Value Profiles for Different Modules of a Computer System

---

[19] See [50]

12. Conclusion

We conclude this brief excursion into the realm of modularity by returning to the question of industry evolution, and commenting on the perils of modularity for incumbent firms.   Modularity in the design of a complex engineering system with high technical potential (high σs in the modules) is likely to be highly disruptive to the pre-existing industry structure. Modularity-in-design allows users or system integrators to mix and match of the best designs within each module category and to incorporate new and improved module designs as they become available. Thus a modular system design requires that a company operate all aspects of its business more efficiently than its competitors. If it is not "the best" in a given module, then competitors will flock to that point of vulnerability.

For example, consider IBM's introduction of the personal computer (PC) in the early 1980s.[20] By this time, IBM had learned the basic lessons of modularity inherent in System/360. Its managers understood how modularity encouraged both innovation and entry on modules. Indeed, the PC was extremely modular-in-design, and IBM leveraged this modularity by outsourcing most of its hardware and software components. But IBM's managers also understood that they needed to protect the company's privileged position within the modular architecture. Thus IBM retained control of what were thought to be the key design and process modules—a chip called the BIOS (Basic Input Output System) and the manufacturing process of the PC itself. By exercising control of these critical and essential "architectural modules," IBM's managers believed that they could manage the rate of innovation in PC's. Their goal was to obtain maximum return from each generation of PC, before going on to the next.

However, the founders of Compaq had a different idea. First, they independently and legally replicated IBM's technical control element, the BIOS. Then they designed a machine that was fully compatible with IBM's published and non-proprietary specifications. They bought the key modules of the PC — the chip and the operating system — and all the other parts they needed from IBM's own suppliers. And they went to market with an IBM-compatible PC built around the newer, faster Intel 386 chip while IBM was still marketing 286 machines. Within a year Compaq had sales of $100 million; by 1990, its revenues were $3 billion and climbing, and IBM was looking to exit from its unprofitable PC business.

However, as everyone knows, the leading player in the PC market today is not Compaq but Dell.[21] In a nutshell, Dell did to Compaq what Compaq did to IBM: it took advantage of the benefits of modularity and designed a line of technologically competitive, yet compatible, and lower-priced PCs. In fact, Dell used *modularity-in-the-design-of-production-processes* more effectively than Compaq in order to arrive at a more efficient, less-asset-intensive business model. Dell outsourced even more of its manufacturing activities than Compaq. It designed its assembly processes to build machines to order (BTO), thereby cutting out most of its inventory. And it sold its

---

[20] This history is recounted in [16].
[21] For detailed comparisons of Compaq's and Dell's business models, see [6, 49].

products directly to consumers, thereby cutting out dealers' margins and more inventory. Compaq simply could not compete. When its top managers saw the handwriting on the wall in PCs in early 1998, they tried to move the company into higher-margin "enterprise computing" through the acquisition of Digital Equipment Corporation, but the transition was not successful. Today, Compaq no longer exists as a separate company—it was acquired by Hewlett Packard Corporation in 2002.

In conclusion, the widespread adoption of modularity-in-design in complex engineering systems with high technical potential can set in motion an uncontrollable process of design and industry evolution. The economic consequences of this process—for good and bad—are depicted in the charts and turnover rates described at the beginning of this chapter. Thus modularity-in-design can open the door to an exciting, innovative, but very Darwinian world in which no one really knows which firms or business models will ultimately prevail. This can be a world of growth, innovation and opportunity. But, as the Internet Bubble and Crash taught us, this world can also fall into periods of extreme value destruction, chaos and inefficiency. Thus in the last analysis, modularity-in-design is neither good or bad. Rather, it is potentially powerful and disruptive, and therefore dangerous to ignore.

Acknowledgments

Our thanks to Datta Kulkarni and Robin Stevenson for many conversations that have informed our work over the past several years, as well as comments and suggestions that greatly improved this paper. Thanks also to Barbara Feinberg, who over many years and countless discussions helped us to develop and refine our ideas.

An earlier version of this talk was prepared as the keynote address for the Opening Conference of the Research Institute of Economy, Trade and Industry (RIETI), "Modularity— Impacts to Japan's Industry," Tokyo, Japan, July 12-13, 2001. Our thanks to the sponsors and participants of the conference, especially Masahiko Aoki and Sozaburo Okamatsu, Nobuo Ikeda, Takahiro Fujimoto, Hiroyuki Chuma, Jiro Kokuryo, Noriyuki Yanagawa, Nobuo Okubo, Shuzo Fujimura, Hiroshi Hashimoto, Hiroshi Kuwahara, and Keiichi Enoki for extremely stimulating discussion and pertinent critiques.

Finally we would like to thank Sarah Woolverton, Joanne Lim and James Schorr for their assistance in compiling our industry database.

We alone are responsible for errors, oversights and faulty reasoning.

References

[1]   Abernathy, William and James Utterback (1978) "Patterns of Industrial Innovation," *Technology Review* 80:41-47.

[2]   Baldwin, Carliss Y. and Kim B. Clark (1992) "Modularity and Real Options: An Exploratory Analysis" Harvard Business School Working Paper #93-026, October.

[3]   Baldwin, Carliss Y. and Kim B. Clark (1997) "Managing in the Age of Modularity," *Harvard Business Review* Sept/Oct: 81-93.

[4]   Baldwin, Carliss Y. and Kim B. Clark (2000). *Design Rules, Volume 1, The Power of Modularity*, MIT Press, Cambridge MA.

[5]   Baldwin, Carliss Y. and Kim B. Clark (2002) "Where Do Transactions Come From? A Perspective from Engineering Design," *Harvard Business School Working Paper* 03-031, Boston, MA.

[6]   Baldwin Carliss Y. and Barbara Feinberg, (1999) "Compaq: The DEC Acquisition," 9-800-199, Harvard Business School Publishing Company, Boston, MA.

[7]   Braha, Dan (2002) "Partitioning Tasks to Product Development Teams," *Proceedings of ASME 2002 International Design Engineering Technical Conferences,* Montreal CN, October.

[8]   Browning, Tyson R. (2001) "Applying the Design Structure Matrix to System Decomposition and Integration Problems: A Review and New Directions," *IEEE Transactions in Engineering Management* 48(3):292-306.

[9]   Browning, Tyson R. (2002) "Process Integration Using the Design Structure Matrix," *Systems Engineering,* 5(3):180-193.

[10]   Browning, Tyson R. and Steven D. Eppinger (2002) "Modeling Impacts of Process Architecture on the Cost and Schedule Risk in Product Development," forthcoming in *IEEE Transactions in Engineering Management.*

[11]   Chandler, Alfred D. (1962) *Strategy and Structure,* MIT Press, Cambridge, MA.

[12]   Chandler, Alfred D. (1977) *The Visible Hand: The Managerial Revolution in American Business,* Harvard University Press, Cambridge, MA.

[13]   Clark, Kim B. (1985) "The Interaction of Design Hierarchies and Market Concepts in Technological Evolution," *Research Policy,* 14(5):235-251.

[14]   Eppinger, Steven D. (1991) "Model-based Approaches to Managing Concurrent Engineering" *Journal of Engineering Design,* 2: 283-290.

[15]   Eppinger, S. D., D.E. Whitney, R.P. Smith, and D.A. Gebala, 1994, "A Model-Based Method for Organizing Tasks in Product Development," *Research in Engineering Design* 6(1):1-13

[16]   Ferguson, Charles H. and Charles R. Morris, *Computer Wars: The Fall of IBM and the Future of Global Technology,* Times Books, NY, 1994.

[17]   Fixson, Sebastian and Mari Sako (2001) "Modularity in Product Architecture: Will the Auto Industry Follow the Computer Industry?" Paper presented at the Fall Meeting of the International Motor Vehicle Program (IVMP).

[18]   Fujimoto, Takahiro (1999) *The Evolution of a Manufacturing System at Toyota,* Oxford University Press, Oxford, UK.

[19]   Fujimoto, Takahiro and Akira Takeishi (2001) Modularization in the Auto Industry: Interlinked Multiple Hierarchies of Product, Production and Supplier Systems, Tokyo University Discussion Paper, CIRJE-F-107, Tokyo, Japan.

[20]   Fujimoto, Takahiro (2002) "Architecture, Capability and Competitiveness of Firms and Industries," presented at the Saint-Gobain Centre for Economic Research 5th Conference, Paris, FR, November.

[21] Garud Raghu and Arun Kumaraswamy (1995) "Technological and Organizational Designs to Achieve Economies of Substitution," *Strategic Management Journal*, 17:63-76, reprinted in *Managing in the Modular Age: Architectures, Networks, and Organizations*, (G. Raghu, A. Kumaraswamy, and R.N. Langlois, eds.) Blackwell, Oxford/Malden, MA.

[22] Gomes, Paulo J. and Nitin R. Joglekar (2003) "The Costs of Organizing Distributed Product Development Processes," Boston University School of Management Working Paper #2002-06, Boston, MA, January.

[23] Holland, John H. (1992) *Adaptation in Natural and Artificial Systems*, 2nd Ed. MIT Press, Cambridge, MA.

[24] Holland, John H. (1996) *Hidden Order: How Adaptation Builds Complexity*, Addison-Wesley Publishing Company, Reading, MA.

[25] Holland, John H. (1999) *Emergence: From Chaos to Order*, Perseus Books, Reading, MA.

[26] Joglekar, Nitin and Steven Rosenthal (2003) "Coordination of Design Supply Chains for Bundling Physical and Software Products," *Journal for Product Innovation Management*, forthcoming.

[27] Klepper, Steven (1996) "Entry, Exit, Growth and Innovation over the Product Life Cycle, *American Economic Review*, 86(30):562-583.

[28] Kusiak, Andrew (1995) *Engineering Design*, Academic Press, New York, NY.

[29] Lindgren, Bernard W. (1968) *Statistical Theory*, Macmillan Publishing Co., New York, NY.

[30] Loch, Christoph H., Christian Terwiesch and Stefan Thomke (2001) "Parallel and Sequential Testing of Design Alternatives," *Management Science*, 45(5):663-678.

[31] Marples, D.L. 1961, "The Decisions of Engineering Design," *IEEE Transactions in Engineering Management*, 2: 55-81

[32] McCord, Kent R. and Steven D. Eppinger, 1993, "Managing the Integration Problem in Concurrent Engineering," MIT Sloan School of Management Working Paper, no. 3594, August.

[33] Mead, Carver and Lynn Conway (1980) *Introduction to VLSI Systems*, Addison-Wesley, Reading, MA.

[34] Merton, Robert C. (1973) "Theory of Rational Option Pricing," *Bell Journal of Economics and Management Science*, 4(Spring): 141-183; reprinted in *Continuous Time Finance*, Basil Blackwell, Oxford, UK, 1990.

[35] Milgrom, Paul and John Roberts (1990) "The Economics of Modern Manufacturing: Technology, Strategy and Organization," *American Economic Review*, 80:511-528.

[36] Parnas, David L. (1972a) "A Technique for Software Module Specification with Examples," *Communications of the ACM* 15(May): 330-36.

[37] Parnas, David L. (1972b) "On the Criteria to Be Used in Decomposing Systems into Modules," *Communications of the ACM* 15(December): 1053-58.

[38] Parnas, David L., P.C. Clements, and D.M. Weiss (1985) "The Modular Structure of Complex Systems," *IEEE Transactions on Software Engineering*, SE-11(March): 259-66.

[39] Sako, Mari (2002) "Modularity and Outsourcing: The Nature of Co-Evolution of Product Architecture and Organization Architecture in the Global Automotive Industry, forthcoming in *The Business of Systems Integration* (A. Prencipe, A. Davies and M. Hobday, eds.) Oxford University Press, Oxford, UK.

[40] Sanchez, Ron (1991) "Strategic Flexibility, Real Options and Product-based Strategy," Ph.D dissertation, Massachusetts Institute of Technology, Cambridge, MA.

[41] Sanchez, R. and Mahoney, J. T. (1996) "Modularity, flexibility and knowledge management in product and organizational design". *Strategic Management Journal*, 17: 63-76, reprinted in *Managing in the Modular Age: Architectures, Networks, and Organizations*, (G. Raghu, A. Kumaraswamy, and R.N. Langlois, eds.) Blackwell, Oxford/Malden, MA.

[42] Sharman, David M., Ali A. Yassine, and Paul Carlile (2002) "Characterizing Modular Architectures," *Proceedings of DETC '02, Design Theory & Methodology Conference*, Montreal, Canada, September.

[43] Sturgeon, Timothy (2002) "Modular Production Networks: A New American Model of Industrial Organization," *Industrial and Corporate Change*, 11(3):451-496.

[44] Sturgeon, Timothy (2003) "Exploring the Benefits, Risks, and Evolution of Value-Chain Modularity in Product-level Electronics," draft, Industrial Performance Center, MIT, February.

[45] Sullivan, Kevin J., William G. Griswold, Yuanfang Cai and Ben Hallen, "The Structure and Value of Modularity in Software Design," University of Virginia Department of Computer Science Technical Report CS-2001-13, submitted for publication to ESEC/FSE 2001.

[46] Sutton, John (1992) *Sunk Costs and Market Structure*, MIT Press, Cambridge, MA.

[47] Topkis, Donald M. (1998) *Supermodularity and Complementarity*, Princeton University Press, Princeton, NJ.

[48] Ulrich, Karl (1995) "The Role of Product Architecture in the Manufacturing Firm," *Research Policy,* 24:419-440, reprinted in *Managing in the Modular Age: Architectures, Networks, and Organizations,* (G. Raghu, A. Kumaraswamy, and R.N. Langlois, eds.) Blackwell, Oxford/Malden, MA.

[49] Wheelwright, Stephen C. and Matt Verlinden, 1998 "Compaq Computer Corporation," 9-698-094, Harvard Business School Publishing Company, Boston, MA

[50] Whitney, Daniel E. (1996) "Why Mechanical Design Cannot Be Like VLSI Design,"http://web.mit.edu/ctpid/www/Whitney/morepapers/design.pdf, viewed April 9, 2001.

Chapter 10

Engineering Complex Systems

Douglas O. Norman
dnorman@mitre.org
Michael L. Kuras
mlk@mitre.org
The MITRE Corporation

1. Introduction

This chapter motivates the need for, and introduces a formal set of processes that constitute the practice of, "Complex Systems Engineering" (CSE). Our experiences and observations strongly suggest Enterprise Engineering is best approached using CSE to engineer and manage the enterprise[1].

Using the current instantiation of the Air and Space Operations Center (AOC[2]), and the desired evolution of it, the AOC is shown to be best thought of as a complex system. Complex Systems are alive and constantly changing. They respond and interact with their environments – each causing impact on (and inspiring change in) the other. We make the case that a traditional systems engineering (TSE) approach does not scale to the AOC; consequently, we don't believe TSE scales to the "enterprise."

We introduce a new set of processes which complement – and do not replace – the processes that constitute traditional systems engineering. The methods for the engineering of complex systems are based on a view of complex systems as having the characteristics of an *ecosystem*, and the use of processes which take advantage of emergence and which deliberately mimic *evolution* to accomplish and manage the engineering outcomes desired.

---

[1] Initially, we appeal to the reader's intuition for the definition of an enterprise. The picture in one's mind should be something like a large collection of independent organizations, loosely associated to achieve something in common.

[2] A special thanks to Col Pete Hoene, USAF, and Col Terry Szanto, USAF, the past and current AOC WS System Program Office Director, respectively, good friends both, for their continuing professional engagement in this topic, and their kind reading of this draft along with their helpful comments. Special thanks also to Col Joe May, USAF (ret) who took the time for healthy debate and discussions both as the Director of Operations for the Air Force's C2 and ISR Center, and subsequently. Joe put the operator's "stink" on the thoughts.

The chapter is structured in four major sections:
- Why Rethink Systems Engineering?
- Complexity and Complex Systems
- Engineering Complex Systems
- Complex Systems Engineering in Practice

We all must come to grips with the non-deterministic nature of enterprises. We hope to extend the concepts and methods of Systems Engineering to complex systems, and to open up the professional dialog so as to codify the engineering and management of complex systems and enterprises.

2. Why Rethink Systems Engineering?

First, we should take stock in everything accomplished. We've designed, built, fielded, and operated two Air and Space Operations Centers (AOC) which provided the tools used to plan, task, and monitor all air operations in both Operation Enduring Freedom in Afghanistan and Operation Iraqi Freedom in Iraq as well as the AOC created for NORAD to manage airspace and combat air patrols in the US after 9/11 (Operation Noble Eagle). Other packagings of the functionality are fielded with Special Operations units, and with the Navy and Marines.

Yet, we are having trouble building, integrating, and modifying large "systems." Our difficulties are legion [e.g., 9]. Struggling with these failures, as a community, we have continued to refine our notions about systems engineering, and how we define, design, and prepare for systems; but – again, as a community – we haven't changed our underlying "mental model" which informs our general (and specific) philosophy and processes. We continue to view Systems Engineering as fundamentally about allocating desired, known functionality among specific elements of a design; all known *a priori* and stable over time. The users of the functionality built often accuse us, the developers and acquirers, of being "late to need," "unresponsive," and "too expensive."

We respond with a lexicon carefully crafted to put the onus back on the users. We say that the users' requirements are unknown or poorly stated; that, if the requirements are known, there is a requirements drift (i.e. modifying the requirements), or requirements creep (i.e. adding additional requirements). We suggest that the user can't (or won't) say what they *really* want, or how they will use that which is to be built and delivered. This situation results in processes that focus on detail: detail of proposed use, detail of environment, detail of design, and detail of planned schedule. The more detail the better.

During the last decade as the world has moved from stand-alone automation that augmented individuals, to networked and shared automation, architectures became the way to deal with the rising complexity. But the early forays into architectural-based engineering didn't seem to pay the dividends anticipated – complaints from users continued, and the most-valued automation tools were ones not built with the same careful attention to detail required by the Architecturally-based Systems Engineering approach practiced. Rather, it seemed much of the architecture-supplied understanding and simplification was implicit. We saw the *technical architectures*, rather than being defined and engineered, were being imposed by the development tools used; and the need for *operational architectures* (what will this system do...and how will it be used) were rendered unnecessary since "experts" were supplying the operational insight directly to the developers. Further, these user/developer collaborations were turning products out quickly; and they were valued by the end-user.

As we attempt to make systems more useful and valued we also start to come face-to-face with the limitations of our current methodologies. For the Air and Space Operations Center, there are two clear forces pressing on it and demanding its evolution: 1) speed of accommodation to new understandings of current missions, or new missions demanding modification of the current AOCs; and 2) application to new doctrinal uses or mission types [1, 2, 3]. In all cases, the manner in which we practice Systems Engineering seems to bog us down; and we're compelled to rethink the practice. This analysis leads us to recognize that there are additional classes of Systems Engineering problems. Each class requires a different qualitative mindset; and consequently, a different set of tools, techniques, and procedures to undertake the task of systems engineering successfully.

Systems engineering, at its simplest, attempts to understand a desired outcome from the interactions among people and things, and is roughly divided into two phases: Analysis and Design. During analysis the context for the desired outcome is explored to uncover the initial conditions and resources available, which can contribute to the desired outcome. Then, with this understanding, processes are designed which are able take the initial conditions and transform them into the desired outcome. As the designing proceeds, roles and activities are allocated among people, hardware, software, and organizations. Flowing out from this allocation are the ancillary activities implied and required by the design. For example, if an activity has been allocated to a person – and that person is expected to perform in a certain manner – then the person must understand their role in the system, and must be trained to perform that role. At all times design boundaries and constraints are examined and evaluated to ensure they are not violated. For example, if precision must be maintained, then cumulative error must be watched; or if certain information flows are required, then there must be connectivity and sufficient bandwidth to satisfy the design, etc.

The preceding certainly seems to make sense. And, the more complex or critical the desired outcome, the more important it would seem to conduct this process with more rigor, more information, and with more detailed plans; after all, we have to get it "right."

So what's the issue? In a nutshell: the mindset, skills, methods, and processes used to develop "systems" in this way seem to fail us when we attempt to craft "Systems of Systems." And, the AOC is certainly a System of Systems.

2.1. Challenges with Engineering the AOC

The AOC is known as a "System of Systems" (SoS). As such, it is envisioned as a system assembled of other systems so as to offer the capabilities needed to perform roles assigned to an AOC. Implicit in this is the expectation that the systems from which the AOC is assembled can be composed into an AOC *System of Systems*. This has proven harder than anticipated, and it provides insight into the challenges of Enterprise Engineering. While still viewed as a "system" by some, the AOC turns out to be an "enterprise" in the small. As in all enterprises, it is composed of different pieces representing different fiefdoms and principalities; (or "tribes" as Gen J. Jumper, AF Chief of Staff likes to say) Listed and discussed below are those characteristics that have proven to add difficulties to the intended composition of systems into an AOC SoS.

The AOC today is assembled from over 80 elements. There are infrastructure elements, communication elements, applications, servers, and databases. The goal is to compose the desired capabilities from the elements found in, or which can be brought into, the AOC. For the most part, today's systems are **not composable**. The systems:

- Don't share a common conceptual basis.
- Aren't built for the same purpose, or for use within specific (AOC) work flows, or for use exclusively at AOCs,
- Share an acquisition environment which pushes them to be "stand alone"[3],
- Have no common control or management,
- Don't share common funding which can be directed to "problems" as required,
- Have many "customers," of which the AOC is only one,
- Evolve at different rates (as do individual system components) subject to different (generally uncoordinated) pressures and needs.

Because of the above, **ensuring integration and interoperability are unbounded, unpredictable engineering activities**. The following observations clarify this further:

Observation 1: The AOC SoS is an opportunistic aggregation, not a design.

- Only the AOC System Program Office (SPO), which has the acquisition responsibility for the AOC SoS, has a strong interest in an overall AOC design, and has no way to enforce such a design on others who supply the component systems of the AOC,
- Since the AOC SPO doesn't spend its money for many of the component systems of an AOC SoS, the component system-owners have little incentive to comply with, or respect, an AOC design,

---

[3] The DoD's acquisition system is built around the concept of a "system" which seeks to separate a given system from every other. This separation extends from the concept through delivery and sustainment. Funds executed on behalf of the system acquisition are, by law, separate from all other monies with Congress carefully monitoring expenditures.

- The SPOs for the component systems in an AOC SoS must remain responsive to customers and users with interests other than the AOC, and,
- The need for, and the appearance of, a specific new capability at an AOC is often driven by a new, immediate need not apparent to or felt by the other customers and users of the component systems in an AOC SoS; and to which the AOC SPO would be unable to satisfy within the time-of-need.

Observation 2: Integration-enabling technologies (*glueware*) are grafted onto the elements (systems) of the AOC, and integration developments are undertaken, after delivery of the component systems to their prime customers,

- Each element in the aggregate is designed and built with its own understanding of the world – around its own set of "conceptual atoms"
  - Integration among these elements requires effort (resources) to understand and bring these potentially disparate "conceptual atoms" in line so they can be composed,
- Integration is a source of work and revenue – using today's dominant business model (employer/contractor) contractors sell engineering hours,
  - "Big Integration" is a potential cash-cow for those who perform it
  - Little incentive to limit the work, or find ways to be more effective
  - Integration of already-developed elements guarantees that the delivery of an integrated, operational AOC will lag behind the availability of the individual elements; however, the expectation from the users is that *general availability* (when the component systems become available to the users) and *integrated* are synonymous. This leads to customer disappointment; and is further compounded by the need to expend additional funds to perform the integration proper.

Observation 3: Funds for integration are limited

- Willingness (and sometimes, the ability) of the user to wait is limited, and accelerating deliver (if even possible) costs additional money;
- Perceived barriers for building automated functionality (in software) are low, setting customer expectations that it's easy, quick, and cheap;
- Integration tends to be built around a defined work flow which implements a specific concept of operation. Integration "glue" which implements the concept of operation binds systems into rigid relationships. This is contrary to achieving "agility" and "net-centricity;" [2, 3]
- The ability to conduct tests has an inverse relation to system size;
  - Resources (organizations, staff, time, money) available to conduct large tests are limited;
  - Test coverage plans for large systems becomes unwieldy;
  - Likelihood of finding incompatibilities during large test rises (due to the possibility of uncovering an unexpected transitive affect);
  - Understanding of how to proceed once a problem is uncovered is an open question;

Note: The process of testing of SoS must be rethought; the goals and the uncovering of problems are good things which must be preserved – just not at system test time; especially operational test.

Observation 4: "Value" assessment is not by those who use the capabilities
- The "marketplace" serviced by the acquisition system is the selling of engineering hours through the promise of future assemblies and creations; and the delivery of these creations; not by the assessment of value or utility by the AOC staff (these aspects are supposedly contained in the formal *requirements*),
- Those who use the creations of the formal system have only an indirect influence; any direct influence being the result of heroic efforts on the part of individuals
    - This tends to bring into being a "black market" of applications and functionality – and the hoarding of local "slush funds" which can be directed by the local commanders to satisfy needs as they arise.

Observation 5: Plans (and Planning) as a primary SoS strategy has problems
- Focuses on the future – but is based in the past
- Tends to fix an early (likely incorrect and incomplete) view
- Activities tend to twist reality (subject to unplanned change) to the plan (static, based on past beliefs)
- Imposes expectations, and dependencies, on partially-interested participants
- Design implied in the plan is based on today's understandings. As things change in the world all the elements to be composed are subject to different pressures and decisions which likely will not align
- Assumes a success based on promises (staying on-plan) – not achievements

Additionally, there are new operational concepts being considered, developed and employed. These include over-arching concepts such as Netcentric Warfare (NCW) [1], and technical concepts such as the Global Information Grid[4]. For the AOC, new operational ideas such as Dynamic Tasking and Effects-Based Operations are taking hold. Supporting this growth and change, at an acceptable rate and at an acceptable cost, is often described as "agile acquisition." Yet, there are few examples of how to achieve this "agile acquisition."

How can these new operational ideas be cast into capabilities, which can then be integrated in the AOC? What works? There seem to be characteristics that militate against success, even when carefully practicing Systems Engineering as we know it. Before these characteristics are introduced and reviewed, and the CSE approaches by which complex systems may be engineered are discussed, we present a quick introduction to systems engineering and complexity as it applies to systems (and systems engineering).

---

[4]Global Information Grid Capstone Requirements Document, 5 JROCM 134-01, August 30, 2001

2.2. What is Systems Engineering?

Traditional systems engineering (TSE) has its foundations in Linear System Theory (LST). Key ideas are proportionality, super-positioning, and the existence of invertible functions (i.e. $x = f^1(f(x))$. There is also the assumption of repeatability. These ideas, coupled to an attention to detail, explain why traditional systems engineering works as well as illuminating the boundaries of its applicability.

Traditional systems engineering begins with the specification of requirements. Closed and complete, precise and fully detailed are the ideals. Systems are then implemented to comply with or to satisfy exactly these requirements.[5] The practice of TSE is the application of a series of linear transformations moving from the statements of the requirements through to a preliminary design, a final design, the actual development, then testing and fielding. A hallmark of the process is the ability to justify everything built in terms of the original requirements. If requirements change it dislodges the careful scaffolding upon which the system rests. Change ripples through everything; and; therefore, approaches which isolate the impacts of change are sought. Since "no change" is the desired (and expected) state of affairs, engineering efforts shift to development efficiencies.

Decomposition and then integration (or assembly) are the bookends for the implementation that follows specification, both of which depend upon the applicability of LST to a given situation. Because of this, there is a strong preference for hierarchy in both implementation activities as well as in the result.[6]

Traditional systems engineering relies on the making of and the fulfilling of predictions. These predictions are more binding for traditional system engineers than any current realities as is seen (for example) in development of PERTs, and in formal testing procedures which are made independent of implementation but that are predicated on the same requirements. As long as predictions and realities diverge, a preference is given to preserving the predictions. Descriptions of traditional systems engineering can be found in many places. A quick "Google"[7] of the term brings many hits (millions), including many universities offering systems engineering curricula and degrees.

Systems Engineering is seen as a professional discipline and, as for other professions, has developed professional associations where the practice itself is codified and socialized. Such a professional organization is the International Council on Systems Engineering (INCOSE – www.incose.org). This (traditional) Systems Engineering definition is taken from the INCOSE web site [18]:

---

[5] Engineering is always an approximation, however. Traditional systems engineering assumes (or asserts) that the ideal result is "closed" and that it can be completely pre-specified. The appearance of Interface Control Documents (ICDs), for example, is an exception that illustrates (or preserves) the validity of the general rule.
[6] Refer to the brief discussion of multiscale analysis. This preference for hierarchy is actually a result of the fact that LST is limited to a uniscale analysis and synthesis of a problem and its solution.
[7] URL = http://www.google.com. Note the irony of using a complex system to help describe complex systems.

"Systems Engineering is an interdisciplinary approach and means to enable the realization of successful systems. It focuses on defining customer needs and required functionality early in the development cycle, documenting requirements, then proceeding with design synthesis and system validation while considering the complete problem:

- Operations
- Performance
- Test
- Manufacturing
- Cost & Schedule
- Training & Support
- Disposal

Systems Engineering integrates all the disciplines and specialty groups into a team effort forming a structured development process that proceeds from concept to production to operation. Systems Engineering considers both the business and the technical needs of all customers with the goal of providing a quality product that meets the user needs."

Figure 1. INCOSE Systems Engineering Definition

What's not described in this definition is the process, previously outlined, to which all of the activities, areas, and disciplines implied in the INCOSE definition are brought to bear to support.

Fundamentally, the practice of TSE seeks to understand the place of an element within the environment, isolate the element under study from the environment, and then treat the environment as a constant.

Reflect on Systems Engineering. Among the characteristics one would require to have a successful, or at least a low risk outcome, there are a few which are absolutely required to ensure success using traditional Systems Engineering. These serve as boundary conditions for applying TSE:

- The specific desired outcome must be known *a priori*, and it must be clear and unambiguous (implied in this is that the edges of the system, and thus responsibility, are clear and known);
- There must be a single, common manager who is able to make decisions about allocating available resources to ensure completion;
- Change is introduced and managed centrally;
- There must be "fungible" resources (that is money, people, time, etc.) which can be applied and reallocated as needed.

Failing to have any of the above raises risk dramatically; and it is unlikely that other mitigation strategies will be possible for the risks introduced. How many of these boundary conditions are found in an enterprise? Our sense is that there seems to be a correlation between small projects – that build stand-alone, fairly simple applications/products, which are under the complete control and management of a single party – and the likelihood of having these boundary conditions satisfied.

Unfortunately, when one considers an *enterprise* every one of the characteristics mentioned above is violated. This isn't too surprising as Systems Engineering has evolved (very successfully) from an industrial, element-manufacturing point of view.

Of note in the INCOSE definition of Systems Engineering is the absence of the concept of the *enterprise*. In fact, the aspects discussed and listed above in the INCOSE definition are appropriate for *elements* of the enterprise – and their manufacture- but not the *enterprise* itself. Any arguments offered wherein one suggests that one does the same things at a grander scale are wrong; they don't scale; they don't work. Our experience stands in stark contrast to this (sometimes implied) assertion. Notwithstanding this seeming omission, INCOSE identifies the changes needed in engineering education and practice to enable engineering on an enterprise basis. They call out for all engineering curricula to be multi-disciplinary. They hint that Systems Engineering is the place where the currently balkanized set of engineering departments can be brought together, and they suggest a difficult interdisciplinary senior challenge problem and "playground" where the students can learn their trade. We agree wholeheartedly with these observations and recommendations. In a real sense, INCOSE is discovering process and methodological elements that fit Complex Systems Engineering, but has yet to name and describe it.

2.3. Can Traditional Systems Engineering be applied to the AOC?

It is clear that processes must be applied where they fit. If boundary conditions for applying a process, or a set of processes, are violated the processes are not really applicable. Is an AOC such a situation?

We can test whether the characteristics we previously argued were required for successful outcomes using TSE fit an AOC:

- *The specific desired outcome must be known* a priori, *and it must be clear and unambiguous (implied in this is that the edges of the system, and thus responsibility, are clear and known);*
    - o   Test Result: Failed.
      There are expectations expressed in documents known as Block Requirements Documents (BRD), which lay out an AOC's planned functionality over time. The fly-in-the-ointment is that the plan implies a convergent set of developments, which would deliver the capabilities found in the AOC BRD. This isn't the case; and it leads to the next characteristic to test.
- *There must be a single, common manager who is able to make decisions about allocating available resources to ensure completion;*
    - o   Test Result: Failed.
      As observed above, there are many component systems, which are managed by many different organizations responding to many constituencies on behalf of a set of users, of which one user community is found at AOCs.
- *Change is introduced and managed centrally;*

o   Test Result: Failed.

It is certainly the case that senior AF management has (and is) attempting to apply centralized management to bring AOCs under control. AOCs have been declared to "Weapon Systems," The senior acquisition authority has asserted personal control over the official configuration, and detailed configuration control processes and measures have been imposed.

To date, these measures have not worked; and we suspect they still won't. The only proximal result is a sense of stasis hovering over the AOC formal definitions. This invites the formation of black markets. Each Combatant Commander has funds that can be spent on their AOCs (the OIF AOC was built on Commander's Initiative funds). Bottom line: they have the means; and when a need surfaces, they can fix their own problems. And they are independent of the corporate AF staff.

Besides, stasis of the AOC definition imposes no stasis on the component systems used to build the AOC, so what does a firm baseline mean in this case?

- *There must be "fungible" resources (that is money, people, time, etc.) which can be applied and reallocated as needed.*
  o   Test Result: Failed.

  As mentioned, few of the total set of resources required to produce an AOC are controlled by, or in a way, that renders them fungible.

The conclusion is pretty straightforward. TSE doesn't lend itself to engineering or managing the engineering of AOCs (and by extension, enterprises). Can one take organizational or management steps to bring the characteristics, which are outside of the boundary conditions back in line? Perhaps one; but they are all violated. A reasonable guess is that it is not likely that there are management or organizational changes which would allow TSE to be applied successfully.

3. Complexity and Complex Systems

To facilitate discussion, terms must be defined. For this discussion there are a few key terms that require definition. While somewhat pedantic, the concepts offered below attempt to define the landscape explored. For those who believe they have a good grasp on the definitions and characteristics of complexity and complex systems, or who have an immediate interest in answering the next logical question with respect to an AOC – is the AOC a complex system? – they might jump ahead to section a.3.4, then return to this point.

"Complex Systems Engineering" contains three terms:
- Complexity
- Systems
- Engineering

Fundamentally, we're talking about an engineering activity centered on complex systems. Yet, to this point the appeal is to a general gut-feel for the concept. But, what is complexity? What are systems? What are complex systems? What is systems engineering? What is complex systems engineering?

3.1. Complexity

"Complexity" as a concept is actually rather slippery. For understanding the difficulties "complex systems" present to engineering and management activities, it's worthwhile taking a few moments and exploring this term "complexity." As we discuss complexity, we will use a "progressive formalism" approach, which initially appeals to intuition, then fills in the intuition with some formal structures. To set complexity in its proper place, we will also use some forward references – i.e. before we define a system, we will use the common notion of a system to help understand "complexity."

"Complexity" does not mean "difficult to understand" (although it might be the case that something complex is difficult to understand). Reaching into the American Heritage Dictionary,

> **Complex** adj.
> **1.a.** Consisting of interconnected or interwoven parts; composite. **b.** Composed of two or more units

This particular definition is not too useful, since every system (using the working definition of *system* below) is "complex" by this definition. The Oxford Dictionary states that something is complex if it is "made of closely connected parts." This definition also does not distinguish between "simple systems" and "complex system." In fact, one could (by simple substitution) quickly create a "simple complex" which seems like an oxymoron.

Bar-Yam [8] suggests that complexity is strongly related to the number of possible states of a collection, or its complete description. That takes the concept closer to a useful understanding for engineering, and borrows in an attractive way from Shannon's Information Theory; but it also seems arbitrary in some ways, as it suggests that a collection becomes more complex when measured with more precision. For example, if one calculates all the possible arrangements of papers on one's desk, the number of discernible possibilities is different depending on the precision of the ruler used. But it's still the same desk and the same set of papers. Arguably, the complexity should be the same. The complexity should not depend on the measuring method. The counter argument is that the use of a different ruler is precisely equivalent to using a different scale; and so finding that the complexity at different scales is different should not be surprising. Nevertheless, the use of a value related to both information theory[8] and entropy[9] remains attractive.

Another aspect to contemplate is the difference between actual number of possibilities and the number of useful possibilities. Consider a spoken language. Is the complexity of a sentence in the language of length 'n' related to the permutations of the number of words in the sentence ($O(n!)$)? Or, is it related to the number of 'useful'

---

[8] Shannon's 10th Theorem:
[9] Second Law of Thermodynamics

arrangements of the 'n' words, which would be significantly less? This is potentially important as one develops metrics to measure complexity and compare complexity levels.

Another view of complexity is Turchin's [14]. He describes complexity in terms of behavior and emergence. He has crafted a theory known as Metasystem Transition Theory, which describes interactions within and among *models of meaning*. The creation of each new level of abstraction and complexity he terms a "quantum of evolution."

Whatever model is used to understand complexity, rendering 'complexity' into a useful engineering concept requires metrics. Return to the statement *"Complexity" does not mean "difficult to understand"* above. Since it is easy to assume that the concepts are synonymous, a concept for "difficult to understand" but is not the concept "complexity" must be found. A candidate term offered is *Intricacy*

Intricacy.

1. Having many complexly arranged elements; elaborate. **2.** Solvable or comprehensible only with painstaking effort.

An example may help drive the point home. There used to be a board game called *Mousetrap* played by children (it may still be played!). In the game, players move their playing pieces (colored mice) around a board and in doing so build a Rube-Goldberg mousetrap which one player ends up using to capture the other player's mouse, thus winning the game.

The advertising copy reads as follows[10]:

> *"Construct a crazy mice-catchin' contraption piece by piece as you race your mice around the track! Once it's built, turn the crank...that kicks the marble...that rolls down the chute...and sets off a zany chain reaction that just might trap a pesky mouse!"*

It's clear that the bizarre mouse-catching device is *intricate*. However, it is not *complex*. It has only one possible configuration, and it results in only one behavior. Each piece is carefully crafted to fit onto the previous structure which sets up the conditions for the subsequent structure. It also doesn't interact at all with its environment. It assembles the same way each time (in fact it must have this characteristic to be a good toy). It is also clear that the mouse-catching system built is *"solvable or comprehensible only with painstaking effort."* That's the appeal of the game. That's why kids enjoy it. That's not a characteristic necessarily appropriate for our military systems.

Measures of complexity and intricacy may serve as good metrics to understand the relative merits of a system, and may be useful for relative comparisons. Mathematical properties of *complexity* and *intricacy* can be shown to relate to specific mathematical characteristics, which we will treat in a subsequent publication. As a precursor to a detailed treatment, it appears that intricacy relates to the number of axes of characterization – i.e. the absolute volume of a hyperspace defined by the axes. Complexity relates to the volume reachable within this hyperspace. Thinking about the mousetrap device, it is a device whose hyperspace has many axes; yet it has a

---

[10] Taken from http://www.areyougame.com

narrow extent along each axis, forming a narrow volume of reachability within this hyperspace. Other models attempt to describe complexity in terms of *variety* and *constraints* [15], which roughly map to Shannon's notion of statistical information entropy and content. Still other formulations of complexity are found in other disciplines.

Alexander [4] offers well-known (architectural) pattern models for considering complexity and emergence in architecture. Emerging from the repeated application of the principles, Alexander's speaks about spaces, homes, towns and cities that are "alive." His concept of "alive" is a reflection of the interactions among the components in the environment and the people, and the support the environment affords to the repeated patterns and events, which make up the peoples' experiences minute-to-minute and day-to-day. He recognizes that there are both patterns formed at higher levels from bottoms-up application of patterns, and there are explicit patterns applied at higher levels – and in this he hints at multiscale analysis. Fundamentally, he is talking about the relations among the entities that interact, and the result of those relations. His work was adopted and interpreted by those who practice "pattern"-oriented approaches to systems and software [13]. These approaches certainly seem to have something to say about complexity, and Complex Systems Engineering.

Alexander's notion of complexity seems to align with the notion of *"order."* The utility of a definition (of order now) depends on its alignment with the informal understanding accorded the term. In the informal sense, *order* is almost always associated with organization as well as with the actions or other forms of direction that lead to this organization (rhyme and reason). Order is not simply a passive thing like color (i.e. a state property). It is dynamic; it is associated with doing something – i.e. both form and function. By focusing on the *relationships among things*, not just the state of the things as a result of the relationships, we can understand the reasons for the molar organization and perhaps understand the implications to change – even infer or deduce state elsewhere which may be out of view.

Formally now, the order (of a system) is a measure. The measure is the set of all of the specific and instant relationships among the parts of a system.[11] In many circumstances, the order (of a system) can be quantized and summarized by the cardinality of such a measure-set.[12] This approach, that of focusing on the relations, not merely the state, will likely provide the most useful characterization of complexity since it characterizes things in an active way.

---

11 A given relationship can vary over time. The "specific and instant" form of a relationship is to be distinguished from these possibilities. A fuller discussion of the meaning of order would elaborate the definition of "relationship" offered here. It would offer that relationships are patterns in attributes, where attributes define the parts of a system (and sets of "values" define attributes). A relationship allows the inference or deduction of the specific values of an attribute of a part of a system based on other attribute values because those attribute values collectively form patterns

12 Random is often used to identify the absence of order. If so, random should not be treated as an exact synonym for stochastic – which actually asserts that there are relationships, but that they are not well enough understood to dependably deduce or infer knowledge of specific parts. In fact, stochastic is a telltale for relationships at multiple levels of scale, something that is taken up briefly below.

3.2. Systems

Here, the term *system* denotes a set of parts that have relationships with one another. This is also the preferred definition for this term. From the American Heritage Dictionary, first definition:

> **1.** A group of interacting, interrelated, or interdependent elements forming a complex whole

Not everyone uses this definition, but it is the definition we use. The major stumbling block some have with this definition is that it requires that the parts have some sort of relationship to one another to constitute a system. Some speak of "systems" using a much looser definition. The American Heritage Dictionary supports a looser definition as well; to wit, definition six:

> **6.** A set of objects or phenomena grouped together for classification or analysis.

The issue to consider is which definition fits one's intuitive notion of "system" better. Number one seems to meet the intuition test: it's alive and active, while definition six is a system of academic or conversational convenience. The definition used here insists that a system has multiple parts, AND that those parts have relationships among them.

INCOSE defines a System [18] in the following way: "...A system can be broadly defined as an integrated set of elements that accomplish a defined objective..." They have also been struggling with scaling up, and the implications. They also note "...It is sometimes confusing as to which elements comprise a system..." offering an example of a broad network with independent databases fused, and a desire to print the results.[ibid] Also noted is the presence of multiple "levels" to a system, and the different roles the same "things" are at different levels[13]. Clearly, the limits of the traditional definition of a *system* are being felt, and it too hints at Multi-scale analyses (discussed later).

3.3. Complex Systems

A Complex System [8, 14, 17, 19] is a system:

- Whose structure and behavior is not deducible, nor may it be inferred, from the structure and behavior of its component parts;
- Whose elements can change in response to imposed "pressures" from neighboring elements (note the reciprocal and transitive implications of this);
- Which has a large number of useful potential arrangements of its elements;
- That continually increases its own complexity given a steady influx of energy (raw resources);
- Characterized by the presence of independent change agents.

A measure of a complex system (for characterization and comparison) might be based on the balance of *complexity* and *intricacy*. Other corollaries of these measurements might be a measurement of the rate at which the complex system's adapts to required/desired change.

---

[13] "Aircraft, automobiles, and homes are other examples of systems at one level, which can be considered elements or subsystems at another level."; Ibid

3.4. Is the AOC a Complex System?

We can test whether AOCS fit the definition of Complex Systems by comparing the two:

A complex system is a system:

- *Whose structure and behavior is not deducible, nor may it be inferred, from the structure and behavior of its component parts;*
    - o   Result: Marginal Pass
      The AOC's desired molar behavior is reasonably well known; even it's desired changes, so this characteristic doesn't necessarily fit. However, if we take a broader view of an AOC as an element of C2 (i.e. the enterprise), then this statement becomes more correct.
- *Whose elements can change in response to imposed "pressures" from neighboring elements (note the reciprocal and transitive implications of this);*
    - o   Result: Pass
      This is certainly the case in the AOC. Independently-introduced applications (through independent agents) such as (for example) ADOCS and Falcon-View cause direct "pressure" on those applications which perform similar roles, or which could potentially act in concert with these introduced applications. As an example of resolving the introduced pressures, TBMCS specifically added certain Information Services to interact with ADOCS without the need to own and control it.
- *Which has a large number of useful potential arrangements of its elements;*
    - o   Result: Pass
      Since the AOC's workflows are numerous, and are in flux due to new missions and doctrine, this fits. IT is also true, though, that it is the people who supply the flexibility; and they often fight the automation present.
- *That continually increases its own complexity given a steady influx of energy (raw resources);*
    - o   Result: Pass
      This also seems to be the case. For example, TBMCS re-architected from a monolith to a set of applications riding on a set of Information Services precisely to increase the number of possible connections and relations, and to allow more independence of creation and use of new clients of the services offered.
- *Characterized by the presence of independent change agents.*
    - o   Result: Pass
      The AOC has upwards of 30 independent agents – in the form of separate Program Elements (PEs). PEs are, by their definition and nature, independent agents.

It seems reasonable to conclude that AOCs are Complex Systems; and, since there is a need to apply a Systems Engineering approach to the AOC which is beyond the traditional Systems Engineering approach (see the earlier discussion in section 2.3), the AOC might benefit from a Complex Systems Engineering approach which acknowledges the differences between the AOC and other more-traditional developments to which TSE can be applied.

4. Engineering Complex Systems

If TSE doesn't scale to AOCs or the enterprise, what does? In the introduction we made the claim that an augmentation of Systems Engineering, to be called Complex Systems Engineering (CSE), should be used to manage and guide the enterprise. As discussed above in some detail, a violation of the boundary conditions required for a favorable application of TSE suggests different tools and approaches are required. In essence, CSE must serve to bring together independent, disparate organizations and entities. It must provide them with a sense of "pressure" that they feel, and a set of processes that can be used to resolve the pressures. CSE must incentivize the partnerships needed; and must compel the engagement of their respective resources to accomplish the integration without resorting to arguments over whose money is being spent, or whether "interoperability" or "integration" is a "requirement" they have.

It's clear that Systems Engineering must extend its philosophic and theoretic foundation to build a consistent (and hopefully complete) framework for the practice of Systems Engineering in general, and Complex Systems Engineering in particular. Within this framework, we describe new roles and responsibilities for new "jobs" which must be performed to do CSE.

Complex Systems Engineering changes the focus from *"...here is the solution designed from the requirements, now go implement it..."* to *"...here are the selective pressures acting on the elements present (likely built using TSE), now resolve or reduce them..."*

CSE does this, and this is the key point, through a deliberate and accelerated mimicry of the processes that drive emergence and natural evolution. Kaufmann [19] noted that complex adaptable systems require both the emergence of novelty and variety as well as selective pressures to account for the richness in ecosystems. We find those characteristics in an AOC. In fact, the AOC (and Command and Control in general) can be thought of as an *Ecosystem*. Bar-Yam [9] has explicitly incorporated this thinking in a recent presentation where he introduces what he calls *Enlightened Evolutionary Engineering*. Using the conceptual models suggested above, one can speculate about niches, selective pressures, competition, adaptation, displacement, etc. It describes a process constantly at work and in line with our daily experience; a process which is alive [16, 17].

Complex systems engineering is NOT a new or renewed attention to detail; it is an attention to overall coherence. The two are obviously and will always be related. However, as the actual order and complexity of a system increases, it becomes humanly impractical to address both from a single perspective – in particular from the perspective of ever increasing detail. What complex systems engineering does is to address overall coherence *without* a direct and immediate attention to detail.[14]

---

[14] This is what baffles those confined to a linear theoretic and uniscale (or reductionist) viewpoint.

Complex Systems Engineering acknowledges the presence and action of "autonomous agents" as important elements of a SoS. These autonomous agents are precisely the effectors, which must be (and are) eliminated to apply TSE. Again, to apply TSE one needs to eliminate the independent agents, or one needs to augment the set of tools for dealing with their continued presence.

4.1. Comparing Traditional and Complex Systems Engineering

Traditional and complex system engineering can be distinguished by contrasting either their methods or the outcomes resulting from the application of those methods. The following briefly contrasts the outcomes that are obtained using traditional and complex system engineering. The term *product* is used to identify the outcome of traditional system engineering; the term *enterprise* is used to identify the outcome of complex system engineering.

Table 1. Comparing TSE and CSE

TSE	CSE
Products are reproducible	No two enterprises are alike.
Products are realized to meet pre-conceived specifications	Enterprises continually evolve so as to increase their own complexity.
Products have well-defined boundaries	Enterprises have ambiguous boundaries
Unwanted possibilities are removed during the realizations of products	New possibilities are constantly assessed for utility and feasibility in the evolution of an enterprise.
External agents integrate products	Enterprises are self-integrating and re-integrating
Development always ends for each instance of product realization	Enterprise development never ends – enterprises evolve
Product development ends when unwanted possibilities are removed and sources of internal friction (competition for resources, differing interpretations of the same inputs, etc.) are removed	Enterprises depend on both internal cooperation and internal competition to stimulate their evolution

Traditional and complex system engineering can (and should) be applied concurrently in the realization and evolution of a complex system. Traditional system engineering is appropriate for managing the decision making processes of individual autonomous agents in a complex system. Complex system engineering must be added when multiple autonomous agents must be a part of any solution and/or when multiscale analysis becomes essential to a sufficiently complete characterization of an evolving problem and its solution.

4.2. The Regimen of Complex Systems Engineering

Complex systems engineering (CSE) operates a bit differently than TSE in that TSE is a practice of direct impact and effects, while CSE tends to be indirect. The goal of CSE is to increase the order of, and the complexity available to, systems. As discussed earlier, there is a practical upper limit to the degree to which this can be done successfully through pre-specification followed by implementation, and all of the other attendant processes of TSE[15].

To engineer a system beyond this limit it is necessary to combine several related activities into a single continuous "regimen" of engineering and development. This regimen is intended to transcend the boundaries of TSE, which have been outlined previously. A regimen is distinguished here from a "recipe," and a recipe here can be understood as shorthand for the cumulative nature of the processes of TSE. A recipe is a tightly and precisely scripted sequence of steps intended to yield reproducible outcomes such as specific kinds of cakes or meat loaves (or cars or planes, or operating systems). In the ideal, every outcome is exactly the same. A regimen is a looser formulation of more generalized steps that can be combined in various ways to yield many different instances of generalized outcomes such as weight loss or increased stamina (or an Air Operations Center, or a Department of Homeland Security). Even in the ideal, there can be no insistence on uniformity, only on acceptability or conformity with broad norms.

The overall regimen of CSE creates and manages an environment[16] in which multiple autonomous agents each address a fraction of the relationships that might be involved in an overall complex system. Autonomous agents (independent development tracks in the context of most engineered systems, especially IT-intensive systems) and their creations both operate in this environment and (continuously) interact to explore the utility and practicality of new or modified relationships.[17] We have some reasonable hope that establishing such an environment is both practical and doable. Holland [17] points to its natural occurrence when the basic elements are present[18].

In terms of today's IT-intensive systems, it is useful to recognize that most of the increases in complexity (or interoperability, or new or expanded relations) are typically associated with the "run-time" of the system, while most of the collaboration (or interoperation or interoperability) that yields these new or modified "run time" relationships occurs among the people who create the "run time" components. This collaboration is said to occur during the "development time" for the system. As a consequence, complex systems engineering for IT-intensive systems needs to establish and manage an environment for "developers" AND for their "run time" creations. These two aspects (scales) of any IT-intensive complex system are not ultimately independent, but they can be discussed separately and then combined to enhance overall understanding.[19]

---

[15] Formally, this can be attributed to the increasing dimensionality of the relational multiscale phase-space that can be used to characterize the actual order and complexity of a system relative to the generally fixed and finite intellectual capacity of any human individual.

[16] There are many analogs to the developmental environment of CSE. See for example Hayek's general thoughts on monetary and trade-cycle theory. An even more familiar analog is the role of "playgrounds" in child development.

[17] As well as to maintain, to modify, or to discard existing relationships.

[18] Holland [17] talks of the basic elements of: aggregation, tagging, nonlinearity, flows, diversity, internal models, and building blocks.

[19] Continued improvement in the engineering of IT-intensive systems will witness the continued convergence of these "separate" development and run times. This can be colloquially summarized as the emergence of self-programming systems. The compelling "evolutionary pressure" for this emergence is the

The following very briefly introduces each of the elements that are combined into the Regimen of CSE. Their combination is discussed after the elements are introduced.

4.2.1. Developmental Environment

An explicit and conscious attention to a developmental environment (including even a pre-specification of its initial form) is the single most important activity underpinning the deliberate development of complex systems.

This developmental environment can be understood as either a separate and distinct environment in which complex systems develop and operate, or – along with that environment – an overall ecosystem that includes both. As such, this first activity focuses on the completeness of the ecosystem relative to supporting the more focused activities that occur within it. Are a sufficient number of the relevant autonomous agents and their creations present? Can new ones be added? Is the means available for these autonomous agents to interact if they so choose? Are resources flowing through the ecosystem? Are the means for supporting both cooperation and competition among the autonomous agents present? Is the flow of resources modulated by cooperation and competition, or is that flow entirely pre-specified? [20] Are there universal signals that can be interpreted locally (and perhaps differently) that are associated with the whole that cannot be entirely explained by any combination of a subset of the parts? [21]

The developmental environment can not be a one-time thing. It must be, nurtured, and managed so it can evolve itself; even after its initial establishment. Attention to this environment must be continuous, deliberate, and it must be available to all the independent agents; this is why it lends itself to being treated as a separate activity.

4.2.2. Outcome spaces

Outcome spaces are identified (or defined) at multiple levels of scale, and from multiple points of view, for a complex system. An outcome space is explicitly distinguished from the many specific outcomes that comprise it. (When very specific outcomes are sought and/or are meant to be exactly reproducible, TSE should be used to achieve them. However, applying TSE becomes increasingly difficult as the number of autonomous agents increases; ideally, only one is involved. All specific outcomes in the outcome space must be viewed as acceptable without there being strong preferences for any of them. [22]

"expense" of human labor in maintaining and expanding IT-intensive systems. Programmers are expensive relative to their programmed creations.

[20] If the flow of resources is entirely pre-specified, then competition cannot operate. The localized decisions of autonomous agents cannot, by definition, influence the flow of resources – although they can still influence the effectiveness of that flow. As a result, complex system development can't occur.

[21] Examples of such signals are the pricing mechanism in a market economy or selective pressures in a biological ecosystem.

[22] This does not mean that outcomes can't be identified as unwanted. Partitioning outcome spaces into wanted and unwanted sub-spaces is one way to do this. This is why outcome spaces are sometimes referred to as the targeted outcome spaces.

When specific outcomes in an outcome space can be realized by individual autonomous agents (or their creations) by themselves, competition is encouraged. When specific outcomes in an outcome space can only be achieved by autonomous agents (or their creations) collectively but not individually, cooperation is encouraged.[23] Network centric operations and Jointness are military domain examples of the latter [1]. Such outcomes are best characterized at a scale in which the autonomous agents (or their creations) are not immediately or directly accessible. Complex systems require both competition and cooperation for sustained development, although competition is almost always more important in the short term.

It is sometimes possible to characterize outcome spaces at multiple levels of scale using identical terminology. This often causes confusion and should be avoided. For example, it may be desired that a complex system achieve a reduced footprint (or reduced power consumption, etc.) and/or it may be desired that individual components of that complex system achieve reduced footprint. These are frequently outcomes at different levels of scale. (It is possible that a complex system could achieve a reduced footprint even though individual components do not – or even increase their footprint.) Such outcome spaces should always be explicitly distinguished (for example, by always and explicitly referring to "component" footprints and the complex system's footprint).

The identification of outcome spaces (vice specific, detailed outcomes) focuses attention on explicitly recognizing sub-spaces and partial volumes in a relational phase-space (vice specific phase points or trajectories).

4.2.3. Rewards

Autonomous agents (independent development tracks in the case of IT-intensive systems) make the decisions that determine the utility and/or the practicality of existing and new relationships within the complex system. Rewards are structured to motivate the autonomous agents to make decisions that cause the complex system to enter the targeted outcome spaces desired.

Rewards shape the decision making processes employed by autonomous agents. The rewards should be clear and should not be dependent on specific processes of the autonomous agents who are subject to rewards. Since one should not assume that these autonomous agent processes are uniform (or that they should be), rewarding based on a specific process, which could be viewed as too invasive by the autonomous agents[24]. Rewards are only one consideration shaping the decisions of the autonomous agents; and should be viewed as incentives only.

In the most general case, rewards are access to the energy flowing through a complex system. In the case of IT-intensive system development and within the context of contemporary acquisition protocols, rewards are almost always associated with access to the money flowing through the entire system development environment. There are other forms of rewards (as well as penalties), however, that

---

[23] There are, of course, many intermediate or blended situations as well in which both cooperation and competition play a part.

[24] Unless the specific outcome space IS a common process. However, insisting on common processes may well stifle innovation and variety needed for evolution.

are neglected only at great peril to the system engineer. Rewards motivate. As long as people are involved, the list of possible rewards is as long as the list of factors that motivate people.

To the degree that rewards can be distributed extra-contractually, that offers motivation to autonomous agents to "keep their eye on" the complex system, even if they are not engaged in a direct manner. Innovations, which these agents can bring to the complex system and that are shown to bring value, should be rewarded. This sets up the potential of new approaches and influences, and avoids stagnation.

4.2.4. Developmental precepts

Developmental precepts constitute the "rules of the game," and by doing so stimulate contextual discovery and interaction among autonomous agents. They do this by establishing (for example) certain constraints on how outcomes are achieved by autonomous agents, or how they interact. It's easy to confuse these precepts with rewards, however they are quite different. Like when playing a board game, even if the next space to promises a great, the rule (developmental precept) says you must move the number of spaces found on the die you throw, you can't move that one space if the number is not "1."

Developmental precepts do not specify specific outcomes or even outcome spaces. In contemporary IT-intensive systems in which development time and run time are treated separately, developmental precepts focus on the interaction of the independent development tracks more so than on such behavior among their creations. In contemporary IT-intensive system acquisition and development, these developmental precepts can be contractually fixed since they can be made as specific as desired.

Developmental precepts are initially difficult for most system engineers to appreciate since they shape autonomous decision making leading to specific outcomes rather than the specific outcomes themselves. An example is useful.

In the case of the Air Force, most IT-intensive systems that contribute to Command and Control (C2) are developed under the supervision of (acquisition) Programs. Many of these Programs are physically housed at the Electronic Systems Center (ESC) on Hanscom AFB, Massachusetts. Each of these C2 systems has explicitly designated military "end users."(For example, an automated mission planning system helps a pilot to plan a route to be flown by an aircraft on a military mission. Pilots are the end users of such an automated system. Pilots do not take delivery of such a system, however. Instead, a Commander responsible for many pilots is the notional end user that takes receipt of such a system at a particular Air Force base on behalf of the Air Force and of the pilots stationed there.) The completion of the acquisition process (the delivery and acceptance of the system) is signified by the signing of a DD-250 form by the end user. The Program uses this signed form to confirm the successful completion of its own (acquisition) mission. Absent that signed form, the mission remains incomplete. Careers depend on the completion of missions, and this is true for the people in Programs as well.

End users (the Commander in this example) take receipt of multiple systems from ESC – often in the same year. The mission planning system is just one such system.

These systems increasingly must interact with one another to fully accomplish their respective purposes. Because such interactions do not fully fall within the scope of any one system, however, the successful realization of such interactions is left to delivery time and is often partially or wholly left to the end users to accomplish.

This is further aggravated by the periodic replacement of an earlier version of a system with a newer (and better) version. This almost always involves moving extensive databases from the earlier to the newer versions of a system to maintain operations. This is frequently left to the end users to accomplish even though improvements in the newer version of a system often involve the reorganization of the existing information in those databases.

This is even further aggravated since many improvements to systems impose new or additional burdens on the infrastructures supporting modern C2 operations (power, bandwidth, connectivity, etc.). New or improved systems are not responsible for augmenting such infrastructures since by definition infrastructures are shared. The end users are however constrained by these infrastructures.

As a result, end users increasingly complain of "drive by" deliveries of systems by ESC Programs and the failure of ESC to deliver "integrated" solutions to their operational needs. Although ESC increasingly talks about an "integrated C2 enterprise," it continues to deliver that enterprise in "kit" form – leaving the hard "integration" part to the non-acquisition community to accomplish.

The traditional response to this complaint (thereby acknowledging its essential truth) has been to attempt to formulate master schedules for the delivery of systems to specific locations and the formulation of detailed integration plans to interconnect the independent systems prior to deliveries in the traditional system engineering fashion. This response has failed, been retried and failed again. The specific reasons explaining each failure is now legion.

Treatment of such overlapping and interdependent deliveries as a complex system development would involve (in part) the formulation of a developmental precept. This precept would alter slightly the mechanism already employed to complete the delivery and acceptance process. It would modify the DD-250 form so that an end user could only take receipt of multiple (say two in the simplest case) systems from ESC at one time. The end user would then have the leverage to compel the acquisition community to address the interconnection of delivered systems. ESC Programs would respond accordingly since a system could no longer be completed by itself. The specifics of which systems to be delivered and how they need to be interconnected, etc. would be left to the Programs (the autonomous agents) to resolve in their own best interests. However, the" global" outcome of more integrated systems from ESC would also be accomplished – even though the specifics of how and when were never explicitly formulated in advance at any "global" level.

The specification and then the enforcement of such a developmental precept would serve to stimulate discovery and interaction among ESC Programs without specifying what the specific outcomes should be. This is the essential characteristic of a developmental precept.

4.2.5. Judging

Judging requires human judgment. Judging associates specific outcomes *achieved* with autonomous agents, and assigns rewards to the autonomous agents accordingly. Rewards are established *prior to* the realization of desired outcomes. Judging, on the other hand, is based on actual outcomes achieved, not before.[25]

Judging for rewards that are associated with outcomes that can be attained directly by autonomous agents (or their creations) is straightforward. Specific outcomes in the targeted outcome spaces are seen to actually occur. They are recognized as such and then the reward is assigned (given) to the autonomous agent(s) responsible.; as quickly as possible following the judgment

Judging for rewards that are associated with specific outcomes that are in outcome spaces that can only be associated collectively with autonomous agents (or their creations) is more demanding. (For example, if a complex system achieves a reduced footprint by virtue of certain components *increasing* their footprint and so allowing others not to be used, or others to decrease their own component footprints so that the net effect is an overall reduction, then judging requires the identification of the autonomous agents responsible and the apportionment of the reward. The actual achievement of network centric operations would be another example.)

4.2.6. Continuous Characterization

Outcome spaces and rewards can (and should) initially be characterized with succinct (even pithy) "bumper sticker" labels.[26] This allows for a maximum role for the autonomous agents in shaping the evolution of the complex system. Because autonomous agents do exactly that – act autonomously – this maximizes the opportunities for inconsistencies in how these characterizations are interpreted. To the extent that consistency matters, outcome spaces, rewards *and the current condition of a complex system* will benefit from continuous and progressively more detailed and complete characterizations. The characterization of the current situation is crucial in this regard because it permits autonomous agents to independently develop metrics to guide their local decision making in ways that will be broadly similar over time. (Autonomous agents, although autonomous, are never wholly dissimilar.) In this regard, the specific outcomes (as distinct from outcome spaces) used as the basis for judgments should be detailed, as should the rationale supporting those judgments.

---

[25] Judging removes an important risk attendant to contemporary acquisition protocols. Contemporary contracts are awarded based on what a successful bidder is predicted (in a source selection) to do in the future. Judging assigns rewards based solely on what actually happens, not on what will happen. This has implications for the nature (size, etc.) of rewards since the independent development tracks must assume greater responsibility for the risks attendant in actually achieving desired outcomes. A discussion of such implications is well beyond the scope of this discussion; but it suggests a wider variety of supported business relations should be explored.

[26] The U.S. Army motivated a tremendous spurt in its evolution with the visionary characterization of a targeted outcome space with the exceptionally pithy expression, "Own the Night."

Figure 2. Continuous Characterization of a Complex System

Consistency can never be guaranteed in complex system development (evolution). (it can be, in theory, with TSE.) As a result, these characterizations can never be made too detailed.[27] Characterization refinement can, however, become less than cost effective.[28] Moreover, consistency in this regard will tend to accelerate complex system evolution but in narrower and narrower directions that are explicitly identified and characterized (see Figure 2). Therefore, as new outcome spaces become apparent and/or attractive, unless they are explicitly added to the characterizations (initially with limited detail), it becomes increasingly unlikely that the new possibilities will be explored even though available. In the extreme, this can even result in stagnation, something that Safety Regulation is used to preclude or at least impede. In any case, deliberately stressing complex system development in this way (very detailed refinement of outcome spaces, etc.) should be carefully weighted.

4.2.7. Safety Regulations

Safety regulations are aimed at preserving the "stability" of a complex system. Their purpose or focus is not on the attainment of desirable outcomes but rather on the continued functioning of the other activities that are intended to do that (defining outcome spaces and developmental precepts, judging, etc.). Their scope is wider than that, however. They are, in short, aimed at preserving the developmental environment. They are indirect measures.

The act of applying safety regulations (or equivalently, safety regulation, and performed by safety regulators) can be thought of as policing a complex system. Although this applies to all levels of scale, in contemporary IT-intensive systems, safety regulation is most important during development time. Safety regulation applies to all developmental activities. Since a list of such activities is open ended, any listing of safety regulations is also open ended.

---

[27] If, however, the detailing of specific outcomes is substituted for outcome spaces (as just one example), the complex system can be caused to collapse even though the complex system appears at first to rapidly accelerate its evolution.
[28] It can even become counter-productive, obscuring with detail rather than further illuminating the essential coherence desired and achieved.

Adding autonomous agents (independent development tracks) to a complex system development is a development-time activity. Vetting such agents (or even preparing them) prior to entry to the developmental environment is an example of safety regulation.

Adding the creations of independent development tracks to the run-time composite of a complex system is roughly analogous to integration in traditional system engineering, except that its goal and its many implementation particulars (defining and refining interfaces, etc.) are the responsibilities of the involved independent development tracks *and not some external integrating agent*. However, progressive steps to regularize this introduction procedure can be formulated and enforced as safety regulations. For example, such "integration" could first be required to happen "offline," and then "online," and finally "inline." Each of these phases would be detailed and enforced by safety regulators. (In this example, offline, online, and inline represent progressively deeper and fuller participation in the run-time composite of the complex system.) The overall purpose of such phases would be to protect the "uninvolved" independent development tracks and their creations from accidental or deliberate rogue behavior by exposing the new components' (admittedly partial) behaviors to independent observers with this goal in mind.

Safety regulation can also be made to apply to the retirement of no-longer-used run-time components in a complex system (when and how this should be done, etc.). This example is cited because it permits attention to be drawn to a role such "soon-to-be retired" components can play as tools themselves in the safety regulation of a system.

In many natural evolutionary complex systems, generations (populations with slightly different capabilities) overlap. Rather than generational replacement, there is gradual displacement with the possibility that older generations (or, equivalently, exact copies of them) can persist. The IT-intensive system analog of this phenomenon is that "older" components remain "on-line" (and in use) while "newer" components are brought "in-line" and then "on-line" as a surety against catastrophic complex system failure. This is illustrative of managed redundancy as a safety regulation in complex systems.

But safety regulation does not have to be entirely *ad hoc*. Safety regulation is about avoiding "collapse" and "stagnation" in overall complex system behavior. These notions can be given rigorous meaning[29] with the use of chaos and catastrophe theory. This meaning can in turn be translated into the specifics of a given complex system's behavior (in terms of its trajectory in its phase-space). These specifics can, in turn, be translated into thresholds used to monitor and control overall system behavior, but only in very specific dimensions. In crude terms, this technique has been in use for a long time in the form of circuit breakers and the like.

Other safety regulations can and should be directed at the detection of the continued presence of both cooperation and competition since *both* are *always* necessary for the sustained operation of any complex system.

---

[29] To do so is well beyond the scope of this system engineering focused narrative.

4.2.8. Duality

It has already been emphasized that a complex system's "development time" can never be fully separated from its "run time." Complex systems continually change as a natural part of their own operation. This can only be fully appreciated using multiscale analysis.

In the context of IT-intensive systems that also utilize people as operators, it is important to apply multiscale analysis in a fashion suggested in Figure 3. Multiscale analysis must be applied to both distinguish between and to understand the relationships among these separate scales in a complex system: the IT-components themselves (application programs, etc.), their developers (groups of people as independent development tracks), and the human operators of the system.

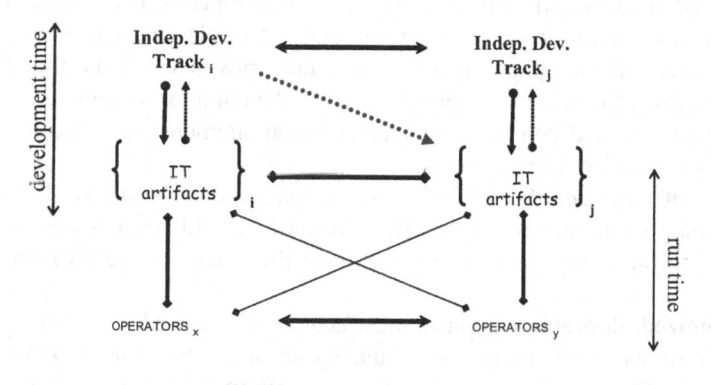

Figure 3. Developmnt Time of IT-intensive systems

The IT components in an IT-intensive complex system do not create themselves (yet). Groups of people do that. These autonomous agent developers interact with one another as well as with their creations. There is only very limited (and often strongly inhibited) interaction between such agents and the creations of other such agents. These interactions are frequently identified as occurring during "development time." The IT components of a complex system also interact with one another. Such interactions can occur during development time, but most of the time such interactions are thought of as an essential element of the "run time" of a system. However, these are not the only interactions that are associated with the run time of a system. In almost every case there is also a strong degree of interaction between these IT components and their human operators. Moreover, these human operators often interact with one another directly – without any intermediate interactions involving the IT components.

To be fully productive in contemporary IT-intensive systems, the definition of outcome spaces is done (at least) at three distinct scales, two corresponding roughly to "run time" and one corresponding to "development time."

Complex system engineering should take all of interactions into account – not just those involving IT components during run time. This is almost never explicitly acknowledged today – although the growing attention to fostering "user/developer" interaction during development is an implicit recognition of this multiscale reality. Duality is the explicit recognition that development cannot be completely separated from operation in the case of a complex system.

4.3. Running the Regimen

There is no single way to characterize how the regimen of complex system engineering unfolds because it is a regimen, not a recipe.

Early on, one should clearly formulate desirable outcome spaces in broad terms (partial volumes and subspaces in the relational phase space for a system). The actual phase points and trajectory of a complex system are determined collectively by the autonomous agents operating within the complex system. In the case of contemporary IT-intensive systems, this is the primary role of the developers – but not exclusively so since operators, for example, can play an important role as well. Recognizing desirable outcomes (actual phase points and trajectories that are in the desired outcome space) when they occur is a primary role of the complex system engineer. In this sense, recognition and continuous characterization augments specification as an engineering activity for complex systems.

Coincident with the identification of outcome spaces must be the publication of rewards available to autonomous agents. These rewards should be expressed in terms that are visible to the autonomous agents – even if the outcomes spaces themselves are not.

Once recognized, desirable outcomes must actually be rewarded. The attainment and recognition of such outcomes does not make this automatic. Human judgment is still required. Complex system engineering can inform this judgment but such judgment must remain the prerogative of those responsible or desirous of the emergent complex system. Once such judgments are made (and rewards and punishments assigned to autonomous agents) the rewards must be restated along with the restatements of desired outcome spaces.

The formulation of desirable outcome spaces should never stop. As new outcomes occur, the desirable outcome spaces need to be restated – along with attendant rewards.

The complex system engineer is responsible for managing the overall developmental (and operational) environment. Key in this regard is formulation of developmental precepts that serve to influence (but not to specify) the decision making of the autonomous agents in the complex system. This requires engineering judgment but judgment that is distinct from the assignment of rewards. These developmental precepts can and should be made "binding" on all developers in the complex system, as should the adherence to safety regulations. Enforcing these precepts and safety regulations are important roles for complex system engineers and are the essential aspects of specifying and managing a developmental environment.

A complex system operates continuously. The complex system engineer is responsible for the overall developmental environment that appropriately mixes operational and developmental contexts. The complex system engineer almost always focuses attention on the developmental aspects of this mixed environment. This includes specifying, operating, maintaining and modifying an infrastructure that supports interactions among autonomous agents and their creations, specifying and enforcing developmental precepts intended to stimulate discovery and interaction among the autonomous agents, and the specification and application of safety regulations.

A complex system operates continuously. In the process of so doing, it is constantly changing and becoming more complex (sic). A complex system engineer continuously characterizes the complex system, emphasizing those aspects that are associated with the order of the system as that order enters targeted outcome spaces. Such system and outcome characterizations become the basis for the complex system engineer's assignment of responsibility for changes in the complex system's order. These assignments become, in turn, the basis for judging – which ultimately assigns rewards to the appropriate autonomous agents. Final judging is always performed by the sponsor or other authority responsible for a complex system.

The complex system engineer assists in the judging process, with the initial formulation of rewards based on targeted outcome spaces, and with their restatements as desirable outcomes are achieved and rewards are assigned.

5. Complex Systems Engineering in Practice

Having presented the motivation and the conceptual basis of CSE in the preceding sections, what works? How would one start to apply CSE? The example of the AOC can serve as a template for other analyses, and we welcome and encourage further description of the edges of TSE and the set of techniques appropriate for CSE and its application.

While not explicitly known as CSE, most people have practical understanding of CSE from common experience. Consider how children are raised, or how large organizations exist and evolve. Even a superficial study of these illustrates how CSE can work in practice.

Consider how children are raised. Upon their birth, we do not set out a detailed set of requirements and a schedule for achieving detailed milestones (of course, some parents try, and end up being rebuffed). Rather, we set out our principles, and help them learn what we, as parents, value, then apply guidance as they grow and mature. They come to find their own way in the world. This might be thought of as engineering through indirection. In that sense, it is a practice of CSE.

Consider the behavior of organizations. Seldom is it the case that detailed control is applied top-down through an organization continuously. Leadership tends to exercise control indirectly by publicizing what their values are, what traits they value in others, what their goals are for the organization, and how the organization should run (precepts). They set context and the desired outcome spaces. Periodically subordinates (autonomous agents) are chosen for promotion (rewards) based on what they've done (outcomes) and how well they fit the valued traits. It is those promoted subordinates who determine the day-to-day activities in most organizations.

5.1.   What are some CS-derived strategies which can be employed within the regimen?

In a real sense, commercial "market places" are complex systems. Commercial practices might be mined for tactics, which lend themselves to CSE. Those with particular relevance are those which make technical change easier; those which transmit "selective pressure" easier; those which permit organizations to collaborate; and those which trim the environment selecting "success" and punishing "failure." Table 2 shows a set of strategies harvested from commercial practices, which map to aspects of the principles outlined for running a CS regimen. We explore each in more detail below.

Table 2. Commercial Practices Embody CSE Principles

	Dev Env	Outcome spaces	Rewards	Dev precepts	Judging	Cont Char	Safety Reg	Duality	Indpnt Agents
½-life Separation	X	X		X			X	X	
Playgrounds	X	X	X	X	X	X	X	X	X
Collaborative Environments	X	X		X		X			X
Partnerships									X
Developers Networks	X	X				X	X	X	X
Branding		X	X	X	X			X	
Co-opetition	X								X
Leveraging others' Investments			X	X		X		X	X
Respect Ricebowls	X	X		X					X
Opportunistic Approach	X	X				X	X	X	X
Advertising and Discovery	X	X	X	X		X		X	X
Value-add business models			X		X	X			X
Experience for test				X	X	X	X	X	

As previously discussed, the essence of CSE is the deliberate modeling of the natural processes found in evolution and ecologies. Evolution and ecologies require interaction among entities, and the ability for the entities to change in response to pressures felt from the environment.

To increase the rate of useful change in the enterprise, entities:

- should be in touch with one another for extended periods of time, and;
- the entities' "pulse" time should be reduced as much as possible;
- "value" must be assessed correctly, and by appropriate parties,
- assessed value must impose the selective pressure.

Enabling *ecological* competition and evolution requires that elements can, in practice, rub against one another and allow respective stresses to be resolved as naturally as possible – based on real value. This requires the "systems" which compose the (for example AOC) SoS to build new connections across each other in useful ways.

5.1.1. Separation of elements based on anticipated half-life

Architects are very familiar with the concepts of layering systems to separate concerns. This is a central principle, and is a good approach. Yet, the way today's "systems" are tightly-bound into monolithic entities prevents the benefit of this to a large extent. Even within each system, limiting the "pulse" of evolution to the slowest changing element causes evolution to move at the slowest pace, rather than at the natural pace of each of the elements.

Evolution also proceeds at a rate strongly dependent on generation time (spiral time, pulse time, etc.). To increase the rate of evolution, one must shorten the generation time. Therefore, in addition to layering based on functionality, one should separate based on likely rate of change.

5.1.2. Playgrounds

How do we come to recognize "goodness," and how is it introduced? Humans, as natural pattern recognizers and problem solvers, learn and innovate through experimentation. We see it everyday among children where they constantly innovate and learn through interaction. "Games" take on an additional dimension when we understand they are using play to prepare for life; they are not merely killing time. "Play" is also a key component of many animals' development. Again, while probably "fun" for the participants, it serves a much more important function. Card [12] reflected on the importance of a playground in his science fiction classic Ender's Game.[30]

Within DoD, "games" are recognized for the powerful tools that they are. They are a key way that leaders and future leaders get to ply their trade. While the word "playground" cannot survive into DoD practice due to its pejorative tone, it serves to make the point by connecting to a truly common understanding, and it is appropriate for these discussions.

Where is the "playground" where technology, doctrine, and *Tactics, Techniques, and Procedures* (TTPs) can come together? Today, the places where technology, doctrine, and TTPs come together are large, carefully-scripted events with carefully (centrally) chosen participants and known answers; they are demonstrations rather than even the experiments they purport to be. Experiments would be fine, but

---

[30] Ender's Game actually is an excellent examination of both the importance of a playground where innovation may be introduced, and an approach to "complex systems engineering" in general.

experiments can result in a negative finding; and, experiments are done on a playground. These are not playgrounds. Where playgrounds exist (e.g. C2 Battle Lab), they are somewhat disconnected from the process that gets new elements into the field where that which they construct could (potentially) provide a qualitative edge. The Air Force has a number of labs, which investigate technologies and operational needs, but there isn't a good connection between them and the formal acquisition process. Where a connection exists it tends to exist because of the heroic efforts of particular persons. The reason for the "potential" rather than "actual" edge is that the new thing has likely been developed assuming all other elements it may impact or influence have themselves remained static; and it remains to be seen whether the potential is realized. Further, the organizational process is for the keepers of the "Systems of Record" to invite an innovator or new-capability provider into the fold. This puts the identification and valuing of innovation into the hands of the organization least likely to welcome it since, by definition, innovation's appearance is disruptive; and the acquisition community is judged, in part, on the smoothness of delivery.

An important point to remember is that among SoS, no element stands alone. Each element exists within the context of those elements around it; and it supplies partial context back to those elements. Thus any change in any element causes a change in context to all elements, which juxtapose the changed element. In this way change flows to neighbors, and they respond, which causes further change, which flows to their neighbors, etc. Generally, the effects and pressures brought by any change can't be predicted, and it occurs independently of any schedule or any *a priori* agreements or expectations, and generally without any insight on how its effects will be felt [11]. For the enterprise, "change" is constant, unplanned, and unpredictable in its complete effect. This is the essence of a Complex System.

Another characteristic of a playground is that "play" tends to be safe. Ideas can be explored without too much risk. There is a (thankfully) natural reluctance to disturb a real-world working process (i.e. the operational system); and if one were to let innovation have free reign within the operational system, the innovation is more likely than not to disturb or interfere with it. This clearly runs counter to the desire to introduce innovation. If one is tempted to suggest that the "new, innovative" element can be developed apart and independently, then introduced into the operational SoS, review the previous paragraph.

A playground is an example of a stigmergic environment [25] which supports innovation. We're unaware of any explicit description of an innovation process appropriate to our domain, so we thought we'd supply one, which can serve as a point of departure for additional discussion.

On the playground, innovation seems to follow four distinct phases:
- Discovery
- Game (compete)
- Codify
- Practice

Discovery. By whatever means it occurs, something new is found and its potential value is envisioned. We *cannot* predict where the new killer idea will emerge. Our goal is to capture the idea, not predetermine or restrict the places where the ideas we accept may gestate. What triggers innovation? This question deserves (and has had) many books of its own. Kuhn has written on this topic extensively describing the sociology surrounding scientific paradigm shifts [21]. Innovation and discovery receive much attention in the scientific world, including labs setup to study it explicitly[31], and the literature dates back to Bacon, Descartes, Leibniz, etc., and continues today. Our use of the *paradigm shift* concept is not so grandiose. From time to time a connection will be made by someone new to the AOC, or by someone finding a new solution to a problem previously constraining their performance.

Game. The idea/insight is reduced to a form, which can be gamed against others who also occupy the same, or close, space. Within an environment where these ideas can be judged against each other, they compete. The new idea might be found to be a qualitative improvement, it might be refined, or it may be rejected. Alternatively, the old idea might be modified and achieve the benefit of "grinding" against the new idea. This must be done in a non-threatening way; and, we can't only celebrate success[32]. Like a playground, consequences that differentiate must exist, and some must fail, but a failure cannot devastate the failed; it must just remove it/them from the game.

Codify. During the gaming, the idea/insight is come to be understood better; as is the area it supposedly improves upon. With this understanding, the insight is reduced to a repeatable process or technique. This allows others to learn and use the innovation.

Practice. Once an innovation is reduced to practice and codified into a TTP, it can be taught and practiced.

To the degree there exists a place to play and innovate, the environment will support evolution at a rate faster than would occur otherwise. If there existed a process for guiding and managing the evolution, then the enterprise can move forward based on demonstrated value rather than future promises of value. The development of capabilities in this manner – through discovery - doesn't require the level of detail and *a priori* planning that a pure engineering approach requires.

5.1.3. Collaborative environments

The ability to work with others is not an altruistic need. It is purely self-serving if done correctly and effectively. And, it's this self-service that sustains useful collaborations. It may be the case that others may have need for that which I produce. Alternatively, I may be able to use that produced by others, allowing me to

---

[31] See, for example, Smithsonian Institute's Lemelson Center for the study of Invention and Innovation
[32] In the commercial world, Venture Capitalists recognize that the one "killer app" often comes after a number of promised, but failed, attempts. Their payback is judged on their portfolio, not on each member of the portfolio.

concentrate on what I do best (and how I add value) rather than expending resources on incidental aspects, which don't discriminate my offering from others.

Additionally, to the degree that what I produce must fit into a bigger whole, if I'm able to easily collaborate with others in that bigger whole, my risk of integration is reduced.

5.1.4. Partnerships

The ability to form and sustain partnerships helps to increase the utility of that which the partners produce and offer. Development risk is spread, and understanding of true need is improved. Successful partnerships seek to reduce overlap, and come to rely on each other to play necessary roles. Partnerships rely on effective collaborations.

5.1.5. Developers networks –creating opportunities for others

As discussed above, to speed up evolution, one must shorten generation time. Among systems, and with regard to systems engineering, this suggests shortening the feedback cycle, and lowering the amount of code which must be written, by allowing interaction among developers; thereby allowing reuse of useful code and connecting better to the run-time context. This is best embodied in the common understanding found in the software marketplace where developers' networks are a common ploy to getting developers to use a specific platform. A short anecdote illustrates the point.

In 2000 at a meeting of industry with the Air Force acquisition leadership, Paul Maritz[33] was asked his opinion about how it is that Microsoft has achieved such an apparently unassailable presence on desktops. His answer came immediately. It was due, he stated, to their commitment to developers[34]. He said Microsoft's strategy was to create opportunities for others; and use the fact of the opportunities created as a force to ensure that small mom-and-pop developer houses would recommend Microsoft products to their clients. Microsoft even extended their development environment (Visual Basic) down into the Microsoft Office suite, about which Maritiz stated that:

> Microsoft really supplied word processing, spreadsheet, and presentation graphics, etc. functionality, which, while bundled as useful office applications, were also available as functional primitives with which developers could provide customized value-added functionality to their customers.

He also noted that Microsoft felt it was necessary to lower the knowledge barriers to developing sophisticated applications.

Enabling integration and interoperability required many developers to be loyal to the Microsoft platform; and winning their loyalty required developer tools and

---

[33] Paul Maritz was a Group Vice President of the Platforms Strategy and Developer Group at Microsoft at the time of his retirement in 2000. He held many different positions there, and was one of Bill Gates trusted advisors.
[34] This was in an unguarded personal conversation between author Norman and Maritz at Lt Gen Kenney's (then Commander of the USAF Electronic Systems Center) first "Presidents Forum" meeting which Norman had defined and helped her put on.

environments, which were attractive and compelling. The resulting mechanism for exchange is Microsoft's Developers Network (MSDN).

This fits the CSE template in that Microsoft was **not** building, or attempting to build, all the functionality themselves, nor were they trying to make a killing on developer tools *per se*. Nor were they trying to get a piece of the action for all the functionality developed using their tools. Instead, they sowed the seeds with their tools, and took advantage of the multitude of applications built on top of their Windows™ platform. They rode the "need" identified by all the thousands of developers who were satisfying their own clients. The developer technical needs also were fed-back to the developer tools developers at Microsoft.

This is not a unique story; it was learned and implemented by others also. For example, Sun Microsystems practiced a similar approach as it brought Java along.

5.1.6. Branding

Branding is an interesting concept that we may be able to apply to bring pressure to coalesce. It is best told with an imaginary example. Suppose a potential supplier shows a General a new capability, which is especially attractive. The General acknowledges the potential value of this new capability and then asks "… have you got the '*Ready for the AOC*' sticker yet?...". The potential supplier responses "no" and the General then sighs his disappointment, and states that, had the potential supplier qualified for the *brand*, the capability offered could be moved into consideration for the AOC immediately. Without it, the capability must be subjected to a long process of evaluation and likely rework to ensure it will be able to integrate; and then is integrated. The General, as a proponent of a capability not carrying the '*Ready for the AOC*' brand, would need to advocate for funds to integrate and sustain the capability for its anticipated lifetime.

Brands can be powerful; and they can influence indirectly.

5.1.7. 'Co-opetition" [35]

This term was coined by Adam Brandenburger of the Harvard Business School and Barry Nalebuff of the Yale School of Management. It described their observation that connectivity among businesses and people require a new way to think about the business environment. Below is an explanation taken from the preface of their book of the same name:

> Co-opetition offers a theory of value. It's a book about creating value and capturing value. There's a fundamental duality here: whereas creating value is an inherently cooperative process, capturing value is inherently competitive. To create value, people can't act in isolation. They have to recognize their interdependence. To create value, a business needs to align itself with customers, suppliers, employees, and many others. That's the way to develop new markets and expand existing ones.

---

[35] Co-opetition http://mayet.som.yale.edu/coopetition/index2.html   (Adam Brandenburger of the Harvard Business School and Barry Nalebuff of the Yale School of Management)

> *But along with creating a pie, there's the issue of dividing it up. This is*
> *competition. Just as businesses compete with one another for market share,*
> *customers and suppliers are also looking out for their slice of the pie.*

Cooperation and competition is well studied. Axelrod [6] wrote about the evolution of cooperation, and showed how it can provide joint benefits. Poundstone [23] also describes mutually-beneficial approaches among autonomous agents.

Co-opetition allows independent parties to cooperate on those elements and aspects, which transcend their individual ability to control, while preserving their ability to compete on demonstrated value in their space.

For C2, various persons from traditional DoD contractors, commercial entities, and DoD officials have envisioned a marketplace where firms may specialize and come to dominate a niche. They maintain their dominance in their niche due to their continuing delivery of valuable and valued goods and services, rather than through contractual dictates.

As currently structured, this is difficult to achieve as there doesn't exist an environment for co-opetition. Consortia are often used as a vehicle for broad cooperation, and these options should be examined for achieving a co-opetive environment.

5.1.8. Leveraging other investments

Partnerships, collaborations, and other instances of cooperation all attempt to use the investments others have made for one's own benefit. An example (besides the obvious ones of using Commercial Off-the-Shelf technologies) is found in the Family of Interoperable Operational Pictures (FIOP) program. The FIOP program was the first to fund TBMCS's desire to build Information Services – which represented a new way for them to build and deploy functionality. FIOP used a relatively small amount of money to leverage the larger development budget TBMCS had. Essentially, FIOP paid for good behavior on TBMCS's part. Both benefited. And, others (who invested nothing) were able to use the Information Services built by TBMCS.

5.1.9. Technical approaches which respect 'ricebowls"

There is no doubt that people come to place great value in that which they are personally involved in and responsible for. These apparently parochial interests are often described as "rice bowls." A great source of resistance to cooperation among independent agents is the thought or impression that others will impose themselves on the independent agents in ways and manners, which they view are inappropriate. After all, each organization has conducted itself according to its needs, and has made decisions according to its assessment of how best to meet the needs. Additionally, there is often fear that one's activity will be subsumed under another's, and one's contributions will be devalued and possibly ignored.

It's clear that cooperation has great value; and those aspects, which interfere with cooperation, cause "innovation drag." Technical approaches, which tend to respect "ricebowls" remove some of the hesitations for forming cooperative partnerships.

Examples of technical approaches, which respect ricebowls, include current developments in Web Services. This technology exposes functionality with the minimum requirements for homogeneity. It presents a "virtual homogeneity" within a heterogeneous world. In this way it offers the potentiality for independent agents to offer their services to others; and thereby permits new associations and relations to be exploited – supporting innovation. It also supports the rise of technical structures and approaches, which permit the agility needed. Assembly moves out towards the end-users further blurring the difference between development and runtime.

5.1.10. Opportunistic approach

An aspect of the usual way in which we do business is to restrict ourselves to a complete capability before fielding. As mentioned earlier this restricts fielding to apparent "complete" sets at a fairly slow rate. If one treated logical sets of users as a unit, and involved them in managing the identification and introduction of functionality and change, then one might be able to be more responsive.

5.1.11. Advertising and Discovery

As currently structured, finding useful capabilities and functionality offered by third-parties is not trivial. Potentially useful functionality is not advertised, and there is no well-known place to go looking. Both for the development and the operational environments, achieving transparency for effective advertising and discovery is critical.

A key enabler for evolving and integrating the enterprise is to create opportunities for *small world* phenomena [7, 26] to emerge. The power of loose connections is clear and convincing. New relations possible are likely discovered in areas not previously explored. This will emerge with in both shared information spaces and with shared behavior. Advertising and discovery technologies are key to enable these results.

5.1.12. Permitting "value-add" business models

A continuing complaint from users in the field is that they don't get "a vote" in what is built for them. This is primarily due to the business models employed in acquisition today. As mentioned earlier, the dominant business model used is employer/contractor. In this model, the employer produces a requirements document, and then various potential contractors propose how they will produce the functionality desired. The market place is contract engineering; the selling and buying of engineering hours (perhaps laced with certain processes which can be argued reduce risk). Those who have successful proposals are those who can tell the story of how they are going to produce a requirements-compliant product for the least risk. It is a promise well told, not a demonstration of specific achievement. Success is measured based on compliance with the requirements, and the maintenance of the cost and schedule negotiated. Success is not directly related to the usefulness of that which is produces. The Government's money is spent for the engineering hours used for development. This is the basis for the complaints.

Assume a by-use payment model. Assume further that there is no *a priori* assumption of the undesirability of redundant functionality[36]. Under such model money flows to those who produce demonstrated utility to the user. The market now shifts to understanding and satisfying real needs rather than the sale of engineering hours. The acquisition organization's role, under this model, shifts to verifying compliance with a set of rules under which functionality is built. Under this model the Government's money pays for demonstrated value.

There are additional aspects to such a model: the emphasis shifts from *cost* to *price*. Suppliers (as opposed to *contractors*) attempt to manage their margins; and they may apply their best and brightest (assuming they can control their intellectual property) and be innovative.

5.1.13. Analysis, simulation and collecting experiences replace full-coverage testing

How does one test a complex adaptable system? Rather than relying on traditional approaches (which attempt to come as close to full-coverage testing as possible), we might collect and catalog things when they go wrong in the field; analyzing these for insight into subtle transitive effects. We also need to employ better testing approaches to develop some sense of belief about the systems we field before fielding. Phadke's[37] Robust Testing™ approach may provide tools for picking better and smaller test cases.

Additionally, the infrastructure should be tested to failure so we know the boundary. Then it should be monitored in the field to shed light on if and when we approach these limits. This permits time to intervene prior to (not after) problems emerge.

6. Summary and Conclusions

The challenge is moving from "things" to "integrated collections of things" which are governed and managed independently. Although we presented the problem as an issue for AOCs, it is not confined to the AF, or to joint forces, or to DoD, or the US Government, or to the US. The observation that one must use methods which respect the characteristics of the enterprise; and which don't require complete control, or complete knowledge in one place.

Our summary would be incomplete if we didn't consider the insights offered by other professions who have faced similar challenges. Architecture has. Christopher Alexander (of architectural fame) talks about the illusion of control; and he observes that attempts to assert control generally has the opposite affect from what is desired; things tend to get worse, more out of control. He notes that the tendency "...to gain 'total design' control of the environment...makes things still worse..."[4, pp. 238] He

---

[36] Ashby's Law of Requisite Variety [5] can be interpreted as explicitly supporting variety as a technique for attempting to supply sufficient potentiality to allow adequate response to selective pressures.
[37] See Phadke Associates' Robust Testing™ at www.phadkeassociates.com/ser/rt.htm

points to the need to construct towns and cities using "patterns" which preserved the correct "nature," and became "alive" in their own right. He calls this the "quality without a name." He's talking about complexity and adaptability.

Traditional Systems Engineering has always attempted to understand and deal with complexity; but the nature of that which was being engineered tended to be stand-alone with well-defined edges. Simple rules could tell what was "in the system" or "out of the system;" and the engineering activities started with, or required that, the requirements were well known, understood, and stable. As systems engineering came to deal with collections and aggregations of elements which were to be integrated into definite, well-understood (and understandable) forms-and-function which were to be stable over time. As we scale this approach up to the enterprise and find ourselves dealing with complex systems, we fall directly into the trap outlined by Alexander: things become worse.

Our traditional systems engineering has been concerned with finding those well-bounded subordinate elements, then (in essence) isolating them so they may be "engineered." From this point one proceeds as if the element is isolated and unmoved by other juxtaposed elements. It's this desire to "divide and conquer" which characterizes our tradition approaches. Is this wrong? No! But it's not always correct; nor is it complete.

Consider the richness now possible due to the potential interconnectivity now available, and the interdependence among elements implied. The forms an ecosystem – where each element responds to its context through some accommodation – potentially evolving to respond (those elements which are "alive" respond and change). Consider further that each element's context is set by the elements, which juxtapose it in almost countless ways (forming a hyperspace of *pressure*). This is certainly an intricate, hard to understand-and-appreciate situation; and in that way, it may be thought of (in the usual vernacular) as *complex*. Using our traditional divide-and-conquer systems engineering (TSE) we would likely measure the external world, then make an assumption of constancy with respect to this external surround. Engineering would proceed on the element from this point of view.

But, the realization that the element under study also forms part of the context for every element which juxtaposes it starts to hint at the limit of the simplifying assumption made to perform the TSE: It imposes a *pressure* (an influence) on its surround in addition to feeling the pressure of the surround.

Note the implications of the transitive nature of these influences. This is what is referred to as *complexity* as opposed to *intricacy* or *difficult to understand*. One could imagine waves and ripples of change flowing through this system of system. Likely, patterns will emerge when viewed from a higher level of abstraction. This is likely where we find ourselves with respect to C2 in a joint and coalition world, and where independent agents can introduce change according to their own agenda and timing.

The big questions: can such an aggregation be engineered at all? Can it even be understood? Will useful patterns present themselves? Should we even bother? We can say that we've failed many times in the past because we've made the simplifying assumptions mentioned earlier.

One of the principles that pops out right away from taking this *complexity* point of view is the need to have tight control over all (or as many) characteristics of a system to be engineered as possible if one wants to apply TSE; e.g., the ability to direct resources to problem areas as they arise. Is this insight new? No; but maybe, where it is impossible to meet that control and authority boundary conditions, there might be other approaches which can be brought to bear. And we believe the codifying of CSE is a step in the right direction.

Understanding complexity, and engineering complex systems is the next step. Systems Engineering is taking the next step. It is maturing past the point where a one-size-fits-all process is what's thought of as "correct." It is finding a new language with which to understand that which it attempts to engineer. This is maturity, and this is its future.

References

[1]   Albert, D., Garstka, J., Stein, F.; 1999; Network Centric Warfare; CCRP
[2]   Alberts, D., Garstka, J., Hayes, R., Signori, D.; 2001; Understanding Information Age Warfare.; CCRP
[3]   Alberts, David S., Hayes, Richard E.; 2003; Power to the Edge;. CCRP
[4]   Alexander, Christopher; 1979; The Timeless Way of Building; Oxford University Press
[5]   Ashby, W.R.; 1956; Introduction to Cybernetics; Chapman & Hall
[6]   Axelrod, R.; 1984; The Evolution of Cooperation; Basic Books.
[7]   Barabasi, L-S.; 2002; Linked: the New Science of Networks; Perseus.
[8]   Bar-Yam, Y.; 1997; Dynamics of Complex Systems; Perseus;.
[9]   Bar-Yam, Y., 2003; When Systems *Engineering Fails---Toward Complex Systems Engineering*; 2003 IEEE International Conference on Systems, Man & Cybernetics, October 5–8 Washington, D.C., USA
[10]   Bonabeau, E.;2003; *Swarming Intelligence*; Swarming Network Enabled C4ISR Conference 13-14 January; McLean, VA
[11]   Breen, P., Case, R., Kazura, A., Norman, D.; 2000; *DSE – A Decision Support Enviroment*; 2000 Command and Control Research and Technology Symposium, Naval Postgraduate School, Monterey, CA, June 26-28, 2000
[12]   Card, O. 1977. Ender's Game; Tor.
[13]   Gamma, E., R. Helm R. Johnson, J, Vlissides; 1994; Design Patterns; Addison-Wesley
[14]   Heylighen F., Joslyn C. & Turchin V. (eds.); 1995; The Quantum of Evolution. Toward a theory of metasystem transitions. ; Gordon and Breach Science Publishers, New York
[15]   Heylighen, F., C. Joslyn; 2001; "Entropy and Information," *Principia Cybernetica Web*, Sep 3,; pespmc1.vub.ac.be/entrinfo.html
[16]   Holland, J.H.; 1992; Adaptation in Natural and Artificial Systems; MIT Press
[17]   Holland, J.H.; 1995; Hidden Order: How Adaptation Builds Complexity.; Reading: Addison-Wesley.
[18]   INCOSE; 2001; Systems Engineering Handbook, v2;
[19]   Kauffman S.A; 1993; The Origins of Order: Self-Organization and Selection in Evolution, Oxford University Press, New York,
[20]   Krygiel, Annette J.;1999; Behind the Wizard's Curtain: An Integration Environment for a System of Systems; CCRP;
[21]   Kuhn, T.S.; 1996; Structure of Scientific Revolutions, , University of Chicago Press,
[22]   Moffat, J.; 2003; Complexity Theory and Network Centric Warfare;. CCRP

[23]   Poundstone, w.; 1992; <u>Prisoner's Dilemma</u>; Anchor Books.
[24]   Van Dyke P.; 1997; *"Go to the Ant": Engineering Principles from Natural Multi-Agent Systems*; <u>Annals of Operations Research 75</u>, pp. 69-101;.
[25]   Van Parunak, A; 2003.; *Making Swarming Happen*; <u>Swarming Network Enabled C4ISR Conference</u> 13-14 January, McLean, VA
[26]   Watts, D.; 2003; <u>Six Degrees: The Science of a Connected Age</u>; Norton.

Douglas O. Norman is the Chief Technologist for USAF Battle Management Capabilities (HQ ESC/AC), and is MITRE Section Leader for Battle Management and C2. Previously, he was the Chief Engineer of the Theater Battle Management Core Systems (TBMCS). He also holds the position of Senior Technical Advisor to the AOC Weapon System Acquisition Group Commander.

Michael L. Kuras is a Principal Systems Engineer in MITRE's USAF Systems Engineering Division with significant interest and experience discovering, defining and applying Complex Systems Engineering principles.

Chapter 11

Negotiation algorithms for collaborative design settings

Mark Klein
Center for Coordination Science
Massachusetts Institute of Technology
Cambridge, MA 02139
m_klein@mit.edu

Peyman Faratin
Laboratory for Computer Science
Massachusetts Institute of Technology
Cambridge MA 02139
peyman@mit,edu

Hiroki Sayama
University of Electro-Communications
Tokyo, Japan
sayama@cx.hc.uec.ac.jp

Yaneer Bar-Yam
New England Complex Systems Institute
Cambridge MA 02138
yaneer@necsi.org

1. Introduction

Work to date on computational models of negotiation has focused almost exclusively on defining 'simple' agreements consisting of one or a few independent issues [1] [2]. These protocols work via the iterative exchange of proposals and counter-proposals. An agent starts with proposal that is optimal for it and makes concessions, in each subsequent proposal, until either an agreement is reached or the negotiation is abandoned because the utility of the latest proposal has fallen below the agents' reservation (minimal acceptable utility) value (Figure 1):

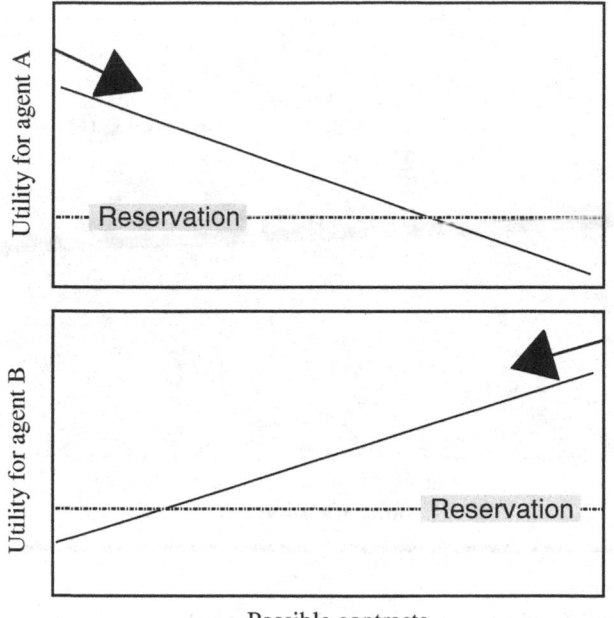

Possible contracts

Figure 1. The proposal exchange model of negotiation, applied to a simple agreement. The Y axis represents the utility of an agreement to each agent. Each point on the X axis represents a possible agreement, ordered in terms of its utility to agent B. Since there is no need to negotiate over issues that both parties agree upon, we only consider issues where improvement for one party represents a decrement for the other. The arrows represent how agents begin with locally optimal proposals, and concede towards each other, with their subsequent proposals, as slowly as possible. Note that we have, for presentation purposes, 'flattened' the agreement space onto a single dimension, but there should actually be one dimension for every issue in the agreement.

This is a perfectly reasonable approach for simple agreements. Since issues are independent, the utility of a agreement for each agent can be calculated as the weighted sum of the utility for each issue. The utility function for each agent is thus a simple one, with a single optimum and a monotonic drop-off in utility as the agreement diverges from that ideal. Such negotiations thus typically progress as follows:

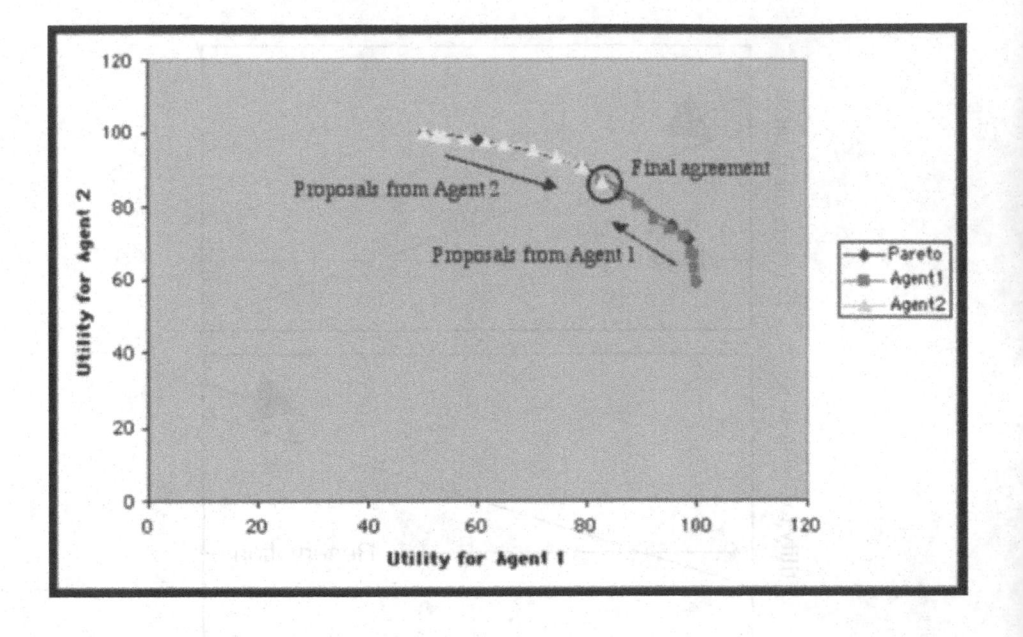

Figure 2. The utilities for the proposals made in a typical simple agreement negotiation. The agreement consisted in this case of 40 binary issues. Each agent starts with a locally optimal proposed agreement (at the extremes of the Pareto frontier) and is required to reduce the Hamming distance (number of issues with different values) between the two agents' proposals, until an agreement is reached. With simple agreements, this results in optimal outcomes. The Pareto frontier, representing the set of optimal agreements, was estimated by applying an annealing optimizer to differently weighted sums of the two agents' utility functions.

As we can see, the proposals from each agent start at their own ideal, and then track the Pareto frontier until they meet in the middle with an optimal agreement. This happens because, with linear utility functions, it is easy for an agent to identify the proposal that represents the minimal concession: the agreement that is minimally worse than the current one is "next" to the current one in the agreement space and can be found by moving in the direction with the smallest downward slope. The simplicity of the utility functions, moreover, makes it feasible for agents to infer enough about their opponents that they can identify concessions that are attractive to each other, resulting in relatively quick negotiations.

Collaborative design negotiations, by contrast, are generally much more complex, consisting potentially of hundreds or even thousands of distinct issues. Even with only 50 issues and two alternatives per issue, we encounter a search space of roughly 10^15 possible agreements, too large to be explored exhaustively. The value of one issue selection to an agent, moreover, will often depend on the selection made for another issue. The value to me of a given couch, for example, depends on whether it is a good match with the chair I plan to purchase with it. Such issue interdependencies lead to *nonlinear* utility functions with *multiple* local optima [3].

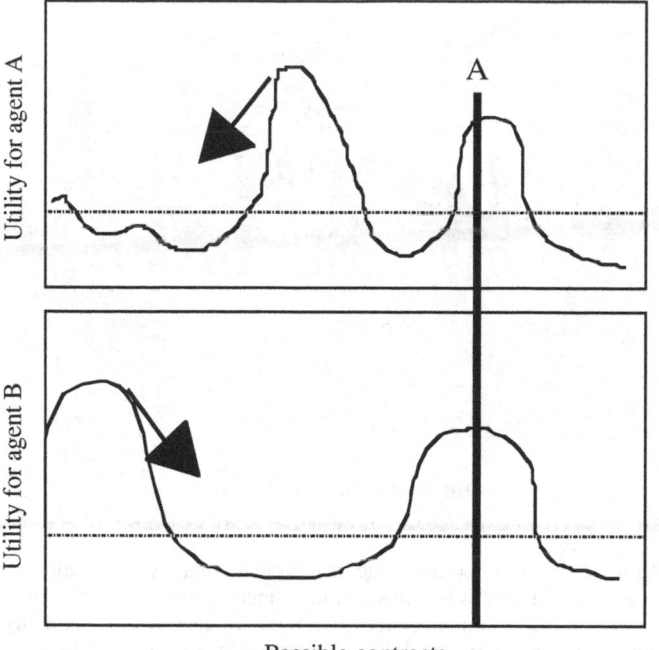

Figure 3. An example of proposal exchange applied to a complex agreement. Because of issue inter-dependencies, the utility functions have multiple optima. The arrows show what happens when each agent begins at a local optimum and concedes towards the other: win-win solutions found elsewhere in the agreement space (e.g. the agreement labeled 'A') can be missed.

In such contexts, an agent identifying a locally optimal design agreement becomes a nonlinear optimization problem, difficult in its own right. Simply conceding toward the other agents' proposals can result in the agents missing agreements that would be superior from both their perspectives (e.g. the agreement labeled "A" in figure 3 above). Standard negotiation techniques thus typically produce the following behavior when applied to complex negotiation:

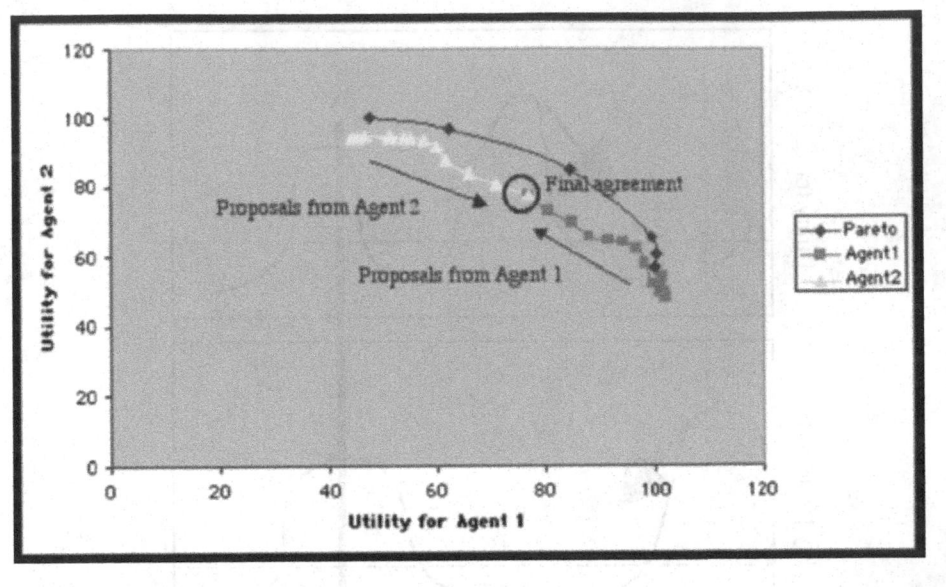

Figure 4. The utilities for the proposals made in a typical complex negotiation. This example differs from figure 2 only in that a nonlinear utility function was used by each agent (details below). As we can see, the minimal concession protocol that works optimally for simple agreements produces outcomes, for complex agreements, that are substantially sub-optimal.

The agents start with an approximation to their ideal design and diverge increasingly from the Pareto frontier as they converge upon an agreement. The degree of sub-optimality depends on the details of the utility function. In our experiments, for example, the final agreements' averaged 94% of optimal. This is a substantial decrement when you consider that the utility functions we used for each agent were, individually, quite easy to optimize: a simple steepest ascent search averaged final utility values roughly 97% of those reached by a nonlinear optimization algorithm. It is striking that such relatively forgiving multi-optima utility functions lead to substantially sub-optimal negotiation outcomes.

These sub-optimal outcomes represent a fundamental weakness with current negotiation techniques. The only way to ensure that subsequent proposals track the Pareto frontier, and thus conclude with a Pareto optimal result, is to be able to identify the proposal that represents the minimal concession from the current one. But in a utility function with multiple optima, that proposal may be quite distant from the current one, and the only way to find it is to exhaustively enumerate all possible agreements. This is computationally infeasible, however, due to the sheer size of the agreement space. Since the utility functions are quite complex, it is in addition no longer practical for one agent to infer the other's utility function and thereby speed the negotiation by well-chosen concessions.

Collaborative design therefore requires different negotiation techniques, ones which allow agents to find 'win-win' agreements in intractable multi-optima search spaces in a reasonable amount of time. In the following sections we describe a family of negotiation protocols that make substantial progress towards achieving these goals. The paper is structured as follows. We begin by describing how a well-known non-linear optimization technique (simulated annealing) can be integrated the mediated single text negotiation protocol to produce an approach that offers near-optimal outcomes for complex negotiations. We reveal the prisoner's dilemma that results from this approach, and propose a refined protocol, based on parity-maintaining annealing mediator, that resolves that problem. We conclude with describing an unmediated version of the negotiation protocol that is also effective at producing near-optimal outcomes with complex agreements.

2. Mediated Single Text Negotiation

A standard approach for dealing with complex negotiations in human settings is the mediated single text negotiation [4]. In this process, a mediator proposes a agreement that is then critiqued by the parties in the negotiation. A new, hopefully better proposal is then generated by the mediator based on these responses. This process continues, generating successively better agreements, until some agreed-upon stopping point (e.g. the reservation utility value is met or exceeded for both parties). We can visualize this process as follows:

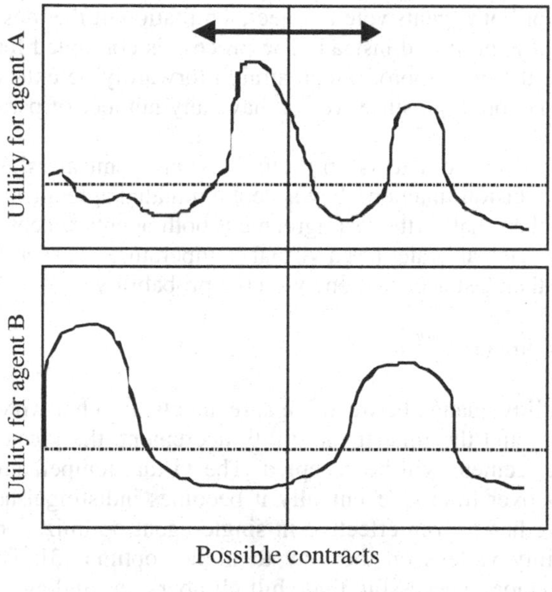

Figure 5. Single text negotiation. The vertical line represents the current proposed agreement, and subsequent proposals move that line in the agreement space.

Here, the vertical line represents the agreement currently proposed by the mediator. Each new agreement moves the line to a different point on the X axis. The goal is to find a agreement that is sufficiently good for both parties.

We defined a simple simulation experiment to help us explore how well this approach actually works. In this experiment, there were two agents negotiating to find a mutually acceptable agreement consisting of a vector S of 100 boolean-valued issues, each issue assigned the value 0 or 1, corresponding to the presence or absence of a given agreement clause. This defined a space of $2^{\wedge}100$, or roughly $10^{\wedge}30$, possible agreements. Each agent had a utility function calculated using its own 100x100 influences matrix H, wherein each cell represents the utility increment or decrement caused by the presence of a given pair of issues, and the total utility of a agreement is the sum of the cell values for every issue pair present in the agreement:

$$U = \sum_{i=1}^{100} \sum_{j=1}^{100} H_{ij} S_i S_j$$

The influence matrix therefore captures the bilateral dependencies between issues, in addition to the value of any individual agreement clause. For our experiments, the utility matrix was initialized to have random values between −1 and +1 in each cell. A different influences matrix was used for each simulation run, in order to ensure our results were not idiosyncratic to a particular configuration of issue inter-dependencies.

The mediator proposes a agreement that is initially generated randomly. Each agent then votes to accept or reject the agreement. If both vote to accept, the mediator mutates the agreement (by randomly flipping one of the issue values) and the process is repeated. If one or both agents vote to reject, a mutation of the most recent mutually accepted agreement is proposed instead. The process is continued for a fixed number of proposals. Note that this approach can straightforwardly be extended to a N-party (i.e. multi-lateral) negotiation, since we can have any number of parties voting on the agreements.

We defined two kinds of agents: 'hill-climbers' and 'annealers'. The hill-climbers use a very simple decision function: they accept a mutated agreement only if its utility to them is greater than that of the last agreement both agents accepted. Annealers are more complicated. Each annealer has a virtual 'temperature' T, such that it will accept agreements worse than last accepted one with the probability:

$$P(accept) = min(1, e^{-\Delta U/T})$$

where ΔU is the utility change between the agreements. In other words, the higher the virtual temperature, and the smaller the utility decrement, the greater the probability that the inferior agreement will be accepted. The virtual temperature of an annealer gradually declines over time so eventually it becomes indistinguishable from a hill-climber. Annealing has proven effective in single-agent optimization, because it can travel through utility valleys on the way to higher optima [3]. This suggests that annealers can be more successful than hill-climbers in finding good negotiation outcomes.

3. The Prisoner's Dilemma

Negotiations with annealing agents did indeed result in substantially superior final agreement utilities, but as the payoff table below shows, there is a catch:

Table 1. The optimality of the negotiation outcomes for different pairings of annealing and hill-climbing agents. The top value in each cell represents how close the social welfare value of the final agreement is to optimal. The pair of values below it represent how close the final agreement is to the optimum for the Agent 1 and Agent 2, respectively.

	Agent 2 hill-climbs	Agent 2 anneals
Agent 1 hill-climbs	.86 .73/.74	.86 .99/.51
Agent1 anneals	.86 .51/.99	.98 .84/.84

As expected, paired hill-climbers do relatively poorly while paired annealers do very well. If both agents are hill-climbers they both get a poor payoff, since it is difficult to find many agreements that represent an improvement for both parties. A typical negotiation with two hill-climbers looks like the following:

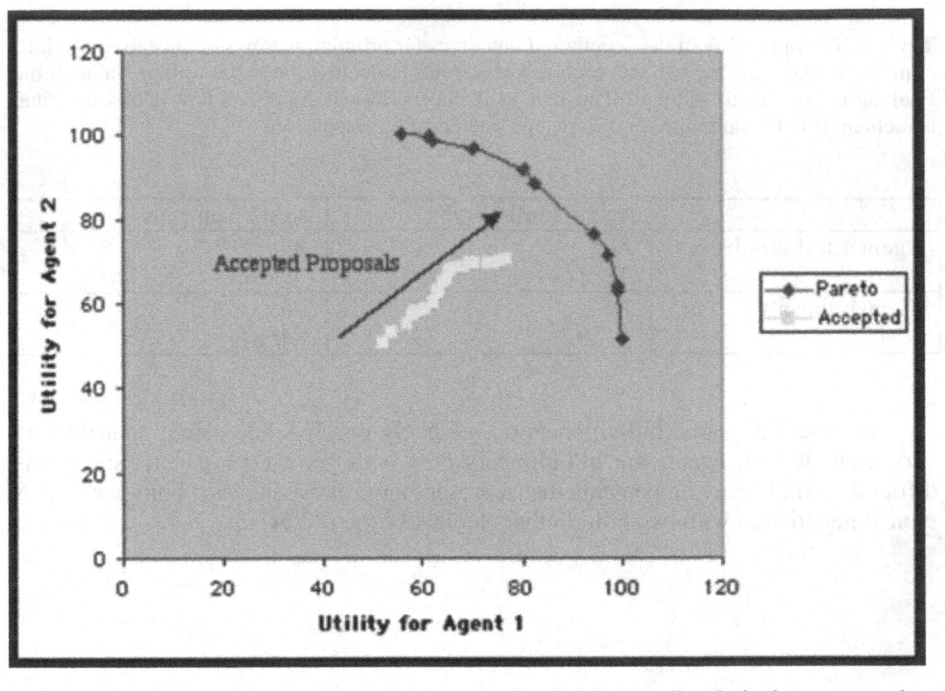

Figure 6. The utilities for the accepted proposals in a typical mediated single text complex negotiation with two hill-climbers. The mediator's initial proposal is at the lower left, and the subsequent accepted proposals move towards higher utilities for both agents.

As we can see, in this case the mediator was able to find only a handful of agreements that increased the utility for both hill-climbers, and ended up with a poor final social welfare.

Near-optimal social welfare can be achieved, by contrast, when both agents are annealers, willing to initially accept individually worse agreements so they can find win-win agreements later on:

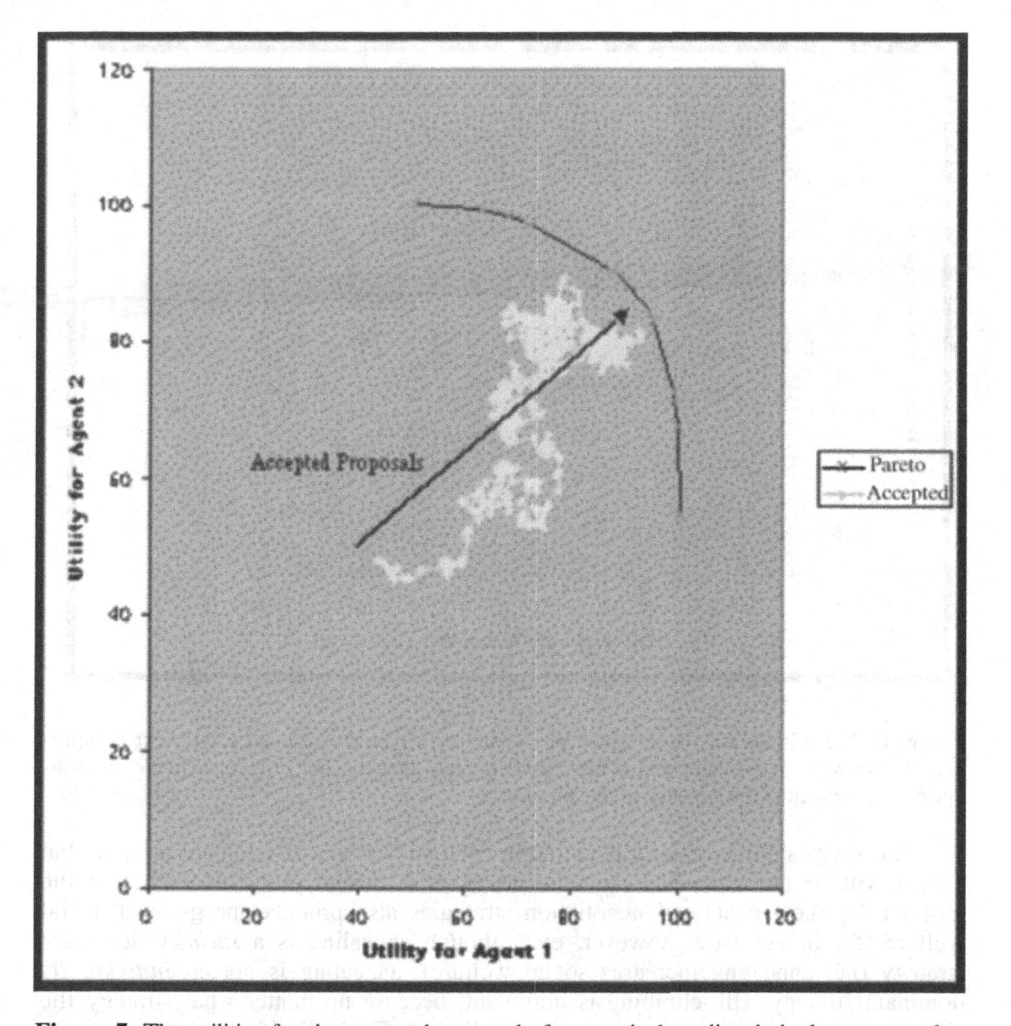

Figure 7. The utilities for the accepted proposals for a typical mediated single text complex negotiation with two annealers. Some of the accepted proposals actually cause utility decrements for one or both agents, but the final result is a near-optimal agreement.

The agents entertain a much wider range of agreements, eventually ending very near the Pareto frontier.

If one agent is a hill-climber and the other is an annealer, however, the hill-climber does extremely well but the annealer fares correspondingly poorly (Figure 8). This pattern can be understood as follows. When an annealer is at a high virtual temperature, it becomes a chronic conceder, accepting almost anything beneficial or not. The hill-climber 'drags' the annealer towards its own local optimum, which is not very likely to also be optimal for the annealer, so the annealer pays a "conceder's penalty":

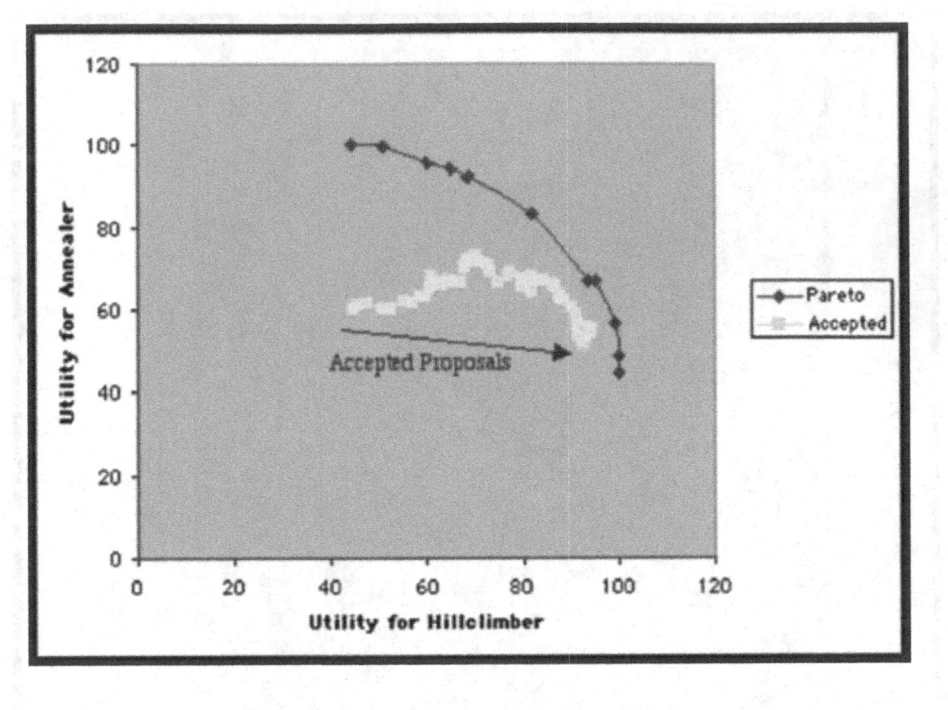

Figure 8. The utilities for the accepted proposals for a typical mediated single text complex negotiation with an annealer and a hill climber. Note that the hill climber achieves a near-optimal agreement at the expense of the annealer.

This reveals a dilemma. In negotiation contexts we typically can not assume that agents will be altruistic, and we must as a result design protocols such that the individually most beneficial negotiation strategies also produce the greatest social welfare [5]. In our case, however, even though annealing is a *socially* dominant strategy (i.e. annealing increases social welfare), annealing is <u>not</u> an *individually* dominant strategy. Hill-climbing is dominant, because no matter what strategy the other agent uses, it is better to be a hill-climber (Table I). If all agents do this, however, then they forego the higher individual utilities they would get if they both annealed. Individual rationality thus drive the agents towards the strategy pairing with the *lowest individual and social welfare.* This is thus an instance of the prisoner's dilemma. It has been shown that this dilemma can be avoided if we assume repeated interactions between agents [6], but we would prefer to have a negotiation protocol that incents socially beneficial behavior without that difficult-to-enforce constraint. Several straightforward approaches to this problem, however, prove unsuccessful. One possibility is to simply reduce the annealer's willingness to make concessions. This can indeed eliminate the conceder's penalty, but at the cost of achieving social welfare values only slightly better than that achieved by two hill climbers. Another option is to have agents switch from being an annealer to a hill-climber if they determine, by observing the proposal acceptance rates of their opponents, that the other agent is being a hill-climber. We found, however, that it takes too long to determine the type of the other agent: by the time it has become clear, much of the agreement utility has been committed, and it is too late to recover from the consequences of having started out as an annealer. See [7] for details.

4. The Annealing Mediator

We were able to define a negotiation protocol that avoids the prisoner's dilemma entirely in mediated single-text negotiation of complex agreements. The trick is simple: rather than requiring that the negotiating agents anneal, and thereby expose themselves to the risk of being dragged into bad agreements, we moved the annealing into the mediator itself. In our original protocol, the mediator would simply propose modifications of the last agreement both negotiating agents accepted. In our refined protocol, the mediator is endowed with a time-decreasing willingness to follow up on agreements that one or both agents rejected (following the same inverse exponential regime as the annealing agents). Agents are free to remain hill-climbers and thus avoid the potential of making harmful concessions. The mediator, by virtue of being willing to provisionally pursue utility-decreasing agreements, can traverse valleys in the agents' utility functions and thereby lead the agents to win-win solutions. We describe the details of our protocol, and our evaluations thereof, below.

In our initial implementations each agent gave a simple accept/reject vote for each proposal from the mediator, but we found that this resulted in final social welfare values significantly lower than what we earlier achieved using annealing agents. In our next round of experiments we accordingly modified the agents so that they provide additional information to the mediator in the form of vote strengths: each agent annotates an accept or reject vote as being *strong* or *weak*. The agents were designed so that there are roughly an equal number of weak and strong votes of each type. This maximizes the informational content of the vote strength annotations. When the mediator receives these votes, it maps them into numeric values (strong accept = 1, weak accept = 0, weak reject = -1, strong reject = -2) and adds them together to produce an aggregate score. A proposal is accepted by the mediator if the score is non-negative, i.e. if both agents voted to accept it, or if a weak reject by one agent is overridden by a strong accept from the other. The mediator can also accept rejected agreements (i.e. those with a negative agrregate score) using the annealing scheme described above. This approach works surprisingly well, achieving final social welfare values that average roughly 99% of optimal despite the fact that the agents each supply the mediator with only two bits of information. We found, in fact, that increasing the number of possible vote weights did not increase final social welfare. This is because the strong/weak vote annotations are sufficient to allow the system to pursue social welfare-increasing agreements that cause a utility decrement for one agent.

5. Incentives for Truthful Voting

Any voting scheme introduces the potential for strategic non-truthful voting by the agents, and our scheme is no exception. Imagine that one of the agents always votes truthfully, while the other exaggerates so that its votes are always 'strong'. One might expect that this would bias negotiation outcomes to favor the exaggerator and this is in fact the case:

Table 2. The optimality of the negotiation outcomes for truth-telling vs. exaggerating agents with a simple annealing mediator. An exaggeration strategy is individually incented, even though it results in outcomes with lower social welfare.

	Agent 2 exaggerates	Agent 2 tells truth
Agent 1 exaggerates	.92 .81/.81	.93 .93/.66
Agent 1 tells truth	.93 .66/.93	.99 .84/.84

As we can see, even though exaggerating has substantial negative impact on social welfare, agents are individually incented to exaggerate, thus re-creating the prisoner's dilemma we encountered earlier. The underlying problem is simple: exaggerating agents are able to induce the mediator to accept all the proposals that are advantageous to them (if they are weakly rejected by the other agent), while preventing the other agent from doing the same. What we need, therefore, is an enhancement to the negotiation protocol that incents truthful voting, preserving equity and maximizing social welfare.

How can this be done? We found that simply placing a limit on the number of strong votes each agent can use does not work. If the limit is too low, we effectively lose the benefit of vote weight information and get the lower social welfare values that result. If the strong vote limit is high enough to avoid this, then all an exaggerator has to do is save all of it's strong votes till the end of the negotiation, at which point it can drag the mediator towards making a series of proposals that are inequitably favorable to it.

Another possibility is to enforce overall parity in the number of "overrides" each agent gets. A override occurs when a agreement supported by one agent (the "winner") is accepted by the mediator over the objections of the other agent. Overrides are what drags a negotiation towards agreements favorable to the winner, so it makes sense to make the total number of overrides equal for each agent. But this is not enough, because exaggerators always win disproportionately more than the truth-teller.

The solution, we found, came from enforcing parity between the number of overrides given to each agent *throughout* the negotiation, so neither agent can get more than a given advantage. This way at least rough equity is maintained no matter when (or whether) either agent chooses to exaggerate. The results of this approach were as follows when the override disparity was limited to 3:

Table 3. The optimality of the negotiation outcomes for truth-telling vs exaggerating agents with parity-enforcing mediator. The parity-enforcing mediator makes truth-telling the rational strategy.

	Agent 2 exaggerates	Agent 2 tells truth
Agent 1 exaggerates	.91 .79/.79	.92 .78/.81
Agent 1 tells truth	.92 .81/.78	.98 .84/.84

When we have truthful agents, we find that this approach achieves social welfare just slightly below that achieved by a simple annealing mediator, while offering a significantly ($p < 0.01$) higher payoff for truth-tellers than exaggerators. We found, moreover, that the same pattern of results holds for a range of exaggeration strategies, including exaggerating all the time, exaggerating at random, or exaggerating just near the end of the negotiation. Truth-telling is thus both the individually dominant and socially most beneficial strategy.

Why does this work? Why, in particular, does a truth-teller fare better than an exaggerator with this kind of mediator? One can think of this procedure as giving agents 'tokens' that they can use to 'purchase' advantageous overrides, with the constraint that both agents spend tokens at a roughly equal rate. Recall that in this case a truthful agent, offering a mix of strong and weak votes, is paired with an exaggerator for whom at least some weak accepts and rejects are presented as strong ones. The truthful agent can therefore only get an override via annealing (see Table 3), and this is much more likely when its vote was a strong accept rather than a weak one. In other words, the truthful agent spends its tokens almost exclusively on agreements that truly offer it a strong utility increase. The exaggerator, on the other hand, will spend tokens to elicit a override even when the utility increment it derives is relatively small. At the end of the day, the truthful agent has spend its tokens more wisely and to better effect.

6. The Unmediated Single Text Protocol

The protocol we have just considered worked well in the contexts studied but suffers from the disadvantage of requiring a mediator. One issue concerns trust. Since the annealing mediator is empowered to selectively ignore agent votes, there is the risk that it may do so in a way that favors one agent over another (though the use of the parity-enforcing token mechanism does somewhat reduce the potential impact of this problem). Another issue concerns how quickly negotiations converge on a result. The annealing mediator generates new proposals by making random mutations to the last provisionally accepted agreement, without taking into account any information about what agreements are preferable or even sensible. As a result, the mediator generates a very high proportion of rejected agreements, which is part of the reason why our experimental runs each involved so many (2500) proposals. The negotiating agents could imaginably provide the mediator with information about their utility functions so that the mediator is able to propose agreements more 'intelligently', but this is problematic for a number of reasons including the typical reluctance of self-interested agents to reveal their utility functions to a party that may or may not be worthy of their trust.

An effective unmediated version of the annealing protocol can, fortunately, be defined. It works as follows. Agents each start with a given number of tokens (2 each, in our experiments) and a mutually agreed-upon starting temperature T. A random agreement is generated, and one of the negotiating agents is selected at random to propose a small (e.g. single-issue) variant thereof, presumably the variant that most increases the utility of the agreement for that agent. The other agent then votes on the proposed variant. The proposing and voting both indicate the strength of their preference for the proposed agreement using the scheme described above (i.e. strong reject, weak reject, weak accept, strong accept). The agreement is provisionally accepted with probability

$$P(\text{accept}) = \min(1, e^{-\Delta U/T})$$

where the aggregate score (ΔU) is calculated as for the annealing mediator, and the outcome is determined using the roll of a fair, mutually observable dice. If the decision to accept a proposal represents the over-ride of one agents' reject vote, the winning agent needs to give one of its' tokens to the over-ridden agent. An over-ride is not permitted if the agent has run out of tokens. The proposer and voter alternate roles thereafter until neither agent can identify any improvements to make to the last accepted agreement. Agents in the proposer role may pass but may not repeat proposals. The temperature T declines at a mutually agreed-upon rate during this process. This protocol thus reproduces the key elements of the annealing mediator protocol – a time-dependent annealing regime plus tokens - without the need for a mediator. Our experiments show that this protocol produces results just as good as the annealing mediator, averaging 99% of optimal, while requiring fewer proposal exchanges (averaging about 200 exchanges per negotiation).

7. Contributions

We have shown that collaborative design negotiation, involving many interdependent issues, has properties that are substantially different from the independent issue case that has been studied to date in the negotiation literature, and requires as a result different protocols to achieve near-optimal outcomes. This work represents, as far as we are aware, the first family of negotiation protocols suited for interdependent issues. While some previous work has studied multi-issue negotiation (e.g. [8], [1] [9] [10]) the issues in these efforts are treated as being independent. Multi-attribute auctions [11] [12] represent another scheme potentially suitable for multiple interdependent issues, wherein negotiators bid on the agreements that they prefer, with the most highly demanded agreement winning. This approach is very unlikely to scale, however, to collaborative design settings where there could easily be trillions of competing agreements up for bid.

The essence of our approach can be summarized simply: conceding early and often (as opposed to little and late, as is typical for independent issue negotiations) is the key to achieving good agreements in contexts like collaborative design. Conceding is not individually rational in the face of agents that may choose not to concede, but this problem can be resolved either by introducing a mediator that stochastically ignores agent preferences, or by introducing dice into the negotiation protocol. In both cases, the exchange of tokens when one agent overrides another can be used to incent the truthful voting that enables win-win outcomes.

8. Next Steps

There are many other promising avenues for future work in this area. The high social welfare achieved by our approach partially reflect the fact that the utility functions for each agent, based as they are solely on binary dependencies, are relatively easy to optimize. Higher-order dependencies, common in many real-world contexts, are known to generate more challenging utility landscapes [13]. We hypothesize that it may be necessary to adapt non-linear optimization techniques such as genetic algorithms into the negotiation context in order to address this challenge. Another possibility involves agents providing limited information about their utility functions to the mediator or to each other in order to facilitate more intelligent search through very large design spaces. Agents can, for example, tell the mediator which issues are heavily dependent upon each other, allowing the mediator to focus its attention within tightly-coupled issue 'clumps', leaving other less influential issues till later. We hypothesize that agents may be incented to tell the truth in order to ensure that negotiations can complete in an acceptable amount of time. Finally, we would like to derive formal incentive compatibility proofs (i.e. concerning when agents are incented to vote truthfully) for our protocols. New proof techniques will probably be necessary because previous results in this area have made strong assumptions concerning the shape of the agent utility functions that do not hold with complex agreements.

Acknowledgements

This work was supported by the DARPA Control of Agent-Based Systems (CoABS) program, and the NSF Computation and Social Systems program.

References

[1]   Faratin, P., C. Sierra, and N.R. Jennings, *Using similarity criteria to make negotiation trade-offs*. Proceedings Fourth International Conference on MultiAgent Systems. IEEE Comput. Soc., 2000.

[2]   Ehtamo, H., E. Ketteunen, and R. Hamalainen, *Searching for Joint Gains in Multi-Party Negotiations*. European Journal of Operational Research, 2001. 1(30): p. 54-69.

[3]   Bar-Yam, Y., *Dynamics of complex systems*. 1997, Reading, Mass.: Addison-Wesley. xvi, 848.

[4]   Raiffa, H., *The art and science of negotiation*. 1982, Cambridge, Mass.: Belknap Press of Harvard University Press. x, 373.

[5]   Rosenschein, J.S. and G. Zlotkin, *Rules of encounter : designing conventions for automated negotiation among computers*. Artificial intelligence. 1994, Cambridge, Mass.: MIT Press. xxi, 229.

[6]   Axelrod, R., *The Evolution Of Cooperation*. 1984: Basic Books.

[7]   Klein, M., P. Faratin, and Y. Bar-Yam, *Using an Annealing Mediator to Solve the Prisoners' Dilemma in the Negotiation of Complex Contracts*, in *Proeedings of the Agent-Mediated Electronic Commerce (AMEC-IV) Workshop*. 2002, Springer-Verlag.

[8]   Kowalczyk, R. and V. Bui, *On Constraint-Based Reasoning in e-Negotiation Agents*, in *Agent-Mediated Electronic Commerce III, Current Issues in Agent-Based Electronic Commerce*, F. Dignum and U. Cortes, Editors. 2001, Springer. p. 31 - 46.

[9]   Jonker, C.M. and J. Treur. *An Agent Architecture for Multi-Attribute Negotiation*. in *Proceedings of IJCAI-01*. 2001.

[10]   Fatima, S.S., M. Wooldridge, and N. R.Jennings. *MultiIssue Negotiation Under Time Constraints*. in *Proceedings of the 2002 International Conference on Autonomous Agents and Multi-Agent Systems (AAMAS-02)*. 2002. Bologna, Italy: ACM.

[11]   Kalagnanam, J. and D. Parkes, *Auctions, Bidding and Exchange Design*, in *Supply Chain Analysis in the eBusiness Area*, Simchi-Levi, Wu, and Shen, Editors. 2003, Kluwer Academic Publishers.

[12]   Bichler, M. and J. Kalagnanam, *Bidding Languages and Winner Determination in Multi-Attribute Auctions*. 2002.

[13]   Kauffman, S.A., *The origins of order: self-organization and selection in evolution*. 1993: Oxford University Press.

Chapter 12

Information Theory — The Bridge Connecting Bounded Rational Game Theory and Statistical Physics

David H. Wolpert

NASA Ames Research Center, Moffett Field, CA, 94035, USA

dhw@email.arc.nasa.gov

A long-running difficulty with conventional game theory has been how to modify it to accommodate the bounded rationality of all real-world players. A recurring issue in statistical physics is how best to approximate joint probability distributions with decoupled (and therefore far more tractable) distributions. This paper shows that the same information theoretic mathematical structure, known as Product Distribution (PD) theory, addresses both issues. In this, PD theory not only provides a principled formulation of bounded rationality and a set of new types of mean field theory in statistical physics. It also shows that those topics are fundamentally one and the same.

1 Introduction

In noncooperative game theory, one has a set of N players, each choosing its strategy x_i independently, by sampling a distribution $q_i(x_i)$ over those strategies. Each player i also has her own utility function $g_i(x)$, specifying how much reward she gets for every possible joint-strategy x of all N players. Let $q_{(i)}(x_{(i)})$ mean the joint probability distribution of all players other than i, i.e., $\prod_{j \neq i} q_j(x_j)$.

Then the "goal" of each player i is to set q_i to so that, conditioned on $q_{(i)}$, the expected value of i's utility is as high as possible.

Conventional game theory assumes each player i is "fully rational", able to solve for that optimal q_i, and that she then uses that distribution. It is primarily concerned with analyzing the such equilibria of the game [15, 7, 27, 4]. In the real world, this assumption of full rationality almost never holds, whether the players are humans, animals, or computational agents [6, 28, 24, 14, 3, 9, 2, 29, 19]. This is due to the cost of computation of that optimal distribution, if nothing else. This real-world **bounded rationality** is one of the major impediments to applying conventional game theory in the real world.

This paper shows how Shannon's information theory [11, 21, 18] provides a principled way to modify conventional game theory to accommodate bounded rationality. This is done by following information theory's prescription that, given only partial knowledge concerning the distributions the players are using, we should use the Maximum Entropy (Maxent) principle to infer those distributions. Doing so results in the principle that the bounded rational equilibrium is the minimizer of a certain set of coupled Lagrangian functions of the joint distribution, $q(x) = \prod_i q_i(x_i)$. This mathematical structure is a special instance of Product Distribution (PD) theory [31, 32, 22, 20, 8, 3].

In addition to showing how to formulate bounded rationality, PD theory provides many other advantages to game theory. Its formulation of bounded rationality explicitly includes a term that, in light of information theory, is naturally interpreted as a cost of computation. PD theory also seamlessly accommodates multiple utility functions per player. It also provides many powerful techniques for finding (bounded rational) equilibria, and helps address the issue of multiple equilibria. Another advantage is that by changing the coordinates of the underlying space of joint moves x, the same mathematics describes a type of bounded rational cooperative game theory, in which the moves of the players are transformed into contracts they all offer one another.

Perhaps the most succinct and principled way of deriving statistical physics is as the application of the Maxent principle. In this formulation, the problem of statistical physics is cast as how best to infer the probability distribution over a system's states when one's prior knowledge consists purely of the expectation values of certain functions of the system's state [17, 18]. For example, this prescription says we should infer that the probability distribution p governing the system is the Boltzmann distribution when our prior knowledge is the system's expected energy. This is known as the "canonical ensemble". Other ensembles arise when other expectation values are added to one's prior knowledge. In particular, if the number of particles in the system is uncertain, but one knows its expectation value, one arrives at the "grand canonical ensemble".

One major difficulty with working with these ensembles is that under them the particles of the system are statistically coupled with one another. For high-dimensional systems, this can make statistical physics calculations very difficult. Accordingly, a large body of work has been produced under the rubric of Mean Field (MF) theory, in which the ensemble is approximated with a distribution in which the particles are independent [26]. In an MF approximation, a product distribution q governs the joint state of the particles — just as a product distribution governs the joint strategy of the players in a game.

MF approximations are usually derived in an ad hoc manner. The principled way to derive a MF approximation (or any other kind) to a particular ensemble is to specify a distance measure saying how close two probability distributions are, and then solve for the q that is closest to the distribution being approximated, p. To do this one needs to specify the distance measure. How best to measure distances between probability distributions is a topic of ongoing controversy and research [33]. The most common way to do so is with the infinite limit log likelihood of data being generated by one distribution but misattributed to have come from the other. This is known as the Kullback-Leibler (KL) distance [11, 12, 21]. It is far from being a metric. In particular, it is not symmetric under interchange of the two distributions being compared.

It turns out that the simplest MF theories minimize the KL distance from q to p. However it can be argued it is the KL distance from p to q that is the most appropriate measure, not the KL distance from q to p. Using that distance, the optimal q is a new kind of approximation not usually considered in statistical physics.

For the canonical ensemble, the type of KL distance arising in simple MF theories turns out to be identical to the maxent Lagrangian arising in bounded rational game theory. This shows how bounded rational (independent) players are formally identical to the particles in the MF approximation to the canonical ensemble. Under this identification, the moves of the players play the roles of the states of the particles, and particle energies are translated into player utilities. The coordinate transformations which in game theory result in cooperative games are, in statistical physics, techniques for more allowing the canonical ensemble to be more accurately approximated with a product distribution.

This identification raises the potential of transferring some of the powerful mathematical techniques that have been developed in the statistical physics community to the analysis of noncooperative game theory. In also suggests translating some of the other ensembles of statistical physics to game theory, in addition to the canonical ensemble. As an example, in the grand canonical ensemble the number of particles is variable, which, after a MF approximation, corresponds to having a variable number of players in game theory. Among other applications, this provides us with a new framework for analyzing games in evolutionary scenarios, different from evolutionary game theory.

In the next section noncooperative game theory and information theory are cursorily reviewed. Then bounded rational game theory is derived, and its many advantages are discussed. The following section starts with a cursory review of the information-theoretic derivation of statistical physics. After that is a discussion of the two kinds of KL distance and the MF theories they induce, and a discussion of coordinate systems. This section also includes a discussion on translating a MF version of the grand canonical ensemble into a new kind of evolutionary game theory.

It should be noted that PD theory is a far-ranging framework for analyzing and controlling distributed systems, with potential applications extending far beyond game theory and statistical physics. In particular, it can be used for distributed optimization and adaptive, distributed control. It also has potential applications in high-dimensional probability sampling, high-dimensional integration, reinforcement learning, and multi-agent systems. See [31, 32, 22, 20, 1, 8] for preliminary investigations of a few of these.

2   PD theory as Bounded Rational Noncooperative Game Theory

This section motivates PD theory as a way of addressing several of the shortcomings of conventional noncooperative game theory.

2.1   Review of noncooperative game theory

In noncooperative game theory one has a set of N **players**. Each player i has its own set of allowed **pure strategies**. A **mixed strategy** is a distribution $q_i(x_i)$ over player i's possible pure strategies. Each player i also has a **utility function** g_i that maps the pure strategies adopted by all N of the players into the real numbers. So given mixed strategies of all the players, the expected utility of player i is $E(g_i) = \int dx \prod_j q_j(x_j) g_i(x)$ [1].

This basic framework can be elaborated to model many interactions between biological organisms, and in particular between human beings. These interactions range from simple abstractions like the famous prisoner's dilemma to iterated games like chess, to international relations [16, 15, 7].

Much of noncooperative game theory is concerned with **equilibrium concepts** specifying what joint-strategy one should expect to result from a particular game. In particular, in a **Nash equilibrium** every player adopts the mixed strategy that maximizes its expected utility, given the mixed strategies of the other players. More formally, $\forall i, q_i = \mathrm{argmax}_{q_i'} \int dx\, q_i' \prod_{j \neq i} q_j(x_j)\, g_i(x)$.

Several very rich fields have benefited from a close relationship with noncooperative game theory. Particular examples are evolutionary game theory (in which the set of N players is replaced by an infinite set of reproducing organisms) and cooperative game theory (in which players choose which **coalitions** of other players to join) [23, 4]. Game theory as a whole is also closely related to economics, in particular the field of mechanism design, which is concerned with how to induce the set of players to do adopt a socially desirable joint-strategy [34, 15, 25, 30].

---

[1]Throughout this paper, the integral sign will be interpreted in the appropriate measure-theoretic terms, e.g., as Lebesgue integrals, point-sums, etc.

2.2   Problems with conventional noncooperative game theory

A number of objections to the Nash equilibrium concept have been resolved. In particular, it was Nash who proved that every game has at least one Nash equilibrium if one expands the realm of discourse to include mixed strategies. (The same is not true for pure strategies.) Other objections have been more or less resolved through numerous **refinements** of the Nash equilibrium concept.

However there are several major problems with the concept that are still outstanding. One of them is the possible multiplicity of equilibria; this multiplicity means the Nash equilibrium concept cannot be used to specify the joint strategy that is actually adopted in a real world game. (Some refinements of the Nash equilibrium concept attempt to address this problem, though none has succeeded.) Another problem is that while calculating Nash equilibria is straightforward in many simple games (e.g., 2 players in a zero-sum game), calculating them in the general case can be a very difficult computational multi-criteria optimization problem. Yet another problem is that there is no general way to extend the concept to allow each player to have multiple utility functions.

However perhaps the major problem with the Nash equilibrium concept is its assumption of **full rationality**. This is the assumption that every player i can both calculate what the strategies $q_{j \neq i}$ will be and then calculate its associated optimal distribution. In other words, it is the assumption that every player will calculate the entire joint distribution $q(x) = \prod_j q_j(x_j)$. If for no other reasons than computational limitations of real humans, this assumption is essentially untenable. This problem is just as severe if one allows statistical coupling among the players [5, 15].

A large body of empirical lore has been generated characterizing the bounded rationality of humans. Similarly much has been learned about the empirical behavior of (bounded rational) machine learning computer algorithms playing games with one another [6, 2]. None of this work has resulted in a full mathematical theory of bounded rationality however.

There have also been numerous theoretical attempts to incorporate bounded rationality into noncooperative game theory by modifying the Nash equilibrium concept. Some of them assume essentially that every player's mixed strategy is its Nash-optimal strategy with some form of noise superimposed [4]. Others explicitly model the humans, typically as computationally limited automata, and assume the automata perform optimally subject to those computational limitations [14]. Both approaches, while providing insight, are very *ad hoc* as models of games involving real-world organisms or real-world (i.e., non-trivial) machine learning algorithms.

The difficulty of calculating equilibria is addressed in the sections below on solving for the distributions of PD theory. The rest of this section shows how information theory can be used to extend game theory to avoid its other shortcomings. Finally, the sections after this one present some other extensions of game theory, in particular to allow for a variable number of players. (Games with variable number of players arise in many biological scenarios as well as economic ones.)

2.3   Review of the maximum entropy principle

Shannon was the first person to realize that based on any of several separate sets of very simple desiderata, there is a unique real-valued quantification of the amount of syntactic information in a distribution $P(y)$. He showed that this amount of information is (the negative of) the Shannon entropy of that distribution, $S(P) = - \int dy\, P(y) ln[\frac{P(y)}{\mu(y)}]$ [2].

So for example, the distribution with minimal information is the one that doesn't distinguish at all between the various y, i.e., the uniform distribution. Conversely, the most informative distribution is the one that specifies a single possible y. Note that for a product distribution, entropy is additive, i.e., $S(\prod_i q_i(y_i)) = \sum_i S(q_i)$.

Say we given some incomplete prior knowledge about a distribution $P(y)$. How should one estimate $P(y)$ based on that prior knowledge? Shannon's result tells us how to do that in the most conservative way: have your estimate of $P(y)$ contain the minimal amount of extra information beyond that already contained in the prior knowledge about $P(y)$. Intuitively, this can be viewed as a version of Occam's razor. This approach is called the maximum entropy (maxent) principle. It has proven extremely useful in domains ranging from signal processing to image processing to supervised learning [21].

2.4   Maxent Lagrangians

Much of the work on equilibrium concepts in game theory adopts the perspective of an external observer of a game. We are told something concerning the game, e.g., its utility functions, information sets, etc., and from that wish to predict what joint strategy will be followed by real-world players of the game. Say that in addition to such information, we are told the expected utilities of the players. What is our best estimate of the distribution q that generated those expected utility values? By the maxent principle, it is the distribution with maximal entropy, subject to those expectation values.

To formalize this, for simplicity assume a finite number of players and of possible strategies for each player. To agree with the convention in other fields, from now on we implicitly flip the sign of each g_i so that the associated player i wants to minimize that function rather than maximize it. Intuitively, this flipped $g_i(x)$ is the "cost" to player i when the joint-strategy is x, rather than its utility then.

Then for prior knowledge that the expected utilities of the players are given by the set of values $\{\epsilon_i\}$, the maxent estimate of the associated q is given by the

---

[2] μ is an *a priori* measure over y, often interpreted as a prior probability distribution. Unless explicitly stated otherwise, in this paper we will always assume it is uniform, and not write it explicitly. See [17, 18, 11].

minimizer of the Lagrangian

$$L(q) \;\equiv\; \sum_i \beta_i[E_q(g_i) - \epsilon_i] - S(q)$$

$$= \;\sum_i \beta_i[\int dx \prod_j q_j(x_j)g_i(x) - \epsilon_i] - S(q) \tag{1}$$

where the subscript on the expectation value indicates that it evaluated under distribution q, and the $\{\beta_i\}$ are Lagrange parameters implicitly set by the constraints on the expected utilities [3].

Solving, we find that the mixed strategies minimizing the Lagrangian are related to each other via

$$q_i(x_i) \propto e^{-E_{q_{(i)}}(G|x_i)} \tag{2}$$

where the overall proportionality constant for each i is set by normalization, and $G \equiv \sum_i \beta_i g_i$ [4]. In Eq. 2 the probability of player i choosing pure strategy x_i depends on the effect of that choice on the utilities of the other players. This reflects the fact that our prior knowledge concerns all the players equally.

If we wish to focus only on the behavior of player i, it is appropriate to modify our prior knowledge. To see how to do this, first consider the case of maximal prior knowledge, in which we know the actual joint-strategy of the players, and therefore all of their expected costs. For this case, trivially, the maxent principle says we should "estimate" q as that joint-strategy (it being the q with maximal entropy that is consistent with our prior knowledge). The same conclusion holds if our prior knowledge also includes the expected cost of player i.

Now modify this maximal set of prior knowledge by removing from it specification of player i's strategy. So our prior knowledge is the mixed strategies of all players other than i, together with player i's expected cost. We can incorporate the prior knowledge of the other players' mixed strategies directly into our Lagrangian, without introducing Lagrange parameters. That **maxent Lagrangian** is

$$L_i(q_i) \;\equiv\; \beta_i[\epsilon_i - E(g_i)] - S_i(q_i)$$

$$= \;\beta_i[\epsilon_i - \int dx \prod_j q_j(x_j)g_i(x)] - S_i(q_i)$$

with solution given by a set of coupled **Boltzmann distributions**:

$$q_i(x_i) \propto e^{-\beta_i E_{q_{(i)}}(g_i|x_i)}. \tag{3}$$

Following Nash, we can use Brouwer's fixed point theorem to establish that for any non-negative values $\{\beta\}$, there must exist at least one product distribution

---

[3]Throughout this paper the terms in any Lagrangian that restrict distributions to the unit simplices are implicit. The other constraint needed for a Euclidean vector to be a valid probability distribution is that none of its components are negative. This will not need to be explicitly enforced in the Lagrangian here.

[4]The subscript $q_{(i)}$ on the expectation value indicates that it is evaluated according the distribution $\prod_{j \neq i} q_j$.

given by the product of these Boltzmann distributions (one term in the product for each i).

The first term in L_i is minimized by a perfectly rational player. The second term is minimized by a perfectly *irrational* player, i.e., by a perfectly uniform mixed strategy q_i. So β_i in the maxent Lagrangian explicitly specifies the balance between the rational and irrational behavior of the player. In particular, for $\beta \to \infty$, by minimizing the Lagrangians we recover the Nash equilibria of the game. More formally, in that limit the set of q that simultaneously minimize the Lagrangians is the same as the set of delta functions about the Nash equilibria of the game. The same is true for Eq. 2.

Eq. 2 is just a special case of Eq. 3, where all player's share the same cost function G. (Such games are known as **team games**.) This relationship reflects the fact that for this case, the difference between the maxent Lagrangian and the one in Eq. 1 is independent of q_i. Due to this relationship, our guarantee of the existence of a solution to the set of maxent Lagrangians implies the existence of a solution of the form Eq. 2.

Typically players aren't close to perfectly self-defeating. Almost always they will be closer to minimizing their expected cost than maximizing it. For prior knowledge consistent with such a case, the β_i are all non-negative.

Finally, our prior knowledge often will not consist of exact specification of the expected costs of the players, even if that knowledge arises from watching the players make their moves. Such other kinds of prior knowledge are addressed in several of the following subsections.

2.5 Alternative interpretations of Lagrangians

There are numerous alternative interpretations of these results. For example, change our prior knowledge to be the entropy of each player i's strategy, i.e., how unsure it is of what move to make. Now we cannot use information theory to make our estimate of q. Given that players try to minimize expected cost, a reasonable alternative is to predict that each player i's expected cost will be as small as possible, subject to that provided value of the entropy and the other players' strategies. The associated Lagrangians are $\alpha_i[S(q_i) - \sigma_i] - E(g_i)$, where σ_i is the provided entropy value. This is equivalent to the maxent Lagrangian, and in particular has the same solution, Eq. 3.

Another alternative interpretation involves **world cost** functions, which are quantifications of the quality of a joint pure strategy x from the point of view of an external observer (e.g., a system designer, the government, an auctioneer, etc.). A particular class of world cost functions are "social welfare functions", which can be expressed in terms of the cost functions of the individual players. Perhaps the simplest example is $G(x) = \sum_i \beta_i g_i(x)$, where the β_i serve to trade off how much we value one player's cost vs. anothers. If we know the value of this social welfare function, but nothing else, then maxent tells us to minimize the Lagrangian of Eq. 1.

2.6   Bounded rational game theory

In many situations we have prior knowledge different from (or in addition to) expected values of cost functions. This is particularly true when the players are human beings (so that behavioral economics studies can be brought to bear) or simple computational algorithms. To apply information theory in such situations, we simply need to incorporate that prior knowledge into our Lagrangian(s).

To give a simple example, say that we know that the players all want to ensure not just a low expected cost, but also that the actual cost doesn't vary too much from one sample of q to the next. We can formalize this by saying that in addition to expected costs, our prior knowledge includes variances in the costs. Given the expected values of the costs, such variances are specified by the expected values of the squares of the cost. Accordingly, all our prior knowledge is in the form of expectation values. Modifying Eq. 3 appropriately, we arrive at the solution

$$q_i(x_i) \propto e^{-E_{q_{(i)}}(\alpha_i(g_i - \lambda_i)^2 | x_i)}.$$

where the Lagrange parameters α_i and λ_i are given by the provided expectations and variances of the costs of the players.

Eq. 4 is our best guess for what the actual mixed strategy of player i is, in light of our prior knowledge concerning that player. Note that this formula directly reflects the fact that player i does not care only about minimizing cost, i.e., maximizing utility. In this, we are directly incorporating the possibility that the player violates the axioms of utility theory — something never allowed in conventional game theory. Other behavioral economics phenomena like risk aversion can be treated in a similar fashion.

A variant of this scenario would have our prior knowledge only give the variances of the costs of the players and not their expected costs. In this cost the Lagrangian must involve a term quadratic in q, in addition to the entropy term and a term linear in q. (See the subsection on multiple cost functions.) More generally, our prior knowledge can be any nonlinear function of q. In addition, even if we stick to prior knowledge that is linear in q, that knowledge can couple the cost functions of the players. For example, if we know that the expected difference in cost of players i and j is ϵ, the associated Lagrange constraint term is $\int dx q(x)[g_i(x) - g_j(x) - \epsilon]$. In this situation our prior knowledge couples the strategies of the players, even though those players are independent. See the discussion on constrained optimization in Sec. Opt.

2.7   Cost of computation

As mentioned above, bounded rationality is an unavoidable consequence of the cost of computation to player i of finding its optimal strategy. Unfortunately, one cannot simply incorporate that cost into g_i, and then presume that the player acts perfectly rationally for this new g_i. The reason is that this cost is associated with the entire distribution $q_i(x_i)$ that player i calculates; it not associated with some particular joint-strategy formed by sampling such a distribution.

How might we quantify the cost of calculating q_i? The natural approach is to use information theory. Indeed, that cost arises naturally in the bounded rationality formulation of game theory presented above. To see how, for each player i define

$$f_i(x, q_i(x_i)) \equiv \beta_i g_i(x) + \ln[q_i(x_i)].$$

Then we can write the maxent Lagrangian for player i as

$$L_i(q) = \int dx \; q(x) f_i(x, q_i(x_i)). \qquad (4)$$

Now in a bounded rational game every player sets its strategy to minimize its Lagrangian, given the strategies of the other players. In light of Eq. 4, this means that we *can* interpret each player in a bounded rational game as being perfectly rational for a cost function that incorporates its computational cost. To do so we simply need to expand the domain of "cost functions" to include probability values as well as joint moves.

Similar results hold for non-maxent Lagrangians. All that's needed is that we can write such a Lagrangian in the form of Eq. 4 for some appropriate function f_i.

2.8   Multiple cost functions per player

Say player i has several different cost functions $\{g_i^j\}$ and wants to choose a strategy that will do well at all of them. In the case of pure strategies we can simply define an aggregate function like $\max_j g_i^j(x)$ or $\sum_j [g_i^j(x)]^2$, and employ that in a conventional, single-cost-function-per-player game theoretic analysis. Player i will perform well according to such a function iff it performs well according to all of the constituent g_i^j.

One might think that for mixed strategies one could just "roll up" the cost functions and say that player i works to minimize an aggregate cost function $\frac{\sum_j g_i^j}{\sum_j 1}$. However especially when player i has many cost functions, it may be that performance according to one or more of the constituent cost functions is quite bad even though the performance according to this average function is good. Similarly, player i can have a low value of the expectation of the minimum of its cost functions, even though the minimum of the expected costs is quite high. More generally, we cannot ensure that $E_q(g_i^j) = \int dx \; g_i^j(x) q_i(x) q_{(i)}(x)$ has a good value for all j by appropriately defining an aggregate g_i. Instead, we must "redefine" expected cost.

Proceeding in analogy to the pure strategy solution, such a redefinition means that player i works to minimize an aggregate *expected* cost function like $\max_j E_q(g_i^j)$ or $\sum_j [E_q(g_i^j)]^2$. Formally, such functions are just Lagrangians of q. If we wish, we can modify them to incorporate bounded rationality, getting Lagrangians like $\sum_j \beta_j [E_q(g_i^j)]^2 + S(q_i)$, where the β_j determine the relative rationalities of player i according to its various cost functions.

These kinds of Lagrangians can also model the process of mechanism design, where there is an external designer who induces the players to adopt a desirable joint-strategy [15]. As an example, "desirable" sometimes means that no single player's expected cost is high. A system that meets this goal fairly well can be modeled with a Lagrangian involving terms like $\sum_i [E_q(g_i)]^2$.

2.9   Uniqueness of equilibria

In general there can be multiple solutions to either the coupled equations of Eq. 2 or those of Eq. 3. This is consistent with the possible multiplicity of Nash equilibria. However say we modify the Lagrangians to be defined for all possible p, not just those that are product distributions. For example the Lagrangian of Eq. 1 becomes

$$L(p) \equiv \sum_i \beta_i \left[\int dx \, g_i(x)p(x) - \epsilon_i \right] - S(p).$$

The first term in this Lagrangian is linear in p. Since entropy is a concave function of the Euclidean vector p over the unit simplex, this means that the overall Lagrangian is a convex function of p over the space of allowed p. This means there is a *unique* minimum of the Lagrangian over the space of all possible legal p. Furthermore, as mentioned previously, for finite β at least one of the derivatives of the Lagrangian is negative infinite at the border of the allowed region of p. This means that the unique minimum of the Lagrangian is interior to that region, i.e., is a legal probability distribution.

In general this optimal p will not be a product distribution, of course. Rather the strategy choices of the players are typically statistically coupled, under this p. Such coupling is very suggestive of various stochastic formulations of non-cooperative game theory. Coupling also arises in cooperative game theory, in which binding contracts couple the moves of the players [16, 4].

Similarly, as in proven in the appendix, the Lagrangian $L(p) = \beta \sum_i [E_p(g_i)]^2 - S(p)$ is concave over the manifold of legal p, assuming non-negative β. So the model of mechanism design introduced in Sec. 2.8 has a unique equilibrium — if we allow the players to be statistically coupled.

2.10   Rationality operators

Often our prior knowledge will not concern expected costs. In particular, this is usually true if our prior knowledge is provided to us before the game is played, rather than afterward. In such a situation, prior knowledge will more likely concern the "intelligences" of the players, i.e., how close they are to being rational. In particular, if we want our prior knowledge concerning player i to be relatively independent of what the other players do, we cannot use i's expected cost as our prior knowledge. Our prior knowledge will often concern how peaked i's mixed strategy is about whichever of its moves minimize its cost (or how peaked we can assume it to be), not the associated minimal cost values.

Formally, the problem faced by player i is how to set its mixed strategy $q_i(x_i)$ so as to maximize the expected value of its **effective cost function**, $E(g_i \mid x_i)$.

Generalizing, what we want is a *rationality* operator $R(U, p)$ that measures how peaked an arbitrary distribution $p(y)$ is about the minimizers of an arbitrary cost function $U(y)$, argmin$yU(y)$.

Formally, we make two requirements of R:

1. If $p(y) \propto e^{-\beta U(y)}$, for non-negative β, then it is natural to require that the peakedness of the distribution — its rationality value — is β.

2. We also need to also specify something of $R(U, p)$'s behavior for non-Boltzmann p. It will suffice to require that of the p satisfying $R(U, p) = \beta$, the one that has maximal entropy is proportional to $e^{-\beta U(y)}$. In other words, we require that the Boltzmann distribution maximizes entropy subject to a provided value of the rationality operator.

As an illustration, a natural choice for $R(U, p)$ would be the β of the Boltzmann distribution that "best fits" p. Information theory provides us such a measure for how well a distribution p_1 is fit by a distribution p_2. This is the **Kullback-Leibler distance** [11, 12]:

$$KL(p_1 \parallel p_2) \equiv S(p_1 \parallel p_2) - S(p_1) \qquad (5)$$

where $S(p_1 \parallel p_2) \equiv - \int dy\, p_1(y) \ln[\frac{p_2(y)}{\mu(y)}]$ is known as the **cross entropy** from p_1 to p_2 (and as usual we implicitly choose uniform μ). The KL distance is always non-negative, and equals zero iff its two arguments are identical.

Define $N(U) \equiv \int dy\, e^{-U(y)}$, the normalization constant for the distribution proportional to $e^{-U(y)}$. (This is called the **partition function** in statistical physics.) Then using the KL distance, we arrive at the rationality operator

$$R_{KL}(U, p) \equiv \text{argmin}_\beta KL(p \parallel \frac{e^{-\beta U}}{N(\beta U)})$$
$$= \text{argmin}_\beta[\beta \int dy\, p(y)U(y) + \ln(N(\beta U))].$$

In the appendix it is proven that R_{KL} respects the two requirements of rationality operators.

The quantity $\ln(N(\beta U))$ appearing in the second equation, when scaled by β^{-1}, is called the **free energy**. It is easy to verify that it equals the Lagrangian $E_p(U) - S(p)/\beta$ if p is given by the Boltzmann distribution $p(y) \propto e^{-\beta U(y)}$.

Say our prior knowledge is $\{\rho_i\}$, the rationalities of the players for their associated effective cost functions. Introduce the general notation

$$[U]_{i,p}(x_i) \equiv \int dx_{(i)} U(x_i, x_{(i)}) p(x_{(i)} \mid x_i),$$

so that $[g_i]_{i,q}$ is player i's effective cost function. Then the Lagrangian for our prior knowledge is

$$L(q) = \sum_i \lambda_i [R([g_i]_{i,q}, q_i) - \rho_i] - S(q). \qquad (6)$$

where the λ_i are the Lagrange parameters. Just as before, there is an alternative way to motivate this Lagangian: if our prior knowledge consists of the entropy of the joint system, and we assume each player will have maximal rationality subject to that prior knowledge, we are led to the Lagrangian of Eq. 6.

It is shown in the appendix that for the Kullback-Leibler rationality operator, we can replace any constraint of the form $R([g_i]_{i,q}, q_i) = \rho_i$ with $E_q(g_i) = \int dx \; g_i(x) \frac{e^{-\rho_i E(g_i | x_i)}}{N(\rho_i g_i)} q_{(i)}(x_{(i)})$. In other words, knowing that player i has KL rationality ρ_i is equivalent to knowing that the actual expected value of g_i equals the "ideal expected value", where q_i is replaced by the Boltzmann distribution of Eq. 3 with $\beta = \rho_i$. This contrasts with the prior knowledge underlying the Lagrangian in Eq. 1, in which we know the actual numerical value of $E_q(g_i)$.

Just as before, we can focus on player i by augmenting our prior knowledge to include the strategies of all the other players. The associated Lagrangian is

$$L_i(q_i) = \lambda_i [R([g_i]_{i,q}, q_i) - \rho_i] - S(q_i). \tag{7}$$

(The prior knowledge concerning the strategies of the other players is manifested in the effective cost function.) It is shown in the appendix that the set of all the Lagrangians in Eq. 7 (one for each player) are minimized simultaneously by any distribution of the form

$$q^g \equiv \frac{\prod_i e^{-\rho_i [g_i]_{i,q}}}{N(\rho_i [g_i]_{i,q})}$$

In addition, since this distribution obeys all the constraints in the Lagrangian in Eq. 6, we know that there exists a minimizer of that Lagrangian. All of this holds regardless of the precise rationality operator one uses.

Note that the Lagrangian L_i of Eq. 7 for player i arises in response to prior knowledge specific to player i. Changing from one player and its Lagrangian to another changes the prior knowledge. (The same is true for the Lagrangians in Eq. 3.) In contrast, the Lagrangian of Eq. 6 arises for a single unified body of prior knowledge, namely the set of all players' rationalities.

For that single body of knowledge, the equilibrium of the game is the solution to a *single*-objective optimization problem. This contrasts with the conventional formulation of full rationality game theory, where the equilibrium is cast as a solution to a multi-objective optimization problem (one objective per player). Furthermore, for finite β, at least one of the derivatives of the Lagrangian is negative infinite at the border of the allowed region of product distributions (i.e., at the border of the Cartesian product of unit simplices). Accordingly, all solutions lie in the interior of that region. This can be a big advantage for finding such solutions numerically, as elaborated below.

2.11   Semi-coordinate systems

Consider a multi-stage game like chess, with the stages (i.e., the instants at which one of the players makes a move) delineated by t. Now strategies are what are set by the players before play starts. So in such a multi-stage game the strategy of player i, x_i, must be the set of t-indexed maps taking what that player has observed in the stages $t' < t$ into its move at stage t. Formally, this set of maps is called player i's **normal form** strategy.

The joint strategy of the two players in chess sets their joint move-sequence, though in general the reverse need not be true. In addition, one can always find a joint strategy to result in any particular joint move-sequence. More generally, any onto mapping $\zeta : x \to z$ that need not be invertible is called a **semi-coordinate system**. The identity mapping $z \to z$ is a trivial example of a semi-coordinate system. Another example is the mapping from joint-strategies in a multi-stage game to joint move-sequences is an example of a semi-coordinate system. So changing the representation space of a multi-stage game from move-sequences z to strategies x is a semi-coordinate transformation of that game.

Typically there is overlap in what the players in chess have observed at stages preceding the current one. This means that even if the players' strategies are statistically independent, their move sequences are statistically coupled. In such a situation, by parameterizing the space of joint-move-sequences z with joint-strategies x, we shift our focus from the coupled distribution $P(z)$ to the decoupled product distribution, $q(x)$. This is the advantage of casting multi-stage games in terms of normal form strategies.

We can perform a semi-coordinate transformation even in a single-stage game. Say we restrict attention to distributions over spaces of possible x that are product distributions. Then changing $\zeta(.)$ from the identity map to some other function means that the players are no longer independent. After the transformation their strategy choices — the components of z — are statistically coupled, even though we are considering a product distribution.

Formally, this is expressed via the standard rule for transforming probabilities,

$$P_z(z) \equiv \zeta(P_x) \equiv \int dx P_x(x) \delta(z - \zeta(x)), \tag{8}$$

where $\zeta(.)$ is the mapping from x to z, and P_x and P_z are the distributions across x-space and z-space, respectively . To see what this rule means geometrically, let $\mathcal{P}$ be the space of all distributions (product or otherwise) over z's. Let $\mathcal{Q}$ be the space of all product distributions over x. Let $\zeta(\mathcal{Q})$ be its image in $\mathcal{P}$. Then by changing $\zeta(.)$, we change that image; different choices of $\zeta(.)$ will result in different manifolds $\zeta(\mathcal{Q})$.

As an example, say we have two players, with two possible strategies each. So z consists of the possible joint strategies, labeled $(1, 1), (1, 2), (2, 1)$ and $(2, 2)$. Have the space of possible x equal the space of possible z, and choose $\zeta(1, 1) = (1, 1)$, $\zeta(1, 2) = (2, 2)$, $\zeta(2, 1) = (2, 1)$, and $\zeta(2, 2) = (1, 2)$. Say that q is given by $q_1(x_1 = 1) = q_2(x_2 = 1) = 2/3$. Then the distribution over joint-strategies z is $P_z(1, 1) = P_x(1, 1) = 4/9$, $P_z(2, 1) = P_z(2, 2) = 2/9$, $P_z(1, 2) = 1/9$. So $P_z(z) \neq P_z(z_1)P_z(z_2)$; the strategies of the players are statistically coupled.

Such coupling of the players' strategies can be viewed as a manifestation of sets of potential binding contracts. To illustrate this return to our two player example. Each possible value of a component x_i determines a pair of possible joint strategies. For example, setting $x_1 = 1$ means the possible joint strategies are $(1,1)$ and $(2,2)$. Accordingly such a value of x_i can be viewed as a set of proffered binding contracts. The value of the other components of x determines which contract is accepted; it is the intersection of the proffered contracts offered by all the components of x that determines what single contract is selected. Continuing with our example, given that $x_1 = 1$, whether the joint-strategy is $(1,1)$ or $(2,2)$ (the two options offered by x_1) is determined by the value of x_2.

Binding contracts are a central component of cooperative game theory. In this sense, semi-coordinate transformations can be viewed as a way to convert noncooperative game theory into a form of cooperative game theory.

While the distribution over x uniquely sets the distribution over z, the reverse is not true. However so long as our Lagrangian directly concerns the distribution over x rather than the distribution over z, by minimizing that Lagrangian we set a distribution over z. In this way we can minimize a Lagrangian involving product distributions, even though the associated distribution in the ultimate space of interest is not a product distribution.

The Lagrangian we choose over x should depend on our prior information, as usual. If we want that Lagrangian to include an expected value over z's (e.g., of a cost function), we can directly incorporate that expectation value into the Lagrangian over x's, since expected values in x and z are identical: $\int dz P_z(z) A(z) = \int dx P_x(x) A(\zeta(x))$ for any function $A(z)$. (Indeed, this is the standard justification of the rule for transforming probabilities, Eq. 8.)

However other functionals of probability distributions can differ between the two spaces. This is especially common when $\zeta(.)$ is not invertible, so the space of possible x is larger than the space of possible z. For example, in general the entropy of a $q \in \mathcal{Q}$ will differ from that of its image, $\zeta(q) \in \zeta(\mathcal{Q})$ in such a case. (The prior probability μ in the definition of entropy only gives us invariance when the two spaces have the same cardinality.) A correction factor is necessary to relate the two entropies.

In such cases, we have to be careful about which space we use to formulate our Lagrangian. If we use the transformation $\zeta(.)$ as a tool to allow us to analyze bargaining games with binding contracts, then the direct space of interest is actually the x's (that is the place in which the players make their bargaining moves). In such cases it makes sense to apply all the analysis of the preceding sections exactly as it is written, concerning Lagrangians and distributions over x rather than z (so long as we redefine cost functions to implicitly pre-apply the mapping $\zeta(.)$ to their arguments). However if we instead use $\zeta(.)$ simply as a way of establishing statistical dependencies among the strategies of the players, it may make sense to include the entropy correction factor in our x-space Lagrangian.

An important special case is where the following three conditions are met: Each point z is the image under $\zeta(.)$ of the same number of points in x-space, n; $\mu(x)$ is uniform (and therefore so is $\mu(z)$); and the Lagrangian in x-space, L_x, is a sum of expected costs and the entropy. In this situation, consider a z-space

Lagrangian, L_z, whose functional dependence on P_z, the distribution over z's, is identical to the dependence of L_x on P_x, except that the entropy term is divided by n [5]. Now the minimizer $P^*(x)$ of L_x is a Boltzmann distribution in values of the cost function(s). Accordingly, for any z, $P^*(x)$ is uniform across all n points $x \in \zeta^{-1}(z)$ (all such x have the same cost value(s)). This in turn means that $S(\zeta(P_x)) = nS(P_z)$ So our two Lagrangians give the same solution, i.e., the "correction factor" for the entropy term is just multiplication by n.

2.12   Entropic prior game theory

Finally, it is worth noting that in the real world the information we are provided concerning the system often will not consist of *exact* values of functionals of q, be those values expected costs, rationalities, or what have you. Rather that knowledge will be in the form of data, D, together with an associated likelihood function over the space of q. For example, that knowledge might consist of a bias toward particular rationality values, rather than precisely specified values:

$$P(D \mid q) \propto e^{-\alpha \sum_i [R_{KL}([g_i]_{i,q}) - \rho_i]^2}.$$

where α sets the strength of the bias.

The extension of the maximum entropy principle to such situations uses the **entropic prior**, $P(q) \propto e^{-\gamma S(q)}$. Bayes' theorem is then invoked to get the posterior distribution [18]:

$$P(q \mid D) \propto e^{-\sum_i \alpha_i [R_{KL}([g_i]_{i,q}) - \rho_i]^2 - \gamma S(q)}.$$

The **Bayes optimal** estimate for q, under a quadratic penalty term, is then given by $E(q \mid D)$. The maxent principle for estimating q is given by this estimate under the limit of all α_i going to infinity. For finite α solving for $E(q \mid D)$ can be quite complicated though. For simplicity, such cases are not considered here.

3   PD theory and statistical physics

There are many connections between bounded rational game theory — PD theory — and statistical physics. This should not be too surprising, given that many of the important concepts in bounded rational game theory, like the Boltzmann distribution, the partition function, and free energy, were first explored in statistical physics. This section discusses some of these connections.

3.1   Background on statistical physics

Statistical physics is the physics of systems about which we have incomplete information. An example is knowing only the expected value of a system's energy

---

[5]For example, if $L_x(P_x) = \beta E_{P_x}(G(\zeta(.))) - S(P_x)$, then $L_z(P_z) = \beta E_{P_z}(G()) - S(P_z)/n$, where P_x and P_z are related as in Eq. 8.

(i.e., its temperature) rather than the precise value of the energy. The statistical physics of such systems is known as the **canonical ensemble**. Another example is the **grand canonical ensemble** (GCE). There the number of particles of various types in the system is also uncertain. As in the canonical ensemble, in the GCE what knowledge we do have takes the form of expectation values of the quantities about which we are uncertain, i.e., the number of particles of the various types that the system contains, and the energy the system.

Traditionally these kinds of ensembles were analyzed in terms of "baths" of the uncertain variable that are connected to the system. For example, in the canonical ensemble the system is connected to a heat bath. In the GCE the system is also connected to a bath of particles of the various types.

Such analysis showed that for the canonical ensemble the probability of the system being in the particular state y is given by the Boltzmann distribution over the associated value of the system's energy, $U(y)$, with β interpreted as the (inverse) temperature of the system: $p(y) \propto e^{-\beta U(y)}$. This result is independent of the details characteristics of the physical system; all that is important is the **Hamiltonian** $U(y)$, and temperature β.

Note that once one knows $p(y)$ and $U(y)$, one knows the expected energy of the system. It is $U(y)$ that is a fixed property of the system, whereas β can vary. Accordingly, specifying β is exactly equivalent to specifying the expected energy of the system.

In the case of the GCE, y implicitly specifies the number of particles of the various types, as well as their precise state. The analysis for that case showed that $p(y) \propto e^{-\beta U(y) - \sum_i \mu_i n_i}$. In this formula β is again the inverse temperature, n_i is the number of particles of type i, and $\mu_i > 0$ is the **chemical potential** of each particle of type i.

Jaynes was the first to show that these results of conventional statistical physics could be derived without recourse to artificial notions like "baths", simply by using the maxent principle. In particular, he used the exact reasoning in Sec. 2.6 to derive the fact that the canonical ensemble is governed by the Boltzmann distribution.

3.2   Mean field theory and PD theory

In practice it can be quite difficult to evaluate this Boltzmann distribution, due to difficulty in evaluating the partition function. For example, in a **spin glass**, y is an N-dimensional vector of bits, one per particle, and $U(y) = \sum_{i,j} H_{i,j} y_i y_j$. So the partition function is given by $\sum_y e^{-\sum_{i,j} H_{i,j} y_i y_j}$, where H is a symmetric real-valued matrix. In general, evaluating this sum for large numbers of spins cannot be done in closed form.

Mean Field (MF) theory is a technique for getting around this problem by approximating the partition function. Intuitively, it works by treating all the particles as independent. It does this by replacing some of the values of the state of a particle in the Hamiltonian by its average state. For example, in the case of the spin glass, one approximates $\sum_{i,j} H_{i,j} [y_i - E(y_i)][y_j - E(y_j)] \approx 0$, where the expectation values are evaluated according to the associated exact Boltzmann distribution, i.e., one assumes that fluctuations about the means are relatively negligible. This then means that

$$U(y) \cong \sum_{i,j} H_{i,j} 2y_i E(y_j) - \sum_{i,j} H_{i,j} E(y_i) E(y_j),$$

The second sum in this approximation cancels out when we evaluate the associated approximate Boltzmann distribution, leaving us with the distribution

$$p^{\beta U}(y) \quad \cong \quad P^{\beta U}(y) \equiv \frac{e^{-\beta \sum_{i,j} H_{i,j} 2y_i E(y_j)}}{\int dy \; e^{-\beta \sum_{i,j} H_{i,j} 2y_i E(y_j)}}$$

$$= \quad \prod_i \frac{e^{-\alpha_i y_i}}{\int dy_i \; e^{-\alpha_i y_i}},$$

where

$$\alpha_i \equiv 2\beta \sum_j H_{i,j} E(y_j).$$

Typically this approximation $P^{\beta U}(y)$ is far easier to work with than the exact Boltzmann distribution, $p^{\beta U}(y) = \frac{e^{-\beta U(y)}}{N(\beta U)}$, since each term in the product is for a single spin by itself. In particular, if we adopt this approximation we can use numerical techniques to solve the associated set of simultaneous equations

$$E(y_i) = \frac{\partial}{\partial \alpha_i} [\int dy_i \; e^{-\alpha_i y_i}] \quad \forall i$$

for the $E(y_i)$. Given those $E(y_i)$ values, we can then evaluate the associated approximate Boltzmann distribution explicitly.

The mean field approximation to the Boltzmann distribution is a product distribution, and in fact is identical to the product distribution q^g of bounded rational game theory, for the team game where $g_i(y) = 2\beta U(y) \; \forall i$. Accordingly, the "mean field theory" approximation for an arbitrary Hamiltonian U can be taken to be the associated team game q^g, which is defined for any U.

This bridge between bounded rational game theory and statistical physics means that many of the powerful tools that have been developed in statistical physics can be applied to bounded rational game theory. They also mean that PD theoretic techniques can be applied in statistical physics. In particular, it is shown elsewhere [32, 22] that if one replaces the identical cost function of each player in a team game with different cost functions, then the bounded rational equilibrium of that game can be numerically found far more quickly. In the context of statistical physics, this means that numerically solving for a MF approximation may be expedited by assigning a different Hamiltonian to each particle.

3.3 Information-theoretic misfit measures

The proper way to approximate a target distribution p with a distribution from a set $\mathcal{C}$ is to first specify a misfit measure saying how well each member of $\mathcal{C}$ approximates p, and then solve for the member with the smallest misfit. This is just as true when $\mathcal{C}$ is the set of all product distributions as when it is any other set.

How best to measure distances between probability distributions is a topic of ongoing controversy and research [33]. The most common way to do so is with the infinite limit log likelihood of data being generated by one distribution but misattributed to have come from the other. This is know as the **Kullback-Leibler distance** [11, 12, 21]:

$$KL(p_1 \parallel p_2) \equiv S(p_1 \parallel p_2) - S(p_1) \tag{9}$$

where $S(p_1 \parallel p_2) \equiv -\int dy\, p_1(y)\ln[\frac{p_2(y)}{\mu(y)}]$ is known as the **cross entropy** from p_1 to p_2 (and as usual we implicitly choose uniform μ). The KL distance is always non-negative, and equals zero iff its two arguments are identical. However it it is far from being a metric. In addition to violating the triangle inequality, it is not symmetric under interchange of its arguments, and in numerical applications has a tendency to blow up. (That happens whenever the support of p_1 includes points outside the support of p_2.)

Nonetheless, this is by far the most popular measure. It is illuminating to use it as our misfit measure. As shorthand, define the "pq distance" as $KL(p \parallel q)$, and the "qp distance" as $KL(q \parallel p$, where p is our target distribution and q is a product distribution. Then it is straightforward to show that the qp distance from q to target distribution $p^{\beta U}$ is just the maxent Lagrangian, up to irrelevant overall constants. In other words, the q minimizing the maxent Lagrangian — the distribution arising in MF theory — is the q with the minimal qp distance to the associated Boltzmann distribution.

However the qp distance is the (infinite limit of the negative log of) the likelihood that distribution p would attribute to data generated by distribution q. It can be argued that a better measure of how well q approximates p would be based on the likelihood that q attributes to data generated by p. This is the pq distance. Up to an overall additive constant (of the canonical distribution's entropy), the pq distance is

$$KL(p \parallel q) = -\sum_i \int dy\, p(y)\ln[q_i(y_i)].$$

This is equivalent to a team game where each coordinate i has the "Lagrangian"

$$L_i^*(q) \quad \equiv \quad -\int dy_i\, p_i(y_i)\ln[q_i(i)],$$

where $p_i(y_i)$ is the marginal distribution $\int dy_{(i)}p(y)$.

The minimizer of this is just $q_i = p_i\ \forall i$, i.e., each q_i is set to the associated marginal distribution of p. So in particular, when our target distribution is the canonical ensemble distribution $p^{\beta U}$, the optimal q according to pq distance is the set of marginals of $p^{\beta U}$. Note that unlike the solution for qp distance, here the solution for each q_i is independent of the $q_{(i)}$. So we don't have a game theory scenario; we do not need to pay attention to the $q_{(i)}$ when estimating each separate q_i.

Another difference between the two kinds of KL distance is how the associated optimal product distributions are typically calculated numerically. The product distribution that optimizes the maxent Lagrangian is usually found via derivative-based traversal of that Lagrangian, or techniques like (mixed) Brouwer updating[32, 22, 8, 20, 1]. In contrast, the integral giving each marginal distribution of p is usually found via adaptive importance sampling of the associated integral, with the proposal distribution for the integral to approximate p_i set adaptively, as $q_{(i)}$[32].

It is possible to motivate yet other choices for the q that best approximates $p^{\beta U}$. To derive one of them, we start with the following lemma, which extends the technique of Lagrange parameters to off-equilibrium points:

Lemma: Consider the set of all vectors leading from $x' \in \mathbb{R}^n$ that are, to first order, consistent with a set of constraints over $\mathbb{R}^n$. Of those vectors, the one giving the steepest descent of a function $V(x)$ is $\vec{u} = \nabla V + \sum_i \lambda_i \nabla f_i$, up to an overall proportionality constant, where the λ_i enforce the first order consistency conditions, $\vec{u} \cdot \nabla f_i = 0 \ \forall i$.

Choose $\mathbb{R}^n$ in the lemma to be the space of real-valued functions over the set of y's (so that n is the number of possible y). Have a single constraint f that restricts us to $\mathcal{P}$, the unit simplex in $\mathbb{R}^n$, i.e., that restricts us to the set of functions that (assuming they are nowhere-negative) are probability distributions. Choose V to be the associated Lagrangian, $L(p) = \beta E_p(U) - S(p)$, p being a point in our constrained submanifold of $\mathbb{R}^n$. Note that this p can be *any* distribution over the y's, including one that couples the components $\{y_i\}$.

Say we are at some current product distribution q. Then we can apply the lemma with the choices just outlined to tell us what direction to move from q in $\mathcal{P}$ so as to reduce the Lagrangian. In general, taking a step in that direction will result in a distribution p' that is not a product distribution. However we can solve for the product distribution that is closest to that p', and move to that product distribution. By iterating this procedure we can define a search over the submanifold of product distributions. We can then solve for the product distribution at which this search will terminate.

To do this, of course, we must define what we mean by "closest". Say that we choose to measure closeness by pq distance. Then the terminating production distribution is the one for which the marginals of $\nabla L + \lambda \nabla f$ all equal 0. For each i, this means that

$$q_i(y_i) \propto \exp\left(-\beta \frac{\int dy_{(i)} U(y)}{\int dy_{(i)} 1}\right). \tag{10}$$

This is akin to the q^g of a bounded rational game, except that each player/particle i sets its distribution by evaluating conditional expected U with a uniform distribution over the $y_{(i)}$, rather than with $q_{(i)}$.

3.4   Semi-coordinate transformations

Let's say there are numerical difficulties with our finding a q that is local minimization of the maxent Lagrangian. That q might still be a poor fit to $p(y)$ if it is far from the global minimizer of the Lagrangian. Furthermore, even the global minimizer might be a poor fit, if $p(y)$ simply can't be well-approximated by a product distribution.

There are many techniques for improving the fit of a product distribution to a target distribution in machine learning and statistics [12]. To give a simple example, say one wishes to approximate the target distribution in $\mathbb{R}^N$ with a product of Gaussians, one Gaussian for each coordinate. Even if the target distribution a Gaussian, if it is askew, then one won't be able to do a good job of approximating it with a product of Gaussians. However one can use Principal Components Analysis (PCA) to find how to rotate one's coordinates so that a product of Gaussians fits the target exactly.

Similar techniques can address both the issue of breaking free of local minima of the Lagrangian, and improving the accuracy of the best product distribution approximation to p. More precisely, identify y with the variables z discussed in Sec. 2.11. Then consider changing the map $\zeta(.) : x \to z$ from the identity map. This will in general change the mapping from P_x to $L_z(\zeta(P_x))$. So if L_z is the Lagrangian we are interested in, the mapping from product distributions over x can be changed by changing $\zeta(.)$, in general.

As an example, consider the case where the space of x's is identical to the space of z's, and consider all possible bijective transformations $\zeta(.)$. Entropy is the same in both spaces for any ζ, i.e., $S(P_z) = S(\zeta(P_x)) = S(P_x)$. So for fixed P_x, the entropy in z-space is independent of $\zeta(.)$. However if we fix P_x and change $\zeta(.)$ the expected values of utilities will change. So $L_z(\zeta(P_x))$ does depend on $\zeta(.)$, as claimed.

This means that by changing $\zeta(.)$ while leaving q_x unchanged, we will in general change whether we are at a local minimum of $L_z(\zeta(q_x))$. Furthermore, such a change will change how closely the global minimizer of $L_z(\zeta(q_x))$ approximates any particular target distribution. Indeed, some such transformation will always transform a team game to have a strictly convex maxent Lagrangian, with only one (bounded rational) equilibrium, an equilibrium that is in the interior of the region of allowed q and that has the lowest possible value of the Lagrangian. In the worst case, we can get this behavior by transforming to the semi-coordinate system in which x is one-dimensional, so that any $p(z)$ — coupling its variables or not — can be expressed as a $q(x) = q_1(x_1)$.

Note that unlike with PCA, semi-coordinate transformations can be used for non-Euclidean semi-coordinates (i.e., when neither x's nor z's are Euclidean vectors). They also can be guided by numerous measures of the goodness of fit to the target distribution (e.g., KL distance), in contrast to PCA's restriction to assuming a Gaussian likelihood.

3.5   Bounded rational game theory for variable number of players

The bridge between statistical physics and bounded rational game theory have many uses beyond the practical ones alluded to the previous subsection. In particular, it suggests extending bounded rational game theory to ensembles other than the canonical ensemble. As an example, in the GCE the number of particles of the various allowed types is uncertain and can vary. The bounded rational game theory version of that ensemble is a game in which the number of players of various types can vary.

We can illustrate this by extending a simple instance of evolutionary game theory [4] to incorporate bounded rationality and allow for a finite total number of players. Say we have a finite population of players, each of which has one of m' possible **types**. (These are sometimes called **feature vectors** in the literature.) Each player i in the population is randomly paired with a different player j, and they each choose a strategy for a two-person game. The set of strategies each of those players can choose among is fixed by its respective attribute vector. In addition the cost player i receives depends on the attribute vectors of itself and of j, in addition to their joint strategy. Finally, to reflect this dependence, we allow each player to vary its strategy depending on the attribute vector of its opponent; we call player i's **meta-strategy** the mapping from its opponent's attribute vector to i's strategy. [6].

We encode an instance of this scenario in an x with a countably infinite number of dimensions. $x_{i,0} \equiv n_i(x)$ specifies the number of players of type i, with $\vec{n}(x)$ being the vector of the number of players of all types. For $1 < j \leq x_{i,0}$, $x_{i,j} \equiv s_{i,j}(x)$ the meta-strategy selected by the j'th player of type i. If its opponent is the j'th player of type T', the cost to the i'th player of type T is $g_{T,i,T',j}(x) \equiv g_{T,i,T',j}(s, s', n_T, n_{T'})$, where s and s' are the two players' respective meta-strategy. To enforce consistency between the index numbers i, j and the associated numbers of players, we set $g_{T,i,T',j}(s, s', \vec{n}) = 0$ if either $i > n_T$ or $j > n_{T'}$.

To start we parallel the GCE, and presume that for each type we know the expected number of players having that type, and the expected cost averaged over all players having that type. Also stipulate that the distribution over x is a product distribution, q. Then our prior information specifies the values of

$$\sum_{k>0} k \, q_{T,0}(k) = \sum_{x_{T,0}} x_{T,0} \, q_{T,0}(x_{T,0})$$

and

---

[6]Note that it is trivial to replace meta-strategies with strategies throughout the analysis below: simply restrict attention to meta-strategies that do not vary with the opponent's attribute vector

$$\sum_{\vec{n}:n_T>0} q(\vec{n}) \sum_{T':n_{T'}>0} [\frac{n_{T'}}{\sum_{T''} n_{T''}}] \sum_{j,k} \int ds_T ds_{T'}$$

$$[1 - \delta_{T,T'} \delta_{j,k}] \frac{q_{T,j}(s_T) q_{T',k}(s_{T'}) g_{T,j,T',k}(s_T, s_{T'}, \vec{n})}{n_T n_{T'}}$$

$$=$$

$$\sum_{x_{1,0}} \cdots \sum_{x_{T,0}>0} \cdots \sum_{x_{m',0}} \sum_{T'} \sum_{j,k} \int dx_{T,j} dx_{T',k}$$

$$\{[1 - \delta_{T,T'} \delta_{j,k}] [\prod_{i=1}^{m'} q_{i,0}(x_{i,0})] \times$$

$$\frac{q_{T,j}(x_{T,j}) q_{T',k}(x_{T',k}) g_{T,j,T',k}(x)}{x_{T,0} \sum_{T''} x_{T'',0}} \}$$

respectively, for all types T. (The sums over j and k all implicitly extend from 1 to ∞, and the delta functions are Kronecker deltas that prevent a player from playing itself.)

We can write these expressions as expectation values, over x, of $2m'$ functions. These functions are the m' functions $n_T(x) = x_{T,0}$ (one function for each T) and the m' functions

$$c_T(x) \equiv \frac{\sum_{T',j,k}\{[1 - \delta_{T,T'} \delta_{j,k}] g_{T,j,T',k}(x)\}}{x_{T,0} \sum_{T''} x_{T'',0}} \Theta(x_{T,0})$$

respectively, where Θ is the Heaviside theta function that equals 1 if its argument exceeds 0, and equals 0 otherwise. Accordingly, the maxent principle directs us to minimize the Lagrangian

$$L(q) = - \sum_T [\mu_T(E(n_T) - N_T) + \beta_T(E(c_T) - C_T)] - S(q)$$

where the integers $\{N_T\}$ and real numbers $\{C_T\}$ are our prior information. In the usual way, the solution for each pair $(i \in \{1, \ldots, m'\}, j \geq 0)$ is

$$q_{i,j}(x_{i,j}) \propto e^{-E([\sum_{T'} \mu_{T'} n_{T'} - \beta_{T'} c_{T'}] \mid x_{i,j})},$$

where the values of the Lagrange parameters are all set by our prior information.

This distribution is analogous to the one in the GCE. As usual, one can consider variants of it by focusing on one variable at a time, having prior knowledge in the form of rationality values, etc. In addition, even if we stay in this random-2-player games scenario, there is no reason for us to restrict attention to prior information paralleling that of the GCE. As with bounded rational game theory with a fixed number of players, our prior information can concern nonlinear functions of q, couple the cost functions, etc.

In particular, in evolutionary game theory we do not know the expected number of players having each type, nor their average costs. In addition, the equilibrium concept stipulates that all players will have type T if a particular condition holds. That condition is that the addition of a player of type other than T to the population results in an expected cost to that added player that is greater than the associated expected cost to the players having type T. This provides a model of the phenotypic interactions underlying natural selection.

We can encapsulate evolutionary game theory in a Lagrangian by appropriately replacing each pair of GCE-type constraints (one pair for each type) with a single constraint. As an example, we could have the (single) constraint for type T be that

$$E(\frac{n_T}{\sum_{T'} n_{T'}}) = E([\frac{\max_{T'} c_{T'} - c_T}{\max_{T'}(c_{T'}) - \min_{T'}(c_{T'})}]^\gamma) \tag{11}$$

for some positive real value γ. For finite γ, the entropy term in the Lagrangian ensures that for no T is the expectation value in the lefthand side of this constraint exactly 0.

In the limit of infinite γ, the distribution minimizing this Lagrangian is non-infinitesimal only for the **evolutionarily stable strategies** of conventional evolutionary game theory. These are the (type, strategy) pairs that are best performing, in the sense that no other pair has a lower cost function value. The distribution for finite γ can be viewed as a "bounded rational" extension of conventional evolutionary game theory. In that extension (type, strategy) pairs are allowed even if they don't have the lowest possible cost, so long as their cost is close to the lowest possible [7].

There is always a solution to this Lagrangian (unlike the case in conventional full rationality evolutionary game theory). The technique of Lagrange parameters provides that solution for each pair $(i \in \{1, \ldots, m'\}, j \geq 0)$ in the usual way:

$$q_{i,j}(x_{i,j}) \propto e^{-E(\sum_{T'} \alpha_{T'} f_{T'}(x) \mid x_{i,j})}$$

where the Lagrange parameters enforce our constraint, and

$$f_{T'}(x) \equiv \frac{n_{T'}}{\sum_{T''} n_{T''}} - [\frac{\max_{T''} c_{T''} - c_{T''}}{\max_{T''}(c_{T''}) - \min_{T''}(c_{T''})}]^\gamma.$$

More general forms of evolutionary game theory allow games with more than two players, and localization via network structures delineating how players are likely to be grouped to play a game. Other elaborations have each player not know the exact attribute vectors of all its opponents, but only an "information structure" providing some information about those opponents' attribute vectors. All such extensions can be straightforwardly incorporated into the current analysis. Many other extensions are simple to make as well. For example, since the cost functions have all components of $\vec{n}$ in their argument lists, they can

---

[7]Many other parameterized constraints will result in this kind of relation between the parameter value and the resultant Lagrangian-minimizing distribution. The one in Eq. 11 was chosen simply for pedagogical clarity.

depend on the total size of the population. This allows us to model the effect on population size of finite environmental resources.

Note that if we change how we encode the number of players of the various types and their joint meta-strategy in x, we change the form of the expectations in Eq. 11. This reflects the fact that by changing the encoding we change the implication of using a product distribution. Formally, such a change in the encoding is a change in the semi-coordinate system. See Sec. 2.11.

4   Appendix

This appendix provides proofs absent from the main text.

4.1 $\beta \sum_i [E_p(g_i)]^2 - S(p)$ is concave over the unit simplex

Proof: Since $S(p)$ is convex over the unit simplex, and the unit simplex is a hyperplane, it suffices to prove that $\sum_i [E_p(g_i)]^2$ is concave over all of Euclidean space. Since a sum of concave functions is concave, we only need to prove that any single function of the form $[\int dx \, p(x)f(x)]^2$ is concave. The Hessian of this function is $2f(x)f(x')$. Rotate coordinates so that f is a basis vector, i.e., so that f is proportional to a delta function. This doesn't change the value of the determinant of our Hessian. After this change though, we can read off that the determinant is non-negative. **QED**

4.2 R_{KL} is a rationality operator

Proof: Since KL distance only equals 0 when its arguments match and is never negative, requirement (1) of rationality operators holds for R_{KL}. Next, since $R_{KL} = \mathrm{argmin}_\beta [\beta \int dy \, p(y)U(y) + \ln(N(\beta U))]$, we know that $E_p(U) = -\frac{1}{N(\beta U)} \frac{\partial N(\beta U)}{\partial \beta}|_{\beta = R_{KL}(U,p)}$. Accordingly, all p with the same rationality have the same expected value $E_p(U)$. Using the technique of Lagrange parameters then readily establishes that of those distributions having the same expected U, the one with maximal entropy is a Boltzmann distribution. Furthermore, by requirement (1), we know that for a Boltzmann distribution the exponent β must equal the rationality of that distribution. **QED**

4.3   Alternative form of a constraint on R_{KL}

Proof: Let $f\{\alpha, v\}$ be any function that is monotonically decreasing in its (real-valued) first argument. Then any constraint $R([g_i]_{i,q}, q_i) - \rho_i = 0$ is satisfied iff the constraint $f\{R([g_i]_{i,q}, q_i), q_{(i)}\} - f\{\rho_i, q_{(i)}\} = 0$ is satisfied. Choose

$$
\begin{aligned}
f\{\alpha, q_{(i)}\} &= -\frac{\partial \ln(N(\beta[g_i]_{i,q}))}{\partial \beta}|_{\beta = \alpha} \\
&= \frac{\int dx_i [g_i]_{i,q} e^{-\alpha[g_i]_{i,q}(x_i)}}{N(\alpha[g_i]_{i,q})}.
\end{aligned}
$$

Differentiating this quantity with respect to α gives the negative of the variance of $[g_i]_{i,q}$ under the Boltzmann distribution $\frac{e^{-\alpha[g_i]_{i,q}}}{N(\alpha[g_i]_{i,q})}$. Since variances are non-negative, this derivative is non-positive, which establishes that f is monotonically decreasing in its first argument.

Evaluating,

$$f\{\rho_i, q_{(i)}\} = \int dx\, g_i(x) \frac{e^{-\rho_i E(g_i|x_i)}}{N_{(\rho_i g_i)}} q_{(i)}(x_{(i)}).$$

In addition, from the equation defining R_{KL}, we know that

$$-\frac{\ln(N(\beta U(x_i)))}{\partial \beta}|_{\beta=R_{KL}(U,q_i)} = \int dx_i q_i(x_i) U(x_i)$$

for any function U. Plugging in $U = [g_i]_{i,q}$, we see that

$$\begin{aligned} f\{R([g_i]_{i,q}, q_i), q_{(i)}) &= \int dx_i q_i(x_i)[g_i]_{i,q}(x_i) \\ &= E_q(g_i). \end{aligned}$$

QED

4.4 q^g minimizes the Lagrangians of Eq. 7

Proof: Following Nash, we can use Brouwer's fixed point theorem to establish that for any non-negative $\{\rho_i\}$, there must exist at least one product distribution given by q^g. The constraint term in all the L_i of Eq. 7 is zero for this distribution. By requirement (2), we also know that given $q^g_{(i)}$ (and therefore $[g_i]_{i,q^g}$), there is no q_i with rationality ρ_i that has lower entropy than q^g_i. Accordingly, no q_i will have a lower value of L_i. Since this holds for all i, q^g minimizes all the Lagrangians in Eq. 7 simultaneously. **QED**

4.5   Constrained steepest descent is along $\nabla V + \sum_i \lambda_i \nabla f_i$

Proof: Consider the set of $\vec{u}$ such that the directional derivatives $D_{\vec{u}} f_i$ evaluated at x' all equal 0. These are the directions consistent with our constraints to first order. We need to find the one of those $\vec{u}$ such that $D_{\vec{u}} g$ evaluated at x' is minimal.

To simplify the analysis we introduce the constraint that $|\vec{u}| = 1$. This means that the directional derivative $D_{\vec{u}} V$ for any function V is just $\vec{u} \cdot \nabla V$. We then use Lagrange parameters to solve our problem. Our constraints on $\vec{u}$ are $\sum_j u_j^2 = 1$ and $D_{\vec{u}} f_i(x') = \vec{u} \cdot \nabla f_i(x') = 0 \ \forall i$. Our objective function is $D_{\vec{u}} V(x') = \vec{u} \cdot \nabla V(x')$.

Differentiating the Lagrangian gives

$$2\lambda_0 u_i + \sum_i \lambda_i \nabla f = \nabla V \ \ \forall i.$$

with solution

$$u_i = \frac{\nabla V - \sum_i \lambda_i \nabla f}{2\lambda_0}.$$

λ_0 enforces our constraint on $|\vec{u}|$. Since we are only interested in specifying $\vec{u}$ up to a proportionality constant, we can set $2\lambda_0 = 1$. Redefining the Lagrange parameters by multiplying them by -1 then gives the result claimed. **QED.**

References

[1] AIRIAU, S., and D. H. WOLPERT, "Product distribution theory and semi-coordinate transformations", Submitted to AAMAS 04.

[2] AL-NAJJAR, N. I., and R. SMORODINSKY, "Large nonanonymous repeated games", *Game and Economic Behavior* **37**, 26-39 (2001).

[3] ARTHUR, W. B., "Complexity in economic theory: Inductive reasoning and bounded rationality", *The American Economic Review* **84**, 2 (May 1994), 406–411.

[4] AUMANN, R.J., and S. HART, *Handbook of Game Theory with Economic Applications*, North-Holland Press (1992).

[5] AUMANN, R. J., "Correlated equilibrium as an expression of Bayesian rationality", *Econometrica* **55**, 1 (1987), 1–18.

[6] AXELROD, R., *The Evolution of Cooperation*, Basic Books NY (1984).

[7] BASAR, T., and G.J. OLSDER, *Dynamic Noncooperative Game Theory*, Siam Philadelphia, PA (1999), Second Edition.

[8] BIENIAWSKI, S., and D. H. WOLPERT, "Adaptive, distributed control of constrained multi-agent systems", Submitted to AAMAS 04.

[9] BOUTILIER, C., Y. SHOHAM, and M. P. WELLMAN, "Editorial: Economic principles of multi-agent systems", *Artificial Intelligence Journal* **94** (1997), 1–6.

[10] CHALLET, D., and N. F. JOHNSON, "Optimal combinations of imperfect objects", *Phys. Rev. Let.* **89** (2002), 028701.

[11] COVER, T., and J. THOMAS, *Elements of Information Theory*, Wiley-Interscience New York (1991).

[12] DUDA, R. O., P. E. HART, and D. G. STORK, *Pattern Classification (2nd ed.)*, Wiley and Sons (2000).

[13] ET AL., G. Korniss, *Science* **299** (2003), 677.

[14] FUDENBERG, D., and D. K. LEVINE, *The Theory of Learning in Games*, MIT Press Cambridge, MA (1998).

[15] FUDENBERG, D., and J. TIROLE, *Game Theory*, MIT Press Cambridge, MA (1991).

[16] GREIF, A., "Economic history and game theory: A survey", *Handbook of Game Theory with Economic Applications*, (R. J. AUMANN AND S. HART eds.) vol. 3. North Holland Amsterdam (1999).

[17] JAYNES, E. T., "Information theory and statistical mechanics", *Physical Review* 106 (1957), 620.

[18] JAYNES, E. T., and G. Larry BRETTHORST, *Probability Theory : The Logic of Science*, Cambridge University Press (2003).

[19] KAHNEMAN, D., "A psychological perspective on economics", *American Economic Review (Proceedings)* 93:2 (2003), 162–168.

[20] LEE, C. Fan, and D. H. WOLPERT, "Product distribution theory and semi-coordinate transformations", Submitted to AAMAS 04.

[21] MACKAY, D., *Information theory, inference, and learning algorithms*, Cambridge University Press (2003).

[22] MACREADY, W., S. BIENIAWSKI, and D.H. WOLPERT, "Adaptive multi-agent systems for constrained optimization", Submitted to AAAI 04.

[23] MESTERTON-GIBBONS, M., and E. S. ADAMS, "Animal contests as evolutionary games", *American Scientist* 86 (1998), 334–341.

[24] NEYMAN, A., "Bounded complexity justifies cooperation in the finitely repeated prisoner's dilemma", *Economics Letters* 19 (1985), 227–230.

[25] NISAN, N., and A. RONEN, "Algorithmic mechanism design", *Games and Economic Behavior* 35 (2001), 166–196.

[26] OPPER, M., and D. SAAD, *Advanced Mean Field Methods: Theory and Practice (Neural Information Processing)*, MIT Press (2001).

[27] OSBORNE, M., and A. RUBENSTEIN, *A Course in Game Theory*, MIT Press Cambridge, MA (1994).

[28] SANDHOLM, T., and V. R. LESSER, "Coalitions among computationally bounded agents", *Artificial Intelligence* 94 (1997), 99–137.

[29] TVERSKY, A., and D. KAHNEMAN, "Advances in prospect theory: Cumulative representation of uncertainty", *Journal of Risk and Uncertainty* 5 (1992), 297–323.

[30] WOLPERT, D., and K. TUMER, "Beyond mechanism design", *International Congress of Mathematicians 2002 Proceedings* (H. G. ET AL. ed.), Qingdao Publishing (2002).

[31] WOLPERT, D. H., "Factoring a canonical ensemble", cond-mat/0307630.

[32] WOLPERT, D. H., "Generalizing mean field theory for distributed optimization and control", Submitted.

[33] WOLPERT, David H., and William MACREADY, "Metrics for sets of more than two points", *Proceedings of the International Conference on Complex Systems, 2004*, (2004), in press.

[34] ZLOTKIN, G., and J. S. ROSENSCHEIN, "Coalition, cryptography, and stability: Mechanisms for coalition formation in task oriented domains", (preprint) (1999).

Chapter 13

Engineering Amorphous Systems, Using Global-to-Local Compilation

Radhika Nagpal

Department of Systems Biology, Harvard Medical School

rad@eecs.harvard.edu

Emerging technologies are making it possible to assemble systems that incorporate myriad of information-processing units at almost no cost: smart materials, self-assembling structures, vast sensor networks, pervasive computing. How does one engineer robust and prespecified global behavior from the local interactions of immense numbers of unreliable parts? We discuss organizing principles and programming methodologies that have emerged from Amorphous Computing research, that allow us to compile a specification of global behavior into a robust program for local behavior.

1   Introduction

Over the next few decades, emerging technologies will make it possible to assemble systems that incorporate myriad of information-processing units at almost no cost. Microelectronic mechanical components have become so inexpensive to manufacture that we can anticipate combining logic circuits, microsensors, actuators, and communications devices, integrated on the same tiny chip to produce particles that could be mixed with bulk materials, such as paints, gels, and concrete. Imagine coating bridges or buildings with smart paint that can sense and report on traffic and wind loads and monitor structural integrity of the bridge. A robot, built of millions of tiny programmable modules, could assemble itself into different shapes, perhaps as a cube for storage and then reconfiguring into a shelter or tool as needed. Already many such novel applications are being envisioned and built [3, 4, 9]. Emerging research in biocomputing will make it possible to harness the many sensors and actuators in cells and program biological cells to function as drug delivery vehicles or chemical factories for the assembly of nanoscale structures [22]. Pervasive computing and sensor networks are creating massive distributed systems at a different scale, from remote habitat monitoring to smart buildings and smart cars [20, 13].

These novel computational environments pose significant challenges, beyond just the manufacturing of parts. Exploiting these new technologies will require tackling two major challenges:

1. How does one engineer robust behavior from immense numbers of unreliable parts?

2. How does one translate *prespecified system-level goals* into the local interactions of vast numbers of identically-programmed parts?

While we can envision producing and deploying vast quantities of individual computing elements, we have few ideas for programming them effectively, without relying on reliable components, precise arrangements of parts and interconnects, and human intervention. The difficulty with engineering such complex systems is that it is often hard to predict how the individual-level activities will interact with each other and whether or not the desired global outcome will be achieved. Hints for how to design robust, complex and predictable collective behavior may come from natural systems, such as biology. Hints for how to specify global goals and manage complexity may come from computer science and other disciplines.

The research in Amorphous Computing has been centered around developing programming methodologies for systems composed of vast numbers of identically-programmed agents[2]. In this article we review some of the work on engineering pattern-formation and self-assembly, using metaphors inspired by biology. We describe two organizing principles that have emerged from this work. First is a common set of well-characterized primitives for achieving robust behavior from locally-interacting agents, inspired by the development in multicellular organisms. Second is constructive programming languages for specifying global goals. Together these allow us to achieve *global-to-local compilation* in the following way: the constructive languages specify complex goals as being constructed from simpler parts, using a fixed set of construction rules. These rules are mapped into local agent behavior using combinations of the primitive behaviors. A global specification of a goal is then *compiled* into a local agent program, using this mapping.

Global-to-local compilation confers many advantages: (1) We can reason about the classes of global shapes and patterns that can and cannot be generated, by analyzing the expressiveness of the language. (2) The local primitives can be made robust by relying on mechanisms inspired by biological systems. (3) The analysis of a complex system becomes tractable because it is built in understood ways from smaller parts. (4) The high level language makes it possible to easily specify complex behavior by programming at the conceptual level, without worrying about the millions of parts that are involved. While the specific examples presented here have to do with pattern and self-assembly, we believe that the same methodology and principles can apply to designing other large multi-agent systems such as sensor networks, distributed robots, and smart materials.

The rest of the article is organized as follows: Section 2 describes the assumptions of the amorphous computing model. Section 3 presents some of the primitives for robust local behavior. Section 4 reviews three amorphous computing languages that have been developed to specify pattern and shape formation and describes how global-to-local compilation is achieved. Section 5 concludes with a discussion of how such an approach can be used to design other complex systems.

2 The Amorphous Computing Model

An amorphous computer consists of massive numbers of identically-programmed and locally-interacting computing agents, embedded in space. We can model this as a collection of "computational particles" sprinkled randomly on a surface or mixed throughout a volume. Some of the characteristics of an amorphous computer are:

- The agents are all programmed identically. Each agent executes its program autonomously and has means for storing local state and generating random numbers.

- Each agent can communicate with a only a few nearby neighbors. We assume that there is a communication radius r, which is large compared with size of individual agents and small compared with the size of the entire area, and that two agents can communicate if they are within distance r.

- The agents are not synchronized, although we assume that they compute at similar speeds and are fabricated by the same process.

- The agents have no a priori knowledge of global position or orientation; however some agents may be started in a special initial state. In general, we assume that access to centralized sources of information is limited and agents must self-organize global information as necessary.

- The agents are possibly faulty, and can die or be replaced at any moment.

- The agents are immobile and we do not have precise control over their placement. In many of the examples here we assume that they are randomly distributed on a two-dimensional plane.

- The agents sense and affect the environment locally.

The massively parallel nature of an amorphous computer resembles, and takes inspiration from, models such as cellular automata. However it relaxes some restrictions such as precise geometric arrangement and synchronicity, while allowing the agents to store more state. The model corresponds to our expectations of the capabilities of individual computing elements and of the constraints of the application environments.

3 Primitives for Robust Local Behavior

The amorphous computing work has relied on a set of simple primitives for local organization, many of which have been inspired by morphogenesis and cell differentiation during the development of multi-cellular organisms [23]. By primitives we mean simple ways in which an agent interacts with its local neighborhood in order to create non-local behavior. Complexity arises by combining these primitives in controlled ways. Here we describe some of the primitives that have been used and analyzed.

Morphogen Gradients: One example of a mechanism common throughout development is the use of gradients of morphogens to determine positional information and polarity. For example, in the *Drosophila* embryo, cells at one end of the embryo emit a morphogen (protein) that diffuses along the length of the embryo. The concentration of this morphogen is used by other undifferentiated cells to determine whether they lie in the head, thorax or abdominal regions [12]. Different morphogens are used in determining the dorsal-ventral axis, in wing and limb development, and even in leg bristle polarity.

We can emulate the concept of a morphogen gradient using a simple agent program. An initial "source" agent, chosen by a cue from the environment or by generating a random value, creates a gradient by sending a message to its local neighborhood with the morphogen name and a value of zero. The neighboring agents forward the message to their neighbors with the value incremented by one and so on, until the morphogen has propagated through the entire population. Each agent stores and forwards only the minimum value it has heard for a particular morphogen name, thus the morphogen value represents the shortest path from the source. The value provides an estimate of distance from the source: a point reached in n steps will be roughly distance nr away. The quality of this estimate depends on the density of the agents and can be theoretically predicted for random distributions [15].

This very simple program can be used in powerful ways. Gradients can be used to create regions of a given size, to provide local orientation towards the source, and to provide routes around obstacles or regions of inactive agents. Gradients are the basis of many different amorphous computing algorithms, but it is also common throughout distributed robotics and sensor network research [13, 7].

Neighborhood Query and Selection: This primitive allows an agent to query its local neighborhood and collect information about their state. For example an agent may collect neighboring values of a gradient for comparison and use this to determine which of its neighbors is closer to the source. This primitive is similar to the cellular automata model of interaction [21]. An agent can also select a particular neighbor by sending a message to that neighbor. However because agents may die and messages may be lost, there is no guarantee that an agent will have received all its neighbors states or that the selection will always work. Instead these primitives are implemented so as to provide correct behavior with high probability.

Local Competition: In many cases there is a need to select a single agent from a region of agents, and this can be done spontaneously through a local process. Each agent picks a random number and counts down. If it reaches zero then it sends a message to inhibit all its neighbors from continuing to compete. If there is a criteria or merit involved, then that can be used to pick the random number such that agents with higher merit choose lower numbers. This distributed leader election process is appropriate for breaking symmetry within small regions and can be analyzed using traditional distributed algorithm techniques [17].

Local Monitoring: This is a more general primitive that states that each agent routinely sends a message to its neighbors indicating that it is alive. This primitive operates in parallel to primitives discussed before and implicitly affects the behavior of an agent by altering its perceived local neighborhood. Agents can also react directly to the disappearance of a particular neighbor.

Contact Neighborhoods: When agents are not fixed on a two dimensional surface but able to move and come into contact with other agents, then the agents in contact seamless become part of each others neighborhood. This primitive also implicitly affects all of the prior primitives by altering the local neighborhood.

These primitives represent building blocks that can then be put together to create more complex agent behavior. These primitives are well-matched to the amorphous setting because they are insensitive to the precise arrangement of the individual agents, so long as the distribution is reasonably dense. If individual agents do not function, or stop broadcasting, the result will not change very much, so long as there are sufficiently many agents. At the same time, it is possible to analyze the behavior of these algorithms, so that we have a solid ground to build on top of. For example, the morphogen algorithm can be thought of as computing a breath first search tree, and the spatial locality of communication gives us a relation from tree depth to distance. These primitives aim to provide a *good-enough answer* with high probability, rather than a perfect answer. This allows the agent behavior to be simple, scalable, and tolerant to variation.

4   Programming Languages for Global Behavior

While biology may provide a means for thinking about organizing local behavior robustly, computer science can provide tools for managing complexity. One such tool is a *programming language*. The ability to think and describe goals in terms of high-level abstractions, make possible a complexity that is almost inconceivable to generate by manipulating 1s and 0s. Yet the final computation does happen as bits, and the *compiler* translates from a language that is natural for expressing how to do something, to a low-level execution model that a computer can interpret [1].

In the amorphous computing setting the goal is similar — we would like to be able to translate complex global goals into local behavior, but in such a way that the translation is not mysterious and not hand crafted for each goal — in other words, a global-to-local compiler. In this section we describe three different amorphous computing languages aimed at pattern formation and self-assembly. In each case the desired goal is described in terms of abstract entities and then compiled to produce the behavior of an agent, such that the identically-programmed agents organize into the prespecified goal. At the end of the section we compare the global and local strategies used by these three systems.

4.1   Growing Point Language

Coore developed a language for forming topological patterns on an amorphous computer on a 2D surface[5]. The growing point language (GPL) can be used to specify topological patterns consisting of lines of various thickness, such as those specifying the interconnect of an electronic circuit. The language represents pattern formation in terms of a botanical metaphor of "growing points" and pheromones. A pheromone is a morphogen with a limited range. A growing point is a locus of activity that modifies the states of agents as it passes through, and it can respond to the gradient of a morphogen by moving towards lower, higher or similar values of morphogens. A pattern is created by writing a program in terms of *abstract entities*: growing points that lay down materials, materials that secrete pheromones, and tropisms that govern the trajectory of the growing point.

The specification is then compiled into an agent program. Initially the agents start out with identical state except for a few agents. As a result of executing the program, the agents "differentiate" into components of the pattern. At the level of the agent, the abstract entities translate to simple local rules. For example, a growing point is simply a piece of state at an agent. The agent collects values of morphogens from its neighbors and uses those value to locally compute which neighboring agent to pass the growing point to. The next agent then repeats the same process to determine where to send the growing point next. Materials are state bits set within an agent.

Figure 1 shows a fragment of a program written in the growing-point language: A growing point process called `make-red-line`, takes one parameter called `length`. This growing point "grows" material called `red-poly` in a band

```
(define-growing-point (make-red-line length)
  (material red-poly)
  (size 1)
  (tropism (and (away-from red-pheromone)
                (keep-constant pheromone-1)
                (keep-constant pheromone-2)))
  (actions
   (secrete 2 red-pheromone)
   (when ((< length 1) (terminate))
         (default
           (propagate (- length 1))))))
```

Figure 1: A program in GPL. This procedure generates a line of specified length that attempts to follow constant values of pheromones 1 and 2.

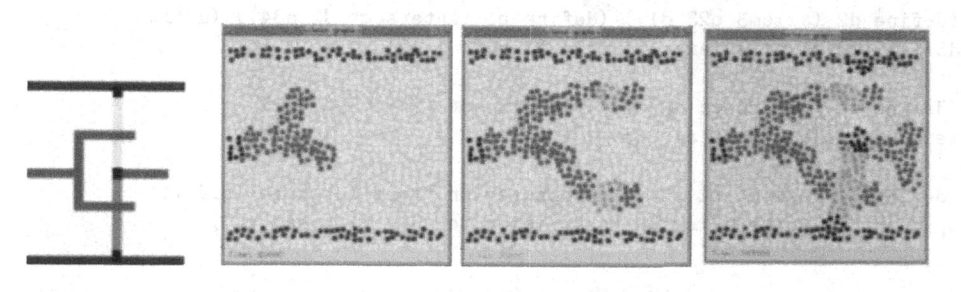

Figure 2: The amorphous surface differentiating to create the inverter pattern. All agents execute the same program, which is compiled directly from the GPL specification of the pattern on the left.

of size 1. This implies that each agent it moves through sets a state bit that will identify the agent as `red-poly`. The growing point moves according to a tropism that directs it away from higher concentrations of `red-pheromone`, in such a way that the concentrations of `pheromone-1` and `pheromone-2` are kept constant. All agents that are `red-poly` secrete `red-pheromone`; consequently, the growing point will tend to move away from the material it has already laid down. The growing point stops when the correct length line has been grown. This procedure is part of a larger GPL program that generates the pattern on the right. This pattern is a caricature of the layout of a CMOS inverter, where the different colored regions represent structures in the different layers of standard CMOS technology: metal, polysilicon and diffusion. Figure 2 shows the agents differentiating to create the inverter pattern. The agents that are part of the top blue rail emit pheromone 1 and the bottom rail emits pheromone 2, thus the code fragment represents the method by which the first red line is drawn parallel to these two rails and away from the edge. The entire program that specifies the shape is only a few paragraphs long, and the resulting state machine for the individual agents requires only about twenty states.

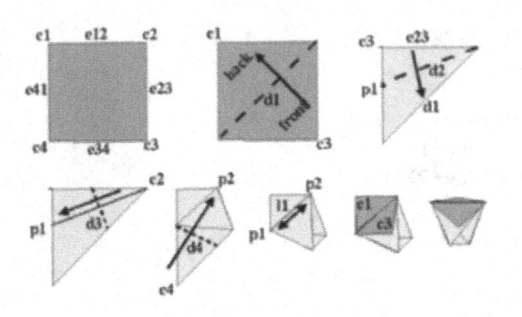

```
;; OSL Cup program ;;---------------------- (define d1 (axiom2 c3
c1)) (define front (create-region c3 d1)) (define back
(create-region c1 d1)) (execute-fold d1 apical c3)

(define d2 (axiom3 e23 d1)) (define p1 (intersect d2 e34)) (define
d3 (axiom2 c2 p1)) (execute-fold d3 apical c2)

(define p2 (intersect d3 e23)) (define d4 (axiom2 c4 p2))
(execute-fold d4 apical c4)

(define l1 (axiom1 p1 p2)) (within-region front (execute-fold l1
apical c3)) (within-region back  (execute-fold l1 basal c1))
```

Figure 3: A program in OSL for folding a square sheet of paper into a cup, as shown in the diagram. At this level of description, there is no notion of agents.

4.2   Origami Shape Language

Nagpal developed a language for shape formation on a simulated foldable sheet [15, 16]. In this case the two dimensional surface of agents represents a sheet with a single layer of randomly but densely distributed agents; a set of agents in a line can coordinate to fold the sheet along that line to create a flat layered structure. The sheet model is inspired by epithelial tissues where a line of epithelial cells can deform to cause the entire tissue to fold along that line, for example during neural tube formation[23]. One could imagine building a programmable reconfigurable sheet composed of such flexible agents.

The shape is specified as folding construction on a continuous sheet, using a programming language called the Origami Shape Language (OSL). The language is based on a set of geometry axioms, described by Humiaki Huzita, to capture the mathematics behind origami paper-folding [8]. A large class of flat folded shapes and line patterns can be constructed using these axioms [11]. OSL builds on these geometry axioms, but also adds concepts such as naming, regions and procedures. Figure 3 shows a diagram for constructing a cup from a blank square sheet of paper, and the corresponding OSL program. The basic elements of the language are points, lines and regions. Initially, the sheet starts out with four

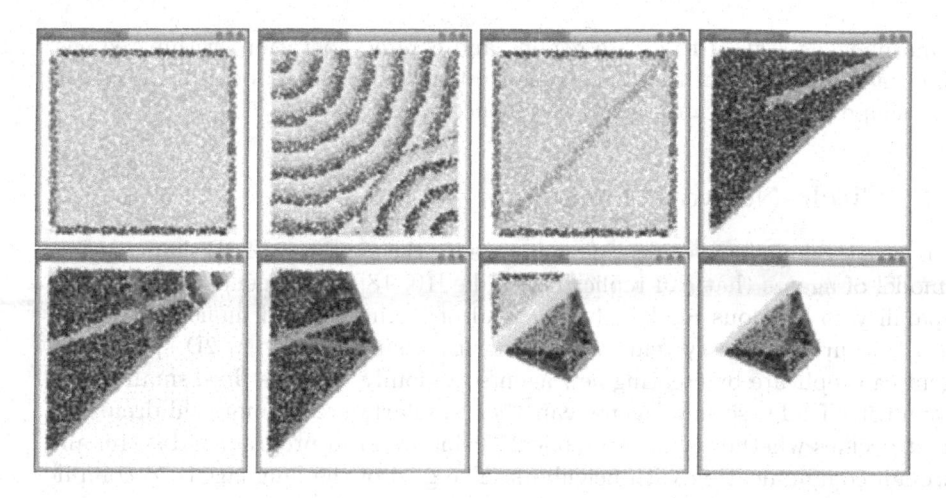

Figure 4: A reconfigurable sheet folding into a cup; each dot represents an agent in the sheet. The agent program is directly compiled from the OSL program in figure /reffig:cup1.

corner points (c1-c4) and four edge lines (e12-e41). The axioms describe how to generate new lines and points from an existing set of lines and points, purely through folding and unfolding paper. For example, the first operation constructs the diagonal d1 from the points c1 and c2 by using axiom 2. Axiom 2 folds the sheet so that c1 lies on c2 and then unfolds the sheet to create a line; this line is tthe perpendicular bisetcor of the line between points c1 and c2. The sheet can be permanently folded flat along a line, hence the structures created by OSL are flat, but layered. Lines can be used to create regions and regions can be used to restrict folds.

The interesting thing about this specification is that it is completely abstract — *there is no notion of morphogens, coordination or even agents!* Rather the programmer thinks in terms of a continuous sheet. The agent program is automatically compiled from this description and is composed from the set of primitives described in the previous section. Figure 4 shows a programmable sheet differentiating to fold into a cup. Initially the surface is mostly homogeneous, with only the agents on the border having special local state. When the agent program is executed by all the agents in the sheet, the sheet is configured into the desired shape. The overall view of this process is very close to what the diagram of the continuous sheet suggests. This is because each global operation is translated into a local agent behavior that emulates the geometry by using the biologically-inspired primitives. For example, in order to implement the first line creation, the agents in regions c1 and c2 create two distinct morphogens. The remaining agents test if the morphogen values are equal; if so then they are part of the new line d1. This is the local rule corresponding to axiom 2. Morphogens also serve as a form of barrier synchronization, so that agents can determine when it is safe to move on to the next fold operation. Selective propagation of morphogens is used to create regions and confine operations within regions. Thus each operation in the OSL language translates into a local rule, and the OSL shape program translates to a sequence of rules.

4.3   Circle-Network Language

Kondacs developed a specification language for the synthesis of 2D shapes, using a model of agents that can replicate and die [10, 18]. The agents are similar in capability to previous models, but instead of having a fixed number of agents on a 2D surface, the system starts out with a single agent in 2D space. An agent can replicate by creating new agents randomly within a fixed small radius around itself. These new agents can then similarly create more children. An agent decides whether or not to replicate using its local program and state, and through communication with neighbors. The goal of the language is to compile a predetermined global shape to produce a program for a seed agent that then "grows" the structure through replication.

The circle-network language (CNL) represents a 2D connected shape as a network of overlapping circles of different sizes. Neighboring circles are linked using local reference points within each circle; a circle can use its internal reference points to triangulate the location of all of its neighboring circles' centers. A rooted directed spanning tree in this network represents a process for constructing the structure starting from the root circle. A key feature of the circle-network representation is that it specifies the formation of the entire structure by recursively executing *only two* high-level operations: creating a circle of a given radius, and determining internal reference points and centers of adjacent circles. Even more importantly, such a specification can be generated *automatically from a graphical description* of an arbitrary connected 2D shape, using techniques borrowed from computer graphics.

Figure 5 shows the circle-network for a cross shape, generated from a drawing program rendition of a cross. The system starts with a single seed agent. The agent program is a set of rules, with rules corresponding to each circle in the network. Agents can assume different roles. For example, an agent with the role of circle A's center induces replication of agents outward to the radius specified by the rule for circle A. It achieves this by creating a morphogen gradient of the correct range, and all agents who recieve non-zero gradient values create children. Reference points within a circle are agents who have taken on the role of "coordinates" of a local grid; these agents create distinct gradient messages that allow other nearby agents (within the circle and neighboring circles) to triangulate their position and determine their roles. Many agents compete to take on the roles of reference points or circle centers. The agent program contains rules that specify an ideal ratio of gradient strengths that a agent should receive in order to be elected for a given role. Agents communicate their fitness to their immediate competing neighbors and the agent with the best fitness is selected as a leader. In this way the system tolerates imperfect agent positions and attempts to locally create the best possible solution. While notions of growth and death

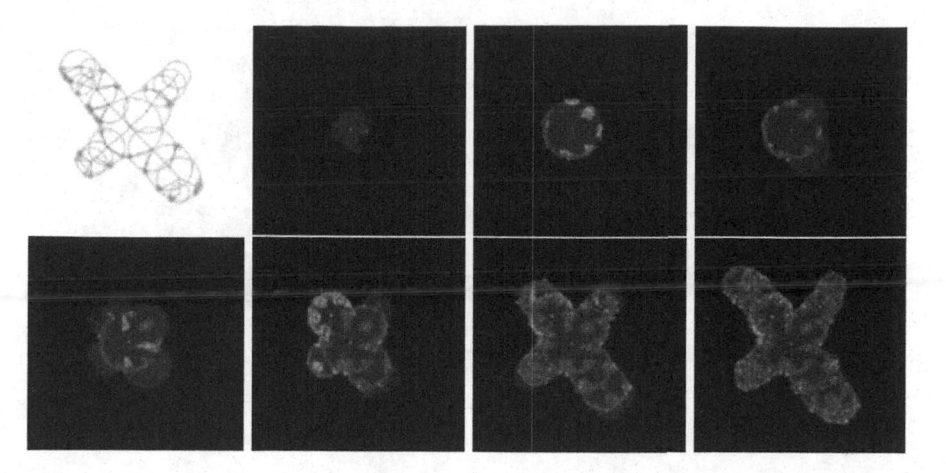

Figure 5: A cross shape represented as a network of linked, overlapping circles. This representation is generated directly from a pixel map of the desired shape. The system starts with a single agent that grows new agents to create the shape, using rules based on the circle-network representation.

are borrowed from living systems, they can be substituted with attachment and detachment in mobile agents.

4.4   Common Properties of the Languages

The three languages presented, GPL, OSL and CNL, specify very different types of patterns and shapes and operate on different models of agent capabilities. Yet they represent a common strategy towards engineering a complex goal — the goal is specified as a construction, using a global language based on a set of construction rules; this specification is then compiled to produce an agent program. The compiler uses a mapping from the elements of the global language to combinations of biologically-inspired local behavior. The *abstractness* of the global specification languages has increased over time. The first language GPL explicitly used gradients and chemotaxis to specify patterns. OSL on the other hand has no notion of gradients or even agents, and reasons in terms of operations on a continuous sheet. Finally the circle-network program is automatically generated from a purely graphical depiction. At the local level, however, the agents rely on roughly the same set of primitives as were first used in GPL, with a few additions.

Global-to-local compilation confers many advantages:

1. We can reason about the classes of structures that can and cannot be generated by analyzing the expressiveness of the language. For example, Coore proved that GPL can generate any prespecified planar graph pattern, up to connection topology, on an amorphous computer. Similarly, OSL can generate any 2D Euclidean construction pattern and all flat folded shapes

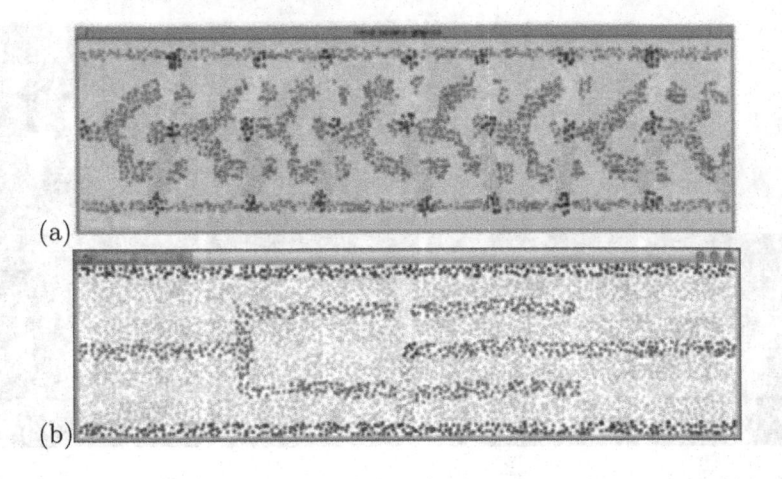

(a)

(b)

Figure 6: Inverter pattern created by (a) GPL (b) OSL, when run on a rectangular surface. The GPL program creates inverters with a fixed scale until the space is full, while OSL creates a inverter that is proportional to the boundary of the surface.

composed of simple folds. The circle-language can generate any 2D connected shape. These results are based on results from geometry, that have nothing to do with multi-agent systems or self-organization.

The languages also encode implicit structural properties. For example, GPL encodes patterns with an *inherent length scale* and can easily describe fractal and space filling structures. The OSL language on the other hand describes *scale-independent* structures, by recursively segmenting space relative to the original boundary. The same pattern created in these two languages behaves differently (but predictably) when the initial conditions are changed. Figure 6 shows an example. When the GPL program for an inverter is executed on a long sheet, it results in a chain of inverters filling the space. When the OSL program for an inverter is executed on a long sheet, it simply results in a stretched inverter pattern. The circle-network implicitly defines structures that can *repair themselves*. Each circle in a the rooted tree is capable of (re) creating its children and can commit apoptosis if its parent circle disappears. Thus any circle-network structure is capable of regeneration, so long as the root circle is not killed.

2. Locally agents interact through a small number of primitives that are robust, general and can be theoretically analyzed. While the three languages described may look very different, they all rely on similar underlying primitives such as gradients and local competition that are general enough to support a variety of global behaviors. These primitives are robust in the face of imprecise placement, asynchronous timing, small amounts of message loss, and random agent death. Building a catalog of simple and general primitives, with appropriate analysis, is key to developing these systems.

3. The analysis of a complex system becomes tractable because it is built in understood ways from smaller parts. We can predict the robustness to various types of faults, by separately analyzing the robustness of primitives such as morphogens, and then analyzing how error accumulates when we combine those primitives. This allows one to predict what densities and numbers of agents are required to satisfactorily achieve a given high-level goal.

   We can also analyze time and space complexity. For example, in each of these languages the space complexity of the agent program is directly proportional to the complexity of the high-level description. By optimizing the high-level description, for example by using procedures to capture regular patterns in GPL/OSL or by reducing the number of circles in the circle-network, one can generate more efficient agent programs. Thus a chain of inverters program need not be more complex than a single inverter pattern. Bounds on formation time can also be inferred directly from the program. In OSL the time taken is proportional to the number of sequential operations. In the circle-network, independent branches can proceed simultaneously so the time is limited by the depth of the tree.

4. The high level language makes it possible to *easily* specify complex behavior, without worrying about the millions of parts that are involved. This is partly because the specification in terms of abstract entities can be very concise and promotes a high level of agent code reuse. For example in OSL, very little of the agent code is modified between one program and the next, the majority of the code implements common primitives and mapping rules. Furthermore the programmer can express complex ideas in terms of languages that are conceptually more closely related to the global goal, rather than in terms of complex local interactions between agents.

One limitation of the languages explored so far is that the desired shapes are globally deterministic and prespecified. An alternate example of desired shapes is presented in [7], where the shape depends on external inputs such as the weight of an object that needs to be lifted or the size of an object that needs to be grasped. There are many cases where a shape may need to satisfy functional and environmental constraints, as opposed to purely geometric or topological ones. Exploring languages that encode such constraints, may allow us to use the same compilation approach to systematically generate agent programs.

5 Discussion

Biology hints that there may be significant power to be achieved from building things out of cheap, imprecise, parts with limited life. The ability to engineer complex but reliable systems from millions of simple and unreliable parts, provides an tantalizing alternative to our current methodologies that rely heavily on making the individual pieces never fail.

Programming such systems to do useful work presents a significant challenge. Decentralized versions of traditionally centralized approaches tend to fail for many reasons, from lack of scalability to the inherent tendency to depend on centralized state such as global clocks or external beacons for global positioning. Planning at the level of individual agents quickly become intractable for large numbers of modules, and more importantly increases the reliance on individual modules never failing [19, 4]. These strategies put pressure on system designers to build complex, precise (and thus expensive) agents rather than cheap, mass-produced, unreliable computing agents that one can conceive of just throwing at a problem. Alternative methods, such as cellular automata and artificial life, have shown the possibility of achieving robust and adaptive behavior from simple individuals [21, 14]. But the approaches have been difficult to generalize; local rules are hand-crafted empirically for each application, without providing a systematic framework for constructing local rules to obtain any desired goal. Evolutionary and genetic approaches are more general, but the local rules are evolved without an understanding of how or why they work. This makes the correctness and robustness of the evolved system difficult to verify and analyze[6].

The amorphous computing approach represents a novel strategy that relies at the individual level on biologically-inspired robust local rules, similar to cellular automata approaches, but then combines those rules using more global frameworks, such as programming languages. One can think of this as constructing a plan at an intermediate level, with cellular automata like local rules that achieve generic elements of the plan. This way there is ability at the individual level to adapt to failure, and the ability at the high level to make guarantees about global results.

So far the work in amorphous computing has focused on languages for pattern and shape formation. However, the desire to achieve global-to-local programming is not unique to amorphous computing. We believe that this programming methodology is applicable to many emerging fields such as reconfigurable robots, self-assembling systems and pervasive computing, where we are trying to program specific global goals from vast numbers of parts. The fact that concepts such as gradients are found everywhere, suggests that there exist general primitives that are fundamentally suited to these types of environments. However high-level programming languages and global-to-local compilation are rare. In the amorphous computing examples, the programming languages made it possible to easily achieve complex and robust desired behavior. We believe that in these other environments, the invention of appropriate global languages could have a similar far-reaching impact.

In order to achieve this we need to (1) find programming languages that are natural for expressing different classes of global goals (2) find ways to translate the primitives, means of combination, and means of abstraction of these global languages, into compositions of robust local rules we already know (3) develop a catalog of robust and general primitives. The work in the amorphous group continues to focus on developing new languages, new primitives, and incorporating new models of agents.

6 Acknowledgements

The Amorphous Computing project was started by Prof. Harold Abelson, Prof. Gerald J. Sussman, and Dr Thomas Knight at the Artificial Intelligence Laboratory at the Massachusetts Institute of Technology. The work discussed here reflects the contributions of many people in the Amorphous Computing group. The author would like to acknowledge discussions with members of the group, especially recent discussions with Daniel Coore, in the development of this chapter. Support for Amorphous Computing research was provided in part by the Advanced Research Project Agency of the Department of Defense, contract number N00014-96-1-1228, and in part by a grant from the National Science Foundation, Division of Experimental and Integrative Activities, contract number EIA-0130391.

References

[1] Abelson and Sussman. *Structure and Interpretation of Computer Programs.* MIT Press, 1996.

[2] H. Abelson, D. Allen, D. Coore, C. Hanson, G. Homsy, T. Knight, R. Nagpal, E. Rauch, G. Sussman, and R. Weiss. Amorphous computing. *Communications of the ACM*, 43(5), May 2000.

[3] A. Berlin. *Towards Intelligent Structures: Active Control of Buckling.* PhD thesis, MIT, Dept of Electrical Eng. and Computer Science, May 1994.

[4] Zack Butler, Sean Byrnes, and Daniela Rus. Distributed motion planning for modular robots with unit-compressible modules. *Proceedings of the Intl Conf. on Intelligent Robots and Systems*, 2001.

[5] D. Coore. *Botanical Computing: A Developmental Approach to Generating Interconnect Topologies on an Amorphous Computer.* PhD thesis, MIT, Dept of Electrical Eng. and Computer Science, February 1999.

[6] S. Forrest and M. Mitchell. What makes a problem hard for a genetic algorithm? *Machine Learning*, 13:285–319, 1993.

[7] Hogg, Bojinov, and Casal. Multiagent control of self-reconfigurable robots. In *4th International Conference on Multi-Agent Systems*, July 2000.

[8] H. Huzita and B. Scimemi. The algebra of paper-folding. In *First International Meeting of Origami Science and Technology*, Ferrara, Italy, 1989.

[9] Kahn, Katz, and Pister. Mobile networking for smart dust. In *MobiCom*, 1999.

[10] A. Kondacs. Biologically-inspired self-assembly of two-dimensional shapes using global-to-local compilation. In *IJCAI*, August 2003.

[11] R. J. Lang. A computational algorithm for origami design. In *Annual Symposium on Computational Geometry*, Philadelphia, PA, 1996.

[12] P. A. Lawrence. *The Making of a Fly: the Genetics of Animal Design.* Blackwell Science, Oxford, U.K., 1992.

[13] Mamei, Zambonelli, and Leonardi. Tuples on the air: a middleware for context-aware computing in dynamic networks. In *Intl. ICDCS Workshop on Mobile Computing Middleware*, 2003.

[14] M. Mataric. Issues and approaches in the design of collective autonomous agents. *Robotics and Autonomous Systems*, 16((2-4)):321–331, December 1995.

[15] R. Nagpal. *Programmable Self-Assembly: Constructing Global Shape using Biologically-inspired Local Interactions and Origami Mathematics.* PhD thesis, MIT, Dept of Electrical Engineering and Computer Science, June 2001.

[16] R. Nagpal. Programmable self-assembly using biologically-inspired multiagent control. In *Autonomous Agents and Multiagent Systems (AAMAS)*, July 2002.

[17] R. Nagpal and D. Coore. An algorithm for group formation in an amorphous computer. In *Proceedings of the 10th IASTED International Conference on Parallel and Distributed Computing and Systems (PDCS'98)*, October 1998.

[18] R. Nagpal, A. Kondacs, and C. Chang. Programming methodology for biologically-inspired self-assembling systems. In *AAAI Spring Symposium: Computational Synthesis*, March 2003.

[19] Pamecha, Ebert-Uphoff, and Chirikjian. Useful metrics for modular robot planning. *IEEE Trans. on Robotics and Automation*, 13(4), August 1997.

[20] Balakrishnan Priyantha, Chakraborty. The cricket location-support system. In *MobiCom*, 2000.

[21] M. Resnick. *Turtles, Termites and Traffic Jams.* MIT Press, Cambridge, MA, 1994.

[22] R. Weiss, G. Homsy, and T. Knight. Toward in vivo digital circuits. In *Dimacs Workshop on Evolution as Computation*, January 1999.

[23] L. Wolpert. *Principles of Development.* Oxford University Press, U.K., 1998.

Chapter 14

A Machine Learning Method for Improving Task Allocation in Distributed Multi-Robot Transportation

Torbjørn S. Dahl
Department of Computing
University of Wales, Newport
torbjorn.dahl@newport.ac.uk
Maja J Matarić and Gaurav S. Sukhatme
Center for Robotics and Embedded Systems
University of Southern California
mataric|gaurav@usc.edu

Machine learning is a means of automatically generating solutions that perform better than those that are hand-coded by human programmers. We present a general behavior-based algorithm that uses reinforcement learning to improve the spatio-temporal organization of a homogeneous group of robots. In this algorithm each robot applies the learning at the level of individual behavior selection. We demonstrate how the interactions within the group affect the individual learning in a way that produces group-level effects, such as lane-formation and task-specialization, and improves group performance. We also present a model of multi-robot task allocation as resource distribution through *vacancy chains*, a distribution method common in human and animal societies, and an algorithm for multi-robot task allocation based on that model. The model predicts the task allocation achieved by our algorithm and highlights its limitations. We present experimental results that validate our model and show that our algorithm outperforms pre-programmed solutions. Last, we present an extension of our algorithm that makes it sensitive to differences in robot performance levels.

1   Introduction

Machine learning (ML) [24] is a means of automatically generating solutions that perform better than those that are hand-coded by human programmers. Such improvement is possible in problem domains where optimal solutions are difficult to identify, i.e., when there are no models available that can accurately relate a system's dynamics to its performance. One such domain is the control of multi-robot systems.

Mobile robots are notoriously difficult to control in a robust, reliable, and repeatable fashion. The challenges stem from uncertainty inherent in physically embodied systems, including in sensors, effectors, and interactions between the system components and the environment. The *behavior-based* (BB) control paradigm [22, 1] provides a means of structuring robot controllers into collections of task-achieving modules or *behaviors*, such as exploration and obstacle avoidance. The modules operate in parallel and interact within the system and also through their effects on the environment. When properly designed, the resulting controller produces robust, repeatable, and reliable overall behavior for the robot.

The BB paradigm takes inspiration from biology and the metaphors used to describe animal behavior. Behaviors also provide a basis for employing ML to improve the robot's performance [23]. ML can be employed at the level of behavior coordination, or at the level of the behaviors themselves. The behaviors typically have a set of *preconditions* used to decide when they are suitable for execution. These preconditions and behaviors together form a tractable search space for improvement through ML.

In Section 4 we present an algorithm that uses ML at the level of behavior selection, applied to the multi-robot control domain. Specifically, we show how a group of robots improved its performance on a cooperative transportation task by using individual reinforcement learning to adjust individual behavior selection polices from feedback related to individual performance.

In Section 5 we present a BB solution to multi-robot task allocation (MRTA) where robots represent tasks internally as behaviors and use ML to automatically improve their individual task selection policies. Through interaction the robots specialize on different tasks and, from these specializations, global task allocation emerges. We also present a model of MRTA called task allocation through *vacancy chains* (TAVC). A vacancy chain [7] is a resource distribution mechanism common in animal and human societies. The TAVC model represents opportunities to improve a group's performance by reallocating tasks, as chains of task-related vacancies to be distributed between the robots. The TAVC model captures the effects of group dynamics in such systems and explains and predicts the task allocations produced by distributed, communication-free MRTA. We demonstrate a TAVC algorithm, in realistic simulations, using a homogeneous group of robots in a dynamic environment where task values change and where robots are susceptible to failure. Our results validate the TAVC model of MRTA and show that our TAVC algorithm improves on pre-programmed solutions to this problem.

In Section 6 we show that our TAVC algorithm also works for heterogeneous groups of robots using a Boltzmann-softmax action selection function. Our TAVC algorithm is a valuable alternative to existing MRTA algorithms as these are typically not sensitive to the complex effects of group dynamics, such as interference and synergy.

2 Motivation

As robot technology becomes increasingly ubiquitous, so will multi-robot systems. Control and coordination of groups of mobile robots is an active field of research; multi-robot task allocation (MRTA) is a key aspect of this problem. MRTA encompasses issues of skill utilization, cooperation, communication, and scheduling. One of the central problems in dealing with these issues is modeling the underlying group dynamics and their effects on the group's performance. We have developed TAVC, a model of MRTA inspired by vacancy chains, a robust and efficient resource allocation mechanism common in biological systems. The advantage of TAVC over current MRTA models is that it is explicit about the effects of group dynamics. Based on the TAVC model, we have developed a novel MRTA algorithm that is sensitive to group dynamics. To achieve this sensitivity, the TAVC algorithm uses distributed RL to make individual task utility estimates. During learning these estimated utilities diverge according to the effects of the prevailing group dynamics. The divergence leads to a general improvement in the group's performance.

2.1 Group dynamics

By the term *group dynamics*, we refer to the interaction among the members of a group of robots. Group dynamics can have positive or negative effects on the group's performance. Interference is a typical negative effect in confined spaces, but positive or *synergistic* effects are also possible when the actions of one robot unintentionally facilitates the actions of another robot, or when robots intentionally cooperate. Existing multi-robot control algorithms and related theory such as *scheduling theory* [6] assume *task independence*. Task are independent when there are no significant interference-related or synergistic effects to be considered. However, in multi-robot systems in general, and in MRTA in particular, such effects are often critical to group performance. When this is the case, understanding and modeling those effects is important for constructing and predicting the quality of the control algorithms. There are major challenges related to modeling group dynamics in general. Each member of a group of robots is a physically embedded system with inherent uncertainty. Interactions among a group of physically embodied systems increase uncertainty to a level where practical accurate models are extremely difficult to construct. Previous work on modeling group dynamics can be divided into *microscopic* (*simulation-based* or otherwise) and *macroscopic* [19]. Microscopic models explicitly model each agent. Simulation-based models simulate the actions taken by each so that properties of the system can be recorded as the agents interact. Macroscopic models describe system properties in terms of abstract features such as the number and general distribution of agents. We briefly overview representative relevant examples from both areas next.

Microscopic Models   Game theory [25] provides an abstract model of interacting agents as *players* in formalized n-person stochastic *games*. Game theory represents players in terms of their *strategies* for playing a given game and their related *payoff matrices* for possible moves. Using these factors, game theory predicts optimality and stability features of games. A game theoretic approach to modeling group dynamics would require a specification of the payoff matrices involved. However, these cannot be known for any but the most trivial multi-robot systems. Littman [20] presented the Minimax-Q algorithm as a means of finding payoff matrices. The algorithm is guaranteed to find the equilibrium of any stochastic game. The exponential growth, n^m, of a payoff matrix for n tasks and m robots also means that it is not feasible to estimate the payoffs for each state/action combination. Bowling *et al.* [5] presented experimental data demonstrating how a reinforcement-learning algorithm based on policy gradient ascent and the WoLF (Win or Learn Fast) principle for adjusting the learning rate can overcome this complexity.

Simulation-Based Models   To learn statistical models of interaction in a space of abstract robot behaviors, Goldberg and Matarić [16] developed Augmented Markov Models (AMMs). AMMs are transition probability matrices with additional temporal information. They used these models to maximize reward in multi-robot foraging/mine collection tasks. Balch *et al.* [3] used live insect colonies to construct three-dimensional histograms of insect presence over a discretized area, with the long-term goal of combining spatio-temporal models with Hidden Markov Models for behavior recognition [18] in order to recognize colony behaviors. Yan and Matarić [30] modeled group behaviors from spatial data describing both human and robot activity. Their models were based on proxemics and considered data related to individuals, pairs, clusters, and the environment. They also used three-dimensional histograms to identify and describe activity distributions produced by underlying behaviors. Seth [27] presented a simulation-based model using genetic algorithms to evolve foraging behaviors for multiple agents in spatially explicit environments. The evolved systems were able to reproduce interference functions previously described in field studies of insects, but not previously reproduced by microscopic models.

Macroscopic Models   Matarić [21] proposed *group density* as a macroscopic model of interference. Group density is defined as the ratio between the agents' *footprints* and the available *interaction space*. An agent's footprint is defined by its sphere of influence, including geometry, motion constraints, and sensor range. When the number and size of the agents are known, a *mean free path* can be computed and used to estimate the *number of expected collisions* for agents executing random walks. Lerman *et al.* [19] studied three different types of models of cooperation and interference in groups of robots: sensor-based simulations, microscopic numerical models, and macroscopic numerical models. The three different models produced corresponding results but the macroscopic model had the advantage of being very fast and independent of the number of robots modeled.

Current macroscopic models are not sophisticated enough for optimizing the control of multi-robot systems with regards to specific problems. Simulation-based models are problem-specific but the time needed to for model-construction makes them unsuitable for real-time control of a multi-robot system. Currently there are no models of group dynamics with the speed, generality, and predictive capabilities needed to specify the effects of interaction on group performance. Simulation-based models are, in general, too slow and macroscopic models make too many simplifying assumptions to be of predictive use.

---

3   Cooperative transportation

Our experiments are related to the problem of *cooperative transportation*, where group dynamics often have a critical impact on performance. In transportation problems, a group of robots traverses a given environment in order to transport items between sources and sinks. We call the time taken to traverse the environment once from a sink via a source and back to a sink again the *traversal time*. We call the time between two arrival events at either source or sink the *target time*. To perform optimally the robots must maximize the number of traversals over time.

The basic transportation problem is one of the sub-problems of foraging [2, 17]. Foraging is reduced to a problem of transportation when the locations of the sources and sinks are known. When there are multiple sources and sinks, the transportation problem becomes the MRTA problem where transportation tasks must be allocated to robots in a way that optimizes overall performance. We refer to a problem as a *prioritized transportation problem* when sources and sinks have (sets of) different values. To optimize performance on prioritized transportation problems, measured in total value of delivered goods over time, a group of robots must strike the correct balance between the positive contribution of high target values and the negative contribution of interference resulting in increased delivery times.

4   Improving performance on cooperative tasks

Taking advantage of the reduction in state/action space made possible by the BB approach, we implemented a general algorithm, based on distributed reinforcement learning [28], that adapts the spatio-temporal properties of each robot's behavior in a way that improves the group's performance on cooperative tasks [9]. The robots learn from performance-related feedback. In homogeneous groups this feedback mainly reflects effects of the prevailing group dynamics and allows the robots to adjust their behavior so as to minimize the negative effects and maximize the positive ones.

In addition to the problem of accurately modeling the effects of group dynamics, multi-robot systems also face the inherent problem of scaling. Algorithms that make decisions based on global system state need complex communication structure to synchronize and update state information and coordinate individual robots. This makes it difficult to scale such centralized algorithms to work efficiently for systems with many robots.

This section presents our distributed algorithm for spatio-temporal organization. The algorithm works purely through *stigmergy*, i.e., implicit communication through the effects of actions on the environment, and hence avoids the scale-related difficulties of communication-dependent algorithms. We have demonstrated the effectiveness of our algorithm and analyzed the qualitative effects it had on individual action selection policies. Our results show that the algorithm changes the spatio-temporal organization of a group of robots according to the effects of the group dynamics. The most interesting result was a clear tendency toward policy differentiation, i.e., specialization. This effect was the motivation for adapting the algorithm as presented in this section for use on MRTA problems, where existing solutions have typically been insensitive to the effects of group dynamics.

4.1   The learning algorithm

Our algorithm for spatio-temporal organization is built on two main principles: *individual learning* and *emergence*. Each robot individually estimates the utility of the available actions and chooses from these in a way that allows it both to keep its estimates current through continuous experimentation and to exploit the actions with high estimated utilities. We used Q-learning for utility estimation.

Our algorithm depends on a reward structure that reflects the effects of the current group dynamics. Q-learning is not sensitive to the frequency of rewards. Hence, the estimated values of actions do not necessarily correspond to the action's contribution to performance over time. In order to use Q-learning to optimize performance over time it is necessary to make the temporal aspect of performance explicit in the reward function. Such a reward function, using the last task processing time, t, and task value, w_i, is given in Equation 1.

$$r = w_i/t \tag{1}$$

Because we want the system to remain adaptive to changes in the environment, the experimentation rate ϵ does not decrease over time, as is common.

Using a reward function that is sensitive to the effects of the group dynamics allows the robots to learn to take actions that minimize these effect with respect to their individual performance. We assume that those individual changes in action selection policies are enough to produce spatio-temporal effects on a group level, such as turn-taking and lanes. The emergence of such phenomena would indicate that the changes in the individual action selection policies can be interpreted as spatio-temporal organization at a group level.

In Section 4.2 we demonstrate how, from individual Q-learning and the effects of the group dynamics, consistent and repeatable group-level effects emerge. In particular we show the effects of the emergent organization on the group's performance.

4.2 Experimental validation of spatio-temporal organization

We hypothesized that it was be possible for recognizable spatio-temporal features of group behavior such as turn-taking, lanes, and territories to emerge in a group of robots using our adaptive algorithm. Each robot was given a set of behaviors that performed the same general task, but did this in ways with different spatio-temporal qualities. We assumed that the spatio-temporal features listed above had genuine positive effects on group performance. If so, these features would represent easily identifiable optima in the problem space of spatially and temporally constrained multi-robot control. As such, these features may emerge from the interaction between robots that use Q-learning to individually estimate each behavior's utility in different locally perceived world states.

To test this hypothesis we choose to the problem of cooperative transportation, a problem that could be given clear spatial and temporal constraints. We also designed a set of transportation-related behaviors with different spatio-temporal qualities.

Simulated Environment For the experiments presented in this section, we simulated five Pioneer 2DX robots with PTZ cameras and SICK laser range-finders in a 6 by 8-meter rectangular area. There was one source and one sink, each placed in the center of one of the short sides of the rectangle. The robots wore color markings recognizable using ActiveMedia's Color-Tracking Software (ACTS), allowing them to perceive each other's orientation. The markings, inspired by ship lanterns which serve the same purpose at sea, were red on the left side, green on the right side, and yellow at the rear. The prototype markings, as they appear on a real Pioneer, are shown in Figure 1.

The source and sink were simulated with unique bar-codes (as shown in Figure 2) made from highly reflective material readable by the SICK laser. Both the recognition of the color markings and the reading of laser bar-codes have been validated on real robot tests.

Figure 1: Prototype Pioneer 2DX robot with color markings

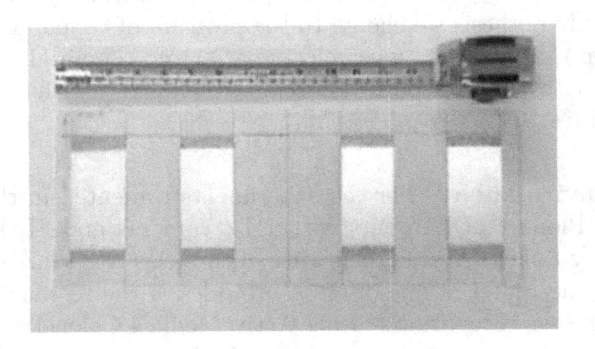

Figure 2: Prototype bar-code readable by the SICK laser

We performed our experiments in simulation, using the Player/Stage [15] software platform. From experience, controllers written for the Stage simulator work with little or no modification on the real Pioneers. A graphical rendering of the simulated environment is presented in Figure 3.

Figure 3: The simulated environment used for experiments on adaptive spatio-temporal organization

Controller architecture   To support flexibility in the spatio-temporal features of the transportation, we implemented three transport-related behaviors:

- **direct target approach:** servoed the robot directly toward the current target.

- **wall following:** followed the boundaries of the simulated environment, eventually arriving at a source or a sink.

- **stopping:** froze the robot in its current position and orientation.

These three behaviors made use of lower level behaviors such as *obstacle avoidance*. The mapping of the transport-related behaviors to input states was done by the learning algorithm presented in Section 4.1.

Learning parameters   The input state presented to the learning algorithm consisted of five bits. The first bit indicated the presence of a source or a sink. The second bit was on for random intervals of time (1-30 seconds), allowing the robot to express certain behaviors for short periods independent of the rest of the input state. For example, instead of always following the walls whenever a target is not visible, the robot could learn to approach the target directly *most of the time* in this state, interspersed with short periods of wall following. The three final bits represented the presence of color-blobs corresponding to the three colored markings on the robots, indicating the presence and orientation of another robot. When several robots were present, these bits reported only on the visible markings of the closest one, distinguished by the height of the visible color-blobs. The five input bits and three behaviors made up a state-action space of size $2^5 * 3 = 96$. Actions were chosen by the adaptive controller using a Boltzmann-softmax action selection function.

Results   We ran a total of five trials. During every trial the five robots each performed around two hundred traversals, resulting in a total of 1000 traversals per trial. The initial individual performance had a mean traversal time, $\hat{\mu}_i$, of 1112.5 seconds, and a standard deviation, s_i, of 1023.7 seconds. These values, and the corresponding values after 50, 100, 150 and 200 traversals, n, are given in columns 2 and 3 of Table 1. The group performance had an initial mean target time, $\hat{\mu}_g$, of 245.2 seconds and a standard deviation, s_g, of 165.3 seconds. Columns 5 and 6 of Table 1 show these values and the corresponding values after 250, 500, 750 and 1000 target arrivals, m.

n	$\hat{\mu}_i$	s_i	m	$\hat{\mu}_g$	s_g
0	1112.5	1023.7	0	245.2	165.3
25	843.9	628.0	125	159.8	102.7
50	759.7	559.8	250	161.4	130.4
75	639.6	445.9	375	133.6	84.4
100	668.3	504.5	500	139.5	114.4
150	537.7	343.0	750	123.1	86.7
200	524.8	296.8	1000	100.0	72.8

Table 1: Converging individual and group performance

The converging of the mean target times over a window of 10 trials is shown graphically in Figure 4.

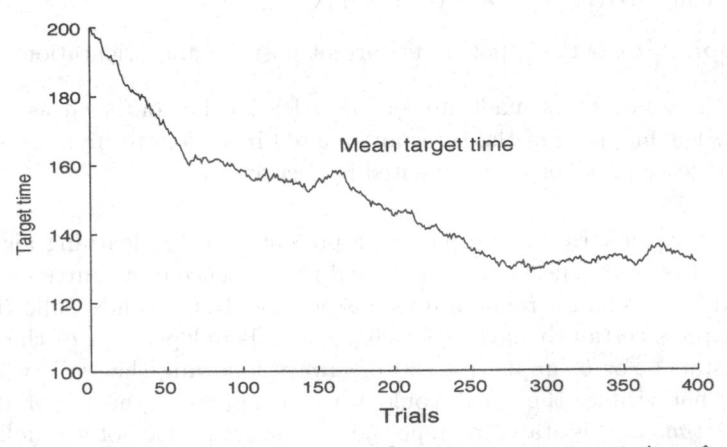

Figure 4: Converging group performance, mean target time, $\hat{\mu}_g$

All mean traversal and target times are lower than the ones produced by the initial random behavior application, at $n = m = 0$, by a statistically significant amount.

Analysis   By analyzing the Q-tables produced during this experiment we found that the robots learned policies dominated by the *direct target approach* behavior, but with significant portions of the *wall-following* behavior. These policies made the robots follow a common circular path from the source to the sink and back, while keeping to the left side to minimize the interference with robots coming from the opposite direction. In most of the experiments, one or two of the robots would learn specialized policies containing disproportionately small amounts of the *wall-following* behavior. When observing the behavior of the specialized robots, it appeared that they effectively cut across the empty center of the common circular path, benefiting from the general dispersion of the rest of the group. The policies show that a structuring of the group's spatial organization is partly responsible for the improvement in performance.

We also found evidence that the robots adapted their temporal organization according to the presence and orientation of other robots. From each robot's Q-table, using only the input states where yellow markings where visible, we calculated the number of times the *stopping* behavior had the highest Q-value. We called this number the *yellow stop-rate* of a robot. The yellow stop-rates were significantly lower than the stop-rate for states in which other robots were visible without displaying any yellow markings, i.e., states where the other robots were seen from the front rather than from the back. The yellow stop-rate was also significantly lower than the corresponding *red* and *green* stop rates. This indicates that the robots in general learned to stop only when they saw a robot without visible yellow markings. Intuitively, visible yellow or rear markings indicate that the other robot is facing, and likely moving, away, and hence not likely to cause interference. Simply put, the robots learned to stop for oncoming traffic only.

4.3   Conclusions

Our analysis of the learned behaviors showed that they had natural interpretations at the group level. The tendency to keep to the left can be seen as the formation of two lanes. The observed utilization of the center space, by one or two of the robots, can be seen as specialization. Our analysis, together with the demonstrated improvements in performance, verify our hypothesis that individual learning is sufficient to produce recognizable forms of group behavior in our domain.

5   Task-allocation through specialization

In Section 4 we demonstrated a distributed learning algorithm that consistently improved a group's performance on a cooperative transportation problem by minimizing the negative effects of the group dynamics. One of the ways in which the performance was improved was the specialization of the members in an initially homogeneous group. The specialization was a result of the algorithm's sensitivity to the effects of the group dynamics. In this section we show how the algorithm can be used for handling MRTA problems.

MRTA is a problem domain that presents a third major challenge, in addition to those we reviewed in Sections 1 and 4. This challenge concerns the fact that if the effects of the group dynamics are considered explicitly, MRTA is an $\mathcal{NP}$-complete problem [10]. This inherent complexity of MRTA implies that it is not feasible to find optimal solutions as the size of the system, i.e., the number of tasks and robots, grows. For such high-complexity problems, solutions must be found through heuristic algorithms that rely on appropriate assumptions about the problem to construct solutions of the necessary quality. Machine learning methods are commonly used to produce the necessary heuristics for such algorithms.

In this section we present the TAVC model of MRTA and the associated TAVC algorithm [11]. Our TAVC model explicitly represents the effects of group dynamics and allowed us to develop the TAVC algorithm, whose main contribution is its sensitivity to those effects. The TAVC algorithm uses Q-learning to make individual estimates of task utilities; in it each robot follows a *greedy* rule for choosing its next task based on its estimated utility. The algorithm goes beyond traditional greedy algorithms [8] in that the utility estimates are sensitive to the effects of the group dynamics produced by the current task allocation.

5.1   Multi-robot task allocation

Multi-robot task allocation (MRTA), is the problem of allocating a set of tasks to the robots in a group in a way that optimizes some group-level performance metric. This problem is receiving growing attention. Recently, Gerkey and Matarić [13, 14] presented a task- and architecture-independent framework in which to study and compare existing empirical solutions. The solution methods include L-ALLIANCE [26], a system for MRTA based on local task allocation decisions only. Robots using L-ALLIANCE individually estimate their own progress and the progress made by other robots, and take over or acquiesce tasks according to those estimates. However, the progress estimation functions are problem-specific and there is no explicit consideration of the effects of interaction.

The M+ algorithm [4] uses a centralized task allocation mechanism based on negotiation through the Contract Net Protocol. The algorithm relies on pre-specified capabilities and does not consider effects of interaction. Also, a distributed negotiation system carries with it a high communication overhead that introduces problems when scaling the system up to handle larger numbers of robots.

Similarly, MURDOCH [12] used a negotiation mechanism based on the Contract Net Protocol for task allocation. However, MURDOCH leaves open a *suitability* parameter to be filled in by the individual robots when bidding for tasks. Though it is was done in their work, the suitability value could reflect the effects of the group dynamics.

The broadcast of local eligibility algorithm [29] defines a similar parameter called *eligibility*. The eligibility value is calculated by the robots individually and used in an inter-robot coordination mechanism based on port-arbitrated behaviors (PABs). PABs are a generalization, for multi-robot systems, of the traditional behavior coordination mechanisms. As in MURDOCH, the locally calculated eligibility could reflect the effects of the group dynamics but was not used in that manner.

Among previous MRTA algorithms, none deal explicitly with the effects of group dynamics. Our TAVC algorithm addresses that deficiency.

5.2 The vacancy chain model

The inspiration for the TAVC algorithm is the vacancy chain resource distribution process [7] common in human and animal societies. The typical example of resource distribution through a vacancy chain is a bureaucracy, where the retirement of a senior employee creates a vacancy to be filled by a less senior employee. The promotion, in turn, creates a second vacancy to be filled, and so on. The initial retirement results in a chain of vacancies linked by the promotions. The resources that are distributed in this example are the positions, and the consumers are the employees. Chase [7] has proposed that major human consumer goods such as cars, boats, and airplanes, also move through vacancy chains and that vacancy chains are common in other species as well.

Inspired by this process we developed a model of MRTA that describes how task values and task processing times combine to produce task allocations among greedy consumers [10]. Our model has the important property of breaking the group performance down into individual robot contributions, a property that allows us to use distributed control algorithms.

According to our TAVC model, any number of robots can be assigned to tasks from a given *class*. When a robot, j, is assigned to a task from a class, i, currently being *serviced* by a set of robots, J, we say that *service-slot*, (i, J, j) is being filled. A particular set of robots, J, servicing the same task, i, will have a task processing frequency, $c_{i,J}$, dependent on the degree to which the robots are able to work concurrently without interference. The difference in the group's task processing frequency before and after robot j joined, together with the task value, w_i, defines robot j's contribution. We call this contribution, which can be negative, the *slot-value*, $s_{i,J,j}$. The formal definition of the slot-value is given in Equation 2.

$$s_{i,J,j} = w_i(c_{i,J\cup\{j\}} - c_{i,J}) \tag{2}$$

Assigning a robot to a task can lead to a decrease in the task processing frequency of the group of robots servicing tasks from that class. The slot-value then becomes negative. When all the available service-slots for a class of tasks have negative slot-values, we say the task is *saturated*.

When the service-slots are allocated optimally, a failure in a robot servicing a high-value task, or the creation of a new class of high-value tasks, will result in empty high-value service-slots that must be re-allocated for the system to perform optimally. If the value of a vacant slot is greater than the value of the occupied service-slots, the vacant slot will have to be filled in order to restore optimal performance. In vacancy chain terminology, a vacant, high-value service-slot is a resource to be distributed among the robots.

We make two simplifying assumptions: the robots are identical and the slot-values decrease monotonically with the number of robots servicing a class of tasks. When these assumptions are true, the task allocation algorithm can be distributed by letting each robot individually optimize the value of the service-slot it occupies. We will relax the first assumption later, in Section 6. Even with these simplifying assumptions, the optimal task allocations cannot be predicted without a model of the group dynamics, and such models are, as discussed in Section 2.1, difficult to construct.

5.3   The vacancy chain algorithm

In the TAVC algorithm, a task corresponds to an *action* in RL terms. Each robot uses a Q-table to keep individual estimates of task utilities and choses its next task using an ϵ-greedy selection function.

The task processing is implemented in terms of pre-programmed high-level behaviors, one per each of the available tasks. The state-space describes what class of tasks the robot last serviced. This state-space allows the robots to learn policies that are not dedicated to one class of tasks; the learned policies can switch between classes of tasks in order to construct optimal task sequences. The action space corresponds to the available classes of tasks.

If a robot consistently occupies a service-slot that is suboptimal due to a high level of interference, the high mean traversal time for that class of tasks will reduce the estimated utility for that class of tasks until it falls below the estimated utility of the class of tasks that contains of the optimal service-slot. This change in estimated utility will attract the robot to the optimal slot. The robots keep a constant level of exploration, allowing them to identify and migrate to new vacancies.

All the robots in our demonstration used the same adaptive, behavior-based controller [22]. However, our vacancy chain task allocation algorithm is independent of the underlying architecture, being defined purely in terms of distributed reinforcement learning of task utilities.

5.4   Experimental validation of TAVC

In this section we present a set of experiments that had two aims. First, the experiments were designed to demonstrate the validity of the TAVC model, i.e., show that a simulated multi-robot system would behave as predicted by the model. Second, they were designed to demonstrate that the TAVC algorithm improved system performance beyond what can be expected from pre-programmed solutions.

The 'initial task distribution' experiment was designed to demonstrate that the interactions between the robots could produce specializations where a number of robots would choose a task with sub-optimal value as a result of the effects of the group dynamics on individual utility estimates.

The 'vacancy from robot breakdown' experiment was designed to demonstrate that it is possible for the TAVC algorithm to recover from individual robot failures. The TAVC model predicts that the failure of a robot servicing tasks from a high-value class of tasks creates a vacant service-slot that is distributed among the robots currently servicing lower-value service-slots. This experiment was designed to show how the individual task utility estimates of the robots change in a way that leads to exactly this kind of migration between classes of tasks by a single robot. This experiment also demonstrates that our TAVC algorithm can perform better than a pre-programmed system.

The 'breakdown without vacancy' experiment was designed as a control experiment, where the failure of a robot leads to a situation where the TAVC model predicts no change. This experiment was designed to show that the TAVC model's prediction was correct by showing that the task allocation in the simulated system did not change.

The 'changing sub-task values' was also a control experiment. Again a vacancy was created, but this time not from a robot failure, but from a change in sub-task values. This experiment was designed to demonstrate that the simulated system produces the patterns predicted by the TAVC model regardless of the underlying reason for a change in the predicted task allocation.

The four experiments above were also designed to show whether the TAVC algorithm could improve the performance of the system to a level significantly above that of a system where task were randomly allocated to robots. Random allocation is a pessimistic approximation of a pre-programmed system, because it might be possible to make universally applicable assumptions, such as 'minimize changes between classes of tasks' or 'distribute robots evenly among the classes of tasks', that would improve the performance of general MRTA algorithms above the performance of random allocation. It is, however, not unreasonable to use a pessimistic approximation since, for most heuristics, one can construct pathological cases where they would perform significantly below the random allocation solution. In such cases, our TAVC algorithm would outperform both the pre-programmed and the randomized solutions.

The simulated environment The experiments presented in this section were done in simulation using six robots identical to those described in Section 4.2. The simulated environment was an 8 by 12-meter rectangular area with two unique laser bar-codes along one of the longer sides of the rectangle, indicating sources, and two along the opposing side indicating sinks. A depiction of the simulated environment is presented in Figure 5; common paths between the sources and sinks are indicated by dashed arrows.

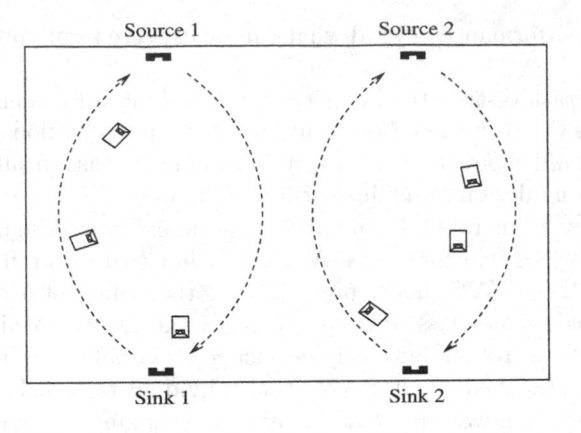

Figure 5: The simulated environment used for experiments on TAVC

Controller architecture   All of the task-related behaviors used by the TAVC algorithm were implemented as a collection of sub-behaviors responsible for different aspects of a safe and efficient task execution.

These sub-behaviors were:

- **obstacle avoidance:** avoided obstacles detected (by laser range finder) in the desired path.

- **visible target approach:** approached the desired target when it was visible (to the laser range finder).

- **target location approach:** approached the location of the desired target when the target itself was not visible.

- **wall following:** followed the walls to the first available target when the desired target was not visible and localization (based on odometry) was deemed to be inaccurate.

The localization was deemed to be inaccurate whenever the desired target was not visible, but should have been so according to the robot's estimated position and orientation. On encountering a target, the localization estimate was reset and again deemed to be accurate.

Learning parameters   The Q-tables ware initialized with random values between -0.1 and 0.1, the learning rate, α, was set to 0.1, and the discount factor γ was set to 0.95. For action selection we used a greedy-ϵ strategy [28], where ϵ was set to 0.1.

We call the circuit with the highest related reward the *high-value* circuit and, correspondingly, the the circuit with the lowest related reward is called the *low-value* circuit. In order to demonstrate task distribution through vacancy chains, we estimated circuit values that would produce an initial task allocation where three robots were allocated tasks from the high-value circuit and three were allocated tasks from the low-value circuit.

In order to produce this task allocation, a situation needed to be set up in which it was less attractive to be one of four robots servicing the high-value circuit than to be one of three servicing the low-value circuit. This constraint on the reward function is presented formally in Equation 3.

$$\forall(x, y).\overline{r}_{x,4} < \overline{r}_{y,3} \tag{3}$$

We also wanted to demonstrate the filling of a vacancy on the high-value circuit. In order for a vacancy in the high-value circuit to be filled, it had to be the case that it was more attractive to be the third robot on that circuit than to be the third robot on the low-value circuit. This is expressed formally in Equation 4, where p denotes the preferred circuit.

$$\forall(x \neq p).\overline{r}_{x,3} < \overline{r}_{p,3} \tag{4}$$

We empirically estimated the relevant mean traversal times. To satisfy the given constraints we chose the circuit values, as defined in Equation 1, to be $w_1 = 2200$ and $w_2 = 2000$. The optimal allocation pattern emerges as each robot's utility estimates converge as a result of individual experience only. The fact that the robots do not know the task values allows new allocation patterns to emerge when external factors, such as the task values or the group size, change.

5.5   Results

For each experiment we defined a convergence period and a stable period according to the stability of the system performance. Our student-t tests for statistical significance were all done at a 90% confidence level.

Initial task distribution   We performed 20 individual trials of 5 hours, each averaging 3000 traversals or 500 traversals per robot. The convergence period was 1.25 hours.

To show the structure that emerged we consider the last target visited by each robot, which yields seven possible system states. We refer to each state using the notation $h : l$, where h is the number of robots whose last target was on the high-value circuit. Correspondingly, l is the number of robots whose last target was on the low-value circuit. The rows labeled $TAVC$ in Table 2 show the mean, $\hat{\mu}$, and standard deviation, s, of the time the system spent in each of the states when controlled by the TAVC algorithm. The values are percentages of the total stable period. The rows labeled R describe the same values for a set of 20 control trials using a group of robots where tasks were allocated at random.

The row labeled T in Table 2 lists the number of different ways to choose a sample of size n from a population of size m, as a percentage of all possible samples, according to Equation 5. The time distribution produced by the six

State		0:6	1:5	2:4	3:3	4:2	5:1	6:0
TAVC	$\hat{\mu}$	0.1	2.8	19.3	44.5	27.6	5.3	0.4
	s	0.1	1.7	5.4	6.0	7.0	2.6	0.4
R	$\hat{\mu}$	1.0	7.4	22.3	33.7	25.3	9.3	1.0
	s	0.6	1.6	1.5	2.2	1.6	1.4	0.6
T		1.6	9.4	23.4	31.2	23.4	9.4	1.6

Table 2: State/time distributions with six robots

random controllers is closely aligned with this theoretical estimate of random distribution, though the differences are statistically significant.

$$T = \frac{m!}{n!(m-n)!2^m} \qquad (5)$$

The two time distributions given in Table 2 are presented as histograms in Figure 6 with the standard deviation indicated by the error bars for each state.

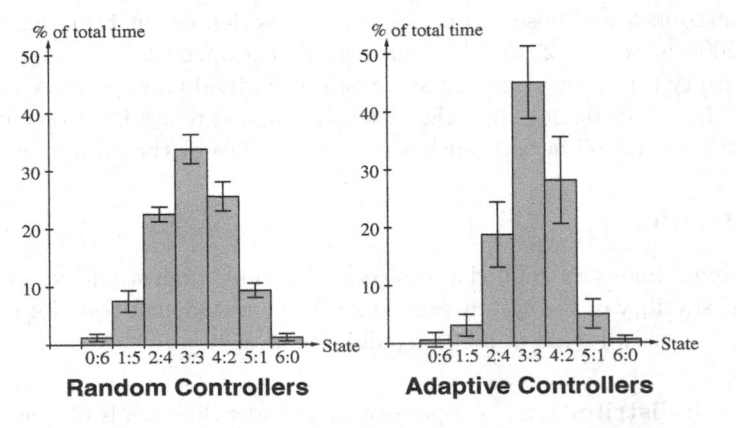

Figure 6: State/time distribution with six robots

The increase in the amount of time spent in state $3 : 3$ is statistically significant. The time the adaptive group spends in state $3 : 3$ is also significantly higher than the time spent in any of the other states.

Figure 7 presents the mean performance of a group of robots controlled by the TAVC algorithm over both the convergence period and the stable period. The TAVC performance is indicated by the thick, solid line. The performance of random allocation is indicated by the dashed line. The values used for the performance plots presented here are 10.0 and 1.0 for the high-value and low-value circuits, respectively.

The performance of a group of robots controlled by the TAVC algorithm is significantly higher than the performance of random allocation.

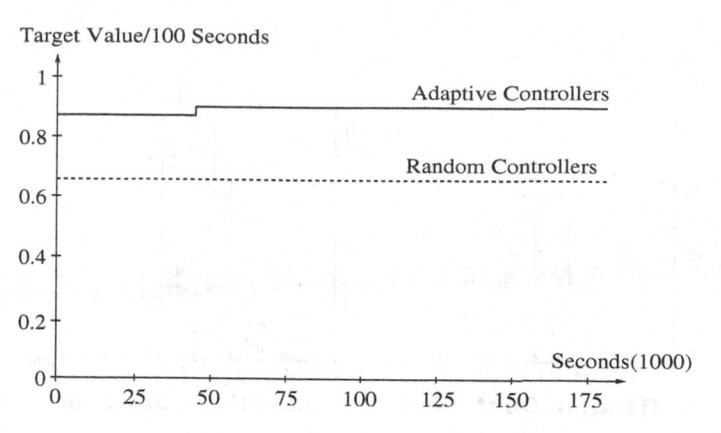

Figure 7: Performance with six robots

Vacancy from robot breakdown This experiment started with the system in the converged state of the 'initial task distribution' experiment. We choose at random one of the robots dedicated to the high-value circuit for removal. According to the MRTA model, that circuit then had a vacancy. We performed 20 experiments of 7.5 hours. The convergence period was 2.5 hours. The values of the time distribution in the stable period are given in Table 3 and a graphical presentation is provided in Figure 8.

State		0:5	1:4	2:3	3:2	4:1	5:0
TAVC	$\hat{\mu}$	0.3	6.5	35.7	46.2	10.4	0.6
	s	0.3	3.8	9.0	8.6	4.7	0.6
R	$\hat{\mu}$	2.2	13.9	31.0	32.9	16.1	2.8
	s	0.8	2.6	2.2	2.3	2.6	0.4
T		3.1	15.6	31.3	31.3	15.6	3.1

Table 3: State/time distribution after breakdown creating a vacancy

The converged controllers kept the system in state 3 : 2 for a significantly larger amount of time than a group of five random controllers. The group adapted its structure from one that promotes the 3 : 3 state to one that promotes the 3 : 2 state. This change implies that a robot from the low-value circuit has filled the vacancy we created in the high-value circuit.

The performance data presented in Figure 9 show that on the removal of a robot from the high-value circuit, the performance drops sharply. After the re-convergence period, however, the performance rises again to a level that is significantly higher than the performance of five random controllers. The new performance level is also significantly higher than the mean performance, over 20 trials, of a group of robots controlled by a static TA algorithm optimized for six robots. The mean performance of the static group is indicated by the thin solid line.

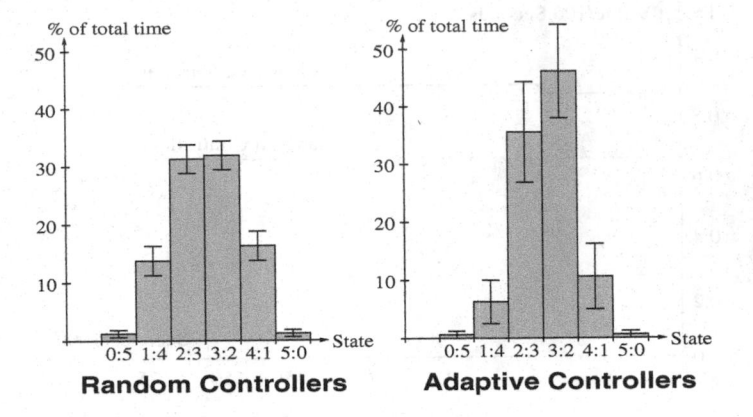

Figure 8: State/time distribution after breakdown creating a vacancy

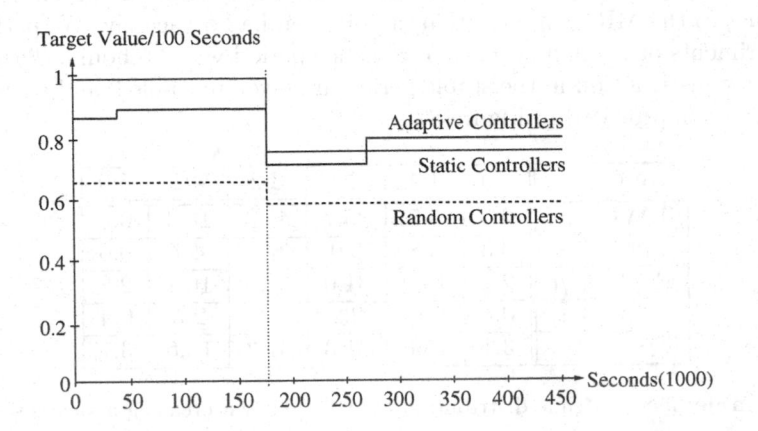

Figure 9: Performance during breakdown creating a vacancy

Breakdown without vacancy This experiment had an initial state identical to the 'vacancy from robot breakdown' experiment. This time, however, we removed a robot from the low-value circuit. We performed 20 trials. The convergence time was 2.5 hours. The time distribution during the stable period of this experiment, presented in Table 4, was not significantly different from the distribution produced during the experiment where a vacancy was created by the removal of a robot.

State	0:5	1:4	2:3	3:2	4:1	5:0
$\hat{\mu}$	0.3	6.7	34.6	47.1	10.5	0.7
s	0.3	3.7	9.2	9.4	3.8	0.4

Table 4: State/time distribution after breakdown not creating a vacancy

As shown in Figure 10, performance fell significantly when the robot was removed, but remained significantly higher than the performance of random allocation among five robots. There was no significant difference in the performance during the stable period of this experiment and the stable period during the experiment where a vacancy was created. Also, there was no significant difference in performance between the convergence period and the stable period. This consistency in performance likely reflects the fact that the group structure remained unchanged.

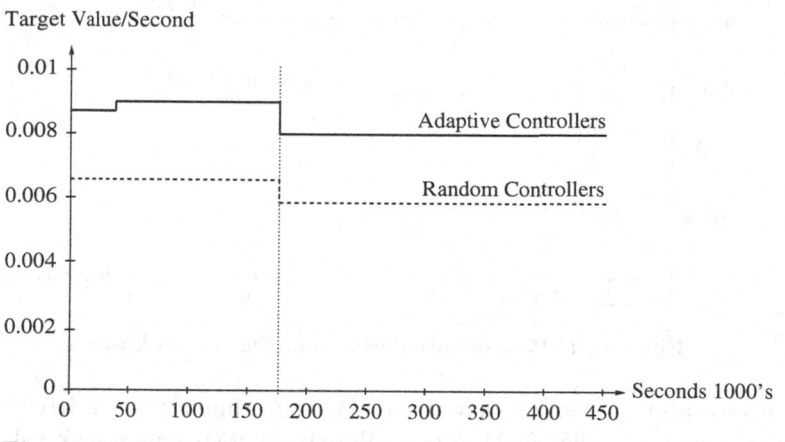

Figure 10: Performance during breakdown not creating a vacancy

This result demonstrates that our algorithm produces the group structure required for VC distribution, independent of which robot should fail.

Changing sub-task values The start state of this experiment was the converged state of the 'vacancy from robot breakdown' experiment. We switched the values of the tasks so that the high-value circuit became the low-value circuit and vice versa. According to the TAVC model, the new setup had a vacancy on the new high-value circuit. We ran 20 individual trials using this setup. The convergence time was 50 minutes.

The time distribution for the stable period, given in Table 5, had significantly higher values for states 3 : 2 and 4 : 1 than the time distribution produced by the experiment where a vacancy was created by the removal of a robot. Compared to the time distribution produced by a group of five robots where tasks were allocated at random, the distribution produced by the TAVC algorithm had a significantly higher value for state 3 : 2 and significantly lower values for *all* other states.

State	0:5	1:4	2:3	3:2	4:1	5:0
$\hat{\mu}$	0.2	5.4	30.3	48.7	14.4	0.9
s	0.5	3.9	5.2	6.1	1.6	0.3

Table 5: State/time distribution after changing sub-task values

When the task values were changed the performance, presented in Figure 11, fell significantly. After the re-convergence period, however, it was back up to a level that was not significantly different from the initial level.

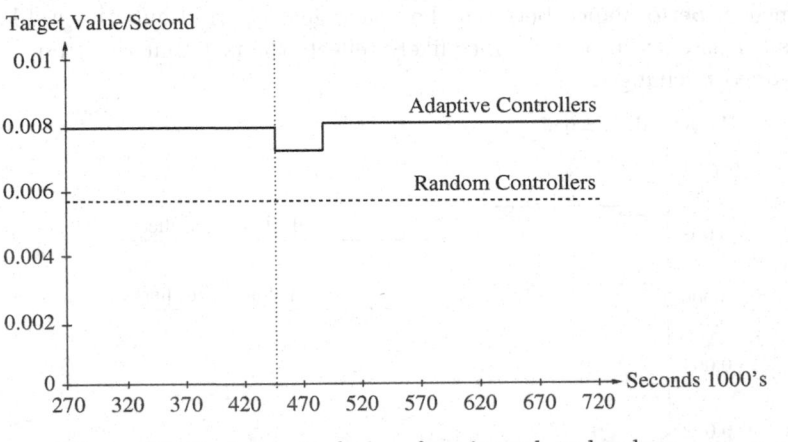

Figure 11: Performance during changing sub-task values

This experiment showed that the TAVC algorithm is not sensitive to how a vacancy is created, whether by robot failure or by a change in task values.

5.6   Conclusions

In the 'initial task distribution' experiment we showed that the performance of the TAVC algorithm was significantly above the performance level of a system where tasks were allocated at random.

In the 'vacancy from robot breakdown' experiment we showed that the performance of the TAVC algorithm after re-convergence was higher than a mis-configured pre-programmed solution. Assuming that a model of the group dynamics is not available in general, a pre-programmed solution must be based on some heuristics that will occasionally lead to sub-optimal task allocations. Our experiment shows that the TAVC algorithm has the potential to improve on pre-programmed solutions.

Together, the experiments validate our TAVC model by reproducing the task allocations predicted by the model. They also demonstrate that the TAVC algorithm outperforms random allocation and can outperform pre-programmed solutions under the assumption that pre-programmed solutions must occasionally make sub-optimal task allocations.

6   Performance-related specialization

The analysis of the TAVC experiments presented in Section 5.4 indicated that persistence with a class of tasks was a determining factor in the estimated utility that class. By persisting with a class of tasks a robot contributes to the level of interference around the related source and sink. This leads to a general fall in the estimated utility of that class of tasks for all the robots in the group. When the fall leads to a situation where one of the robots involved estimates another class of tasks to have a higher utility, that robot will migrate to the class of tasks. We refer to this effect of the group dynamics as *forced migration through persistence*.

A severe limitation of our TAVC algorithm as presented in Section 5 is that it is limited to homogeneous groups of robots. To remove this limitation, we modified it to be sensitive to the individual performance levels of the robots in a *heterogeneous* group of robots.

We hypothesized that this could be achieved by using the perceived ability of robots to force migration through persistence. If performance could be related to persistence in a way that made robots with high performance levels more persistent, then the observed effect of forced migration through persistence would allocate robots with high performance levels to high-value classes of tasks and robots with low performance levels to low-value classes of tasks. This would produce a completely communication-free algorithm for MRTA that is sensitive to individual robot performance levels as a result of being sensitive to the general effects of the group dynamics.

In this section we present action selection through a Boltzmann-softmax function as a way of relating a robot's performance level to its persistence. We present experimental evidence showing that the TAVC algorithm, when using a Boltzmann-softmax action selection function, is sensitive to the performance level of the individual robots to a degree which produces a significant difference in system performance over a system that allocates tasks to robots at random.

6.1   Action Selection

The TAVC algorithm used a greedy-ϵ action selection function. The problem with using this function is that all the tasks, apart from the one with the highest estimated utility, have equal probability, ϵ, of being explored. With a Boltzmann-softmax action selection function however, the probability of trying a task is correlated with the relative estimated utility of that task. The probability of selecting action, a, according to a Boltzmann-softmax function in given in Equation 6.

$$P(a) = \frac{e^{Q_t(a)/\tau}}{\sum_{b=1}^{n} e^{Q_t(b)/\tau}} \tag{6}$$

A robot that on average has task processing time of $\bar{p}$ will have a difference in estimated task utility that correlates with the expression $\frac{w_h - w_l}{\bar{p}}$ where w_h and w_l are the values of high- and low-value tasks respectively. A robot with a lower mean task processing time will have a correspondingly higher difference in estimated utility of high- and low-value tasks. This greater difference in estimated utility implies a greater persistence on high-value tasks in terms of a higher probability of servicing these tasks. This higher persistence can hypothetically lead to fast robots forcing the slow robots to migrate from high-value tasks to low-value tasks.

The experimental results presented in this section demonstrate that the ability of the Boltzmann-softmax action selection function to differentiate between suboptimal actions can indeed be made to work on an inter-robot level, as a mechanism for allocating high-value tasks to high-performance robots without explicit communication.

6.2   Experimental Validation

We designed an experiment to test our hypothesis that a Boltzmann-softmax action selection rule could relate a robot's individual performance level to its persistence and that a difference in persistence would lead to high-value tasks being allocated to robots with high performance levels and low-value tasks to robots with low performance levels.

The simulated environment was identical to the one used in the experiments presented in Section 5.4.

Control architecture   We made one change to the system architecture presented in Section 5.4. The original group of six robots was made heterogeneous by dividing it into two sub-groups of three. One sub-group was made to operate at a default speed of 300 mm/sec while the other was made to operate at a default speed of 200 mm/sec. These speeds were chosen as equidistant points from the default operating speed of 250 mm/sec used in the homogeneous robot experiments. The equidistance was intended to preserve the allocation of three robots to the high-value circuit and three to the low-value circuit.

The general difference in default speed of operation encompasses more specific differences in robot morphology and task competence. Hence, the results presented here generalize to heterogeneous multi-robot systems where the differences between the participating robots can be expressed in terms of differences in task processing speed.

Learning parameters   The Q-tables ware initialized with random values between -0.1 and 0.1, the learning rate, α, was set to 0.1, and the discount factor γ was set to 0.95.

For the Boltzmann-softmax function, the temperature parameter, τ, was set empirically to 0.005. Because we wanted the system to remain adaptive to changes in the environment we did not decrease τ over time, as is common. The reward structure was identical to the one presented in Section 5.4.

6.3   Results

We defined a convergence period of 15 hours based on the stability of the system performance. To examine the performance we consider which of the robots visited the high-value circuit last. We used three fast and three slow robots, yielding fifteen possible system states. We refer to each state using the notation $f : s$, where f is the number of fast robots whose last target was on the high-value circuit. Correspondingly, s is the number of slow robots whose last target was on the high-value circuit. The columns with subscript VC in Table 6 show the mean and standard deviation of the time the system spent in each of the states while running the TAVC algorithm. The values are percentages of the total stable period. The columns with subscript R describe the same values for a set of 15 trials using randomly tasks allocations.

$f : s$	$\hat{\mu}_{VC}$	s_{VC}	$\hat{\mu}_R$	s_R	T	$\hat{\mu}_{VC} - \hat{\mu}_R$	$\frac{\hat{\mu}_{VC} - \hat{\mu}_R}{\hat{\mu}_R}$
0:0	0.2	0.2	1.4	0.5	1.5	-1.3	-0.88
0:1	1.5	1.2	3.4	1.7	4.7	-3.2	-0.67
0:2	2.0	2.6	5.0	1.4	4.7	-2.7	-0.58
0:3	0.5	0.7	2.0	1.1	1.5	-1.1	-0.70
1:0	2.9	1.6	3.6	1.4	4.7	-1.8	-0.38
1:1	12.1	4.9	13.5	3.3	14.1	-2.0	-0.14
1:2	14.4	6.1	14.4	2.4	14.1	0.3	0.02
1:3	4.1	2.7	4.0	1.4	4.7	-0.6	-0.13
2:0	7.0	5.0	5.1	1.2	4.7	2.3	0.48
2:1	19.4	7.4	15.7	2.3	14.1	5.30	0.37
2:2	18.5	5.0	14.7	3.0	14.1	4.4	0.3
2:3	5.7	3.6	4.4	2.4	4.7	1.0	0.20
3:0	2.7	4.1	1.7	0.6	1.5	1.2	0.78
3:1	5.2	4.6	5.2	1.6	4.7	0.5	0.10
3:2	3.3	2.6	4.2	1.2	4.7	-1.4	-0.29
3:3	0.7	0.7	1.5	0.8	1.5	-0.9	-0.56

Table 6: State/time distribution for six heterogeneous robots

The column labeled T lists the probability of choosing a sample of size f from a population of $g = 3$ fast robots as well as choosing a sample of size s from a population of $h = 3$ slow robots according to Equation 7. The time distribution produced by the six random controllers is closely aligned with this theoretical estimate, though the differences are statistically significant.

$$T = \frac{100g!h!}{f!(g-f)!s!(h-s)!2^g2^h} \tag{7}$$

The difference between the state-time distribution produced by the TAVC algorithm and the distribution produced by random task allocation is presented in the column labeled $\hat{\mu}_{VC} - \hat{\mu}_R$. This difference is presented as a percentage of the mean of the random distribution, $\hat{\mu}_R$, for each state in the last column, labeled $\frac{\hat{\mu}_{VC}-\hat{\mu}_R}{\hat{\mu}_R}$.

The time distributions for the random allocation and the TAVC algorithm, i.e., columns 2 and 4 of Table 6, are presented graphically in Figure 12.

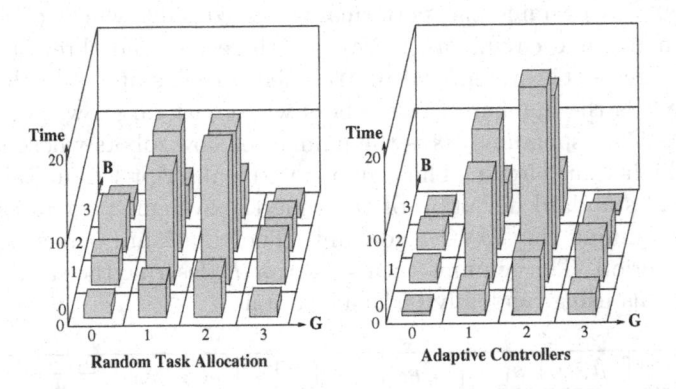

Figure 12: State/time distributions for six homogeneous robots

The differences between the distributions produced by the TAVC algorithm and the random task allocation, i.e., the last two columns, are also presented graphically in Figure 13.

Over 15 experiments, the difference in time spent in state 0 : 3 is statistically significant. The second histogram in Figure 13 shows the differences relative to an underlying random distribution. The optimal state, 0 : 3 stands out as the state with the highest relative increase. This validates the TAVC model by confirming that the group behavior has converged to promote the state predicted by the TAVC model. The performance data shows that the TAVC algorithm produces a total target value throughput of $0.81 * 10^{-2}$ per second. This is significantly higher than the performance of a group of robots where tasks are allocated randomly, $0.78 * 10^{-2}$ per second. Together, the time distribution data and the performance data show that the TAVC algorithm improves the group's performance by promoting the dedicated service structure predicted by the TAVC model. The transition functions also indicate this. The fast robots, on average, have higher estimated utilities for servicing the high-value circuit

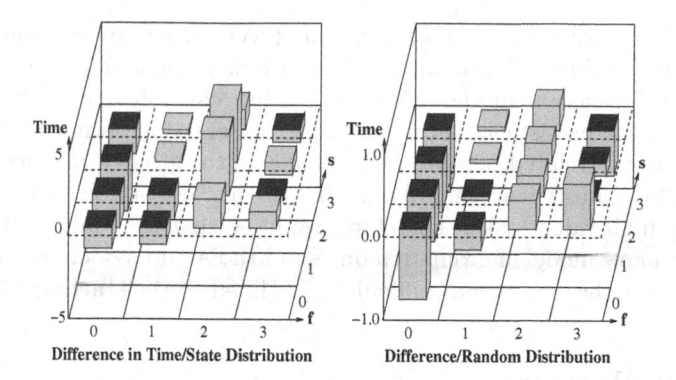

Figure 13: Difference in the state/time distributions produced by TAVC and random task allocation

while the slow robots have higher estimated utilities for servicing the low-value circuit. Finally, as predicted, the mean difference between the estimated utilities are greater for the fast robots than for the slow robots.

6.4   Conclusions

Our results demonstrate that when using a Boltzmann-softmax action selection function, there is an observable migration through persistence effect. This validates our hypothesis that a performance-related persistence affects the group dynamics in a way that promotes task allocations where high-value tasks are allocated to high-performance robots and low-value tasks are allocated to low-performance robots.

Our results demonstrated a significant improvement in group performance over a control group using random task allocation. This shows that the TAVC algorithm is able to produce task allocations that are sensitive to the underlying group dynamics, including differences in individual robot performance levels.

7   Summary and conclusions

We have discussed three main challenges in producing high-quality solutions to MRTA problems: modeling group dynamics, problem complexity, and communication overhead. We then presented a learning algorithm that used individual learning and feedback that reflected the effects of the prevailing group dynamics. We presented experimental evidence that the algorithm produces effects that improve group performance. We then showed how one such effect, specialization, can be used to construct a MRTA algorithm based on a model of MRTA as TAVC. The TAVC algorithm is an important contribution to existing MRTA algorithms in that it is demonstrably sensitive to the effects of group dynamics, including the effects of different individual performance levels among the robots in a group. We presented experimental evidence that validated the

TAVC model of MRTA and showed that the TAVC algorithm can out-perform pre-programmed solutions. Finally we showed how using a Boltzmann-softmax action selection function in the TAVC algorithm can relate a robot's performance level to its persistence level. We also presented experimental evidence that a performance-related persistence level affects the group dynamics and produces task allocations where the robots with the highest performance levels are allocated the high-value tasks. Therefore, we hope that the main contributions of this work, a new model of group dynamics in MRTA and associated algorithm, will help address the three main difficulties outlined at the start.

Acknowledgments

This work was supported in part by a Department of Energy (DOE) Robotics and Intelligent Machines (RIM) grant DE-FG03-01ER45905 and in part by an Office of Naval Research (ONR) Defense University Research Instrumentation Program (DURIP) grant 00014-00-1-0638.

The Player/Stage robot device server and simulator was developed originally by Brian Gerkey, Kasper Støy, Andrew Howard and Richard Vaughan at the Center for Robotics and Embedded Systems (CRES) at the University of Southern California. The laser bar-code technology was developed by Andrew Howard. These technologies are freely available under the GNU General Public License.

References

[1] ARKIN, Ronald C., *Behaviors-based robotics*, MIT Press Cambridge, Massachusetts (1998).

[2] BALCH, Tucker R., "The impact of diversity on performance in multi-robot foraging", *The proceedings of the Third International Conference on Autonomous Agents (Agents'99)* (Seattle, Washington, May 1 - 5, 1999) (O. ETZIONI, J. P. MÜLLER, AND J. M. BRADSHAW eds.), ACM Press, 92–99.

[3] BALCH, Tucker R., Zia KHAN, and Manuela M. VELOSO, "Automatically tracking and analyzing the behavior of live insect colonies", *Proceedings of the Fifth International Conference on Autonomous Agents (Agents'01)* (Montreal, Canada, May 31 - June 4, 2001) (J. P. MÜLLER, E. ANDRE, S. SEN, AND C. FRASSON eds.), ACM Press, 521–528.

[4] BOTELHO, Silvia, and Rachid ALAMI, "M+ : a scheme for multi-robot cooperation through negotiated task allocation and achievemen", *Proceedings of the 1999 IEEE International Conference on Robotics and Automation (ICRA'99)* (Detroit, Michigan, May 10-15, 1999), IEEE Press, 1234–1239.

[5] BOWLING, Michael, and Manuela M. VELOSO, "Rational and convergent learning in stochastic games", *Proceedings of the 17th International Joint Conference on Artificial Intelligence (IJCAI'01)* (Seattle, Washington, August 4 - 10, 2001) (B. NEBEL ed.), Morgan Kaufmann, 1021–1026.

[6] BRUCKER, Peter, *Scheduling Algorithms* second ed., Springer Verlag Berlin, Germany (1998).

[7] CHASE, Ivan D., Marc WEISSBURG, and Theodore H. DEWITT, "The vacancy chain process: a new mechanism of resource distribution in animals with application to hermit crabs", *Animal Behavior* **36** (1988), 1265–1274.

[8] CORMEN, Thomas H., Charles E. LEISERSON, Ronald L. RIVEST, and Clifford STEIN, *Introduction to algorithms* second ed., MIT Press Cambridge, Massachusetts (2001).

[9] DAHL, Torbjørn S., Maja J. MATARIĆ, and Gaurav S. SUKHATME, "Adaptive spatio-temporal organization in groups of robots", *Proceedings of the 2002 IEEE/RSJ International Conference on Intelligent Robots and Systems (IROS'02)* (Lausanne, Switzerland, September 30 - October 4, 2002), IEEE Press, 1044–1049.

[10] DAHL, Torbjørn S., Maja J MATARIĆ, and Gaurav S. SUKHATME, "Scheduling with group dynamics: A multi-robot task allocation algorithm based on vacancy chains", *Tech. Rep. no. CRES-002-07*, Center for Robotics and Embedded Systems, University of Southern California, Los angeles, CA (2002).

[11] DAHL, Torbjørn S., Maja J. MATARIĆ, and Gaurav S. SUKHATME, "Multi-robot task-allocation through vacancy chains", *Proceedings of the 2003 IEEE International Conference on Robotics and Automation (ICRA'03)* (Taipei, Taiwan, September 9 - 14, 2003), IEEE Press, 2293–2298.

[12] GERKEY, Brian P., and Maja J MATARIĆ, "Sold!: Auction methods for multi-robot coordination", *IEEE Transactions on Robotics and Automation* **18**, 5 (October 2002), 758–768.

[13] GERKEY, Brian P., and Maja J MATARIĆ, "A framework for studying multi-robot task allocation", *Multi-Robot Systems: From Swarms to Intelligent Automata, Volume II, Proceedings of the 2nd International Naval Research Laboratory Workshop on Multi-Robot Systems* (Washington, DC, March 17 - 19, 2003) (A. C. SCHULTZ ET AL. eds.).

[14] GERKEY, Brian P., and Maja J. MATARIĆ, "A formal analysis and taxonomy of task allocation in multi-robot systems", *International Journal of Robotics Research* (2004), to appear.

[15] GERKEY, Brian P., Richard T. VAUGHAN, Kasper STØY, Andrew HOWARD, Gaurav S. SUKHATME, and Maja J MATARIĆ, "Most valuable

player: A robot device server for distributed control", *Proceedings of the 2001 IEEE/RSJ International Conference on Intelligent Robots and Systems (IROS'01)* (Wailea, Hawaii, October 29 - November 3, 2001), IEEE Press, 1226–1231.

[16] GOLDBERG, Dani, and Maja J MATARIĆ, "Learning multiple models for reward maximization", *Proceedings of the 17th International Conference on Machine Learning (ICML'00)* (Stanford, California, June 29 - July 2, 2000) (P. LANGLEY ed.), Morgan Kaufmann, 319–326.

[17] GOLDBERG, Dani, and Maja J MATARIĆ, "Design and evaluation of robust behavior-based controllers for distributed multi-robot collection tasks", *Robot Teams: From Diversity to Polymorphism*, (T. R. BALCH AND L. E. PARKER eds.). A K Peters Ltd (2001), pp. 315–244.

[18] HAN, Kwun, and Manuela M. VELOSO, "Automated robot behavior recognition applied to robotic soccer", *Proceedings of the 9th International Symposium on Robotics Research (ISRR'99)* (Snowbird, Utah, October 9 - 12, 1999), 199–204.

[19] LERMAN, Kristina, Asram. GALSTYAN, Alcherio MARTINOLI, and Auke J. IJSPEERT, "A macroscopic analytical model of collaboration in distributed robotic systems", *Artificial Life* **7**, 4 (2001), 375–393.

[20] LITTMAN, Michael L., "Markov games as a framework for multi-agent reinforcement learning", *Proceedings of the 11th International Conference on Machine Learning (ICML'94)* (New Brunswick, New Jersey, July 10 - 13, 1994) (W. W. COHEN AND H. HIRSH eds.), Morgan Kaufmann, 157–193.

[21] MATARIĆ, Maja J., *Interaction and Intelligent Behavior*, PhD thesis Massachusetts Institute of Technology Cambridge, Massachusetts (1994).

[22] MATARIĆ, Maja J., "Behavior-based control: Examples from navigation, learning, and group behavior", *Journal of Experimental and Theoretical Artificial Intelligence, special issue on Software Architectures for Physical Agents* **9**, 2-3 (1997), 323–336.

[23] MATARIĆ, Maja J., "Reinforcement learning in the multi-robot domain", *Autonomous Robots* **4**, 1 (1997), 73–83.

[24] MITCHELL, Tom M., *Machine Learning*, McGraw-Hill New York (1997).

[25] MYERSON, Roger B., *Game Theory: Analysis of Conflict*, Harvard University Press Cambridge, Massachusetts (1991).

[26] PARKER, Lynne E., "L-ALLIANCE: Task-Oriented Multi-Robot Learning in Behaviour-Based Systems", *Advanced Robotics, Special Issue on Selected Papers from IROS'96* **11**, 4 (1997), 305–322.

[27] SETH, Anil K., "Modelling group foraging: Individual suboptimality, inter-ference, and a kind of matching", *Adaptive Behavior* **9**, 2 (2001), 67–91.

[28] SUTTON, Richard S., and Andrew G. BARTO, *Reinforcement learning: an introduction*, MIT Press Cambridge, Massachusetts (1998).

[29] WERGER, Barry B., and Maja J MATARIĆ, "Broadcast of local eligibility for multi-target observation", *Distributed Autonomous Robotic Systems 4, Proceedings of the 5th International Symposium on Distributed, Autonomous Robotic Systems (DARS'00)* (Knoxville, Tennessee, October 4-6, 2000) (L. E. PARKER, G. BEKEY, AND J. BARHEN eds.), Springer, 347–356.

[30] YAN, Helen, and Maja J MATARIĆ, "General spatial features for analysis of multi-robot and human activities from raw position data", *Proceedings of the 2002 IEEE/RSJ International Conference on Intelligent Robots and Systems (IROS'02)* (Lausanne, Switzerland, September 30 - October 4, 2002), IEEE Press, 2770–2775.

Chapter 15

Towards Pro-active Embodied Agents: On the Importance of Neural Mechanisms Suitable to Process Time Information

G. de Croon
IKAT, Universiteit Maastricht
P.O. Box 616 - 6200 MD - Maastricht - The Netherlands
e-mail: g.decroon@cs.unimaas.nl
S. Nolfi
Institute of Cognitive Sciences and Technologies, CNR
Viale Marx, 15 - 00137 - Rome - Italy
e-mail: s.nolfi@istc.cnr.it
E.O. Postma
IKAT, Universiteit Maastricht
P.O. Box 616 - 6200 MD - Maastricht - The Netherlands
e-mail: postma@cs.unimaas.nl

In Embodied Cognitive Science, many studies have focused on reactive agents, i.e. agents that have no internal state and always respond in the same way to the same stimulus. However, this particular focus is not due to a rejection of the importance of internal states. Rather, it is due to the difficulty of developing pro-active embodied and situated agents, that is agents able to: (a) extract internal states by integrating sensorymotor information through time and, (b) later use these internal states to modulate their motor behaviour according to the current environmental circumstances. In this chapter we will focus on how pro-active agents can be developed and, more specifically, on which are the neural mechanisms that might favour the development of pro-active agents. By comparing the results of five sets of evolutionary experiments in which simulated robots are provided with different types of recurrent neural networks, we gain insight into the relation between the robots' capabilities and the characteristics of their neural controllers. We show how special mechanisms for processing information in time facilitate the exploitation of internal states.

1 Introduction

A new research paradigm, that has been called Embodied Cognitive Science [18], has recently challenged the traditional view according to which intelligence is an abstract process that can be studied without taking into consideration the physical aspects of natural systems. In this new paradigm, researchers tend to stress situatedness, i.e., the importance of studying systems that are situated in an external environment [3, 4], embodiment, i.e., the importance of study systems that have bodies, receive input from their sensors and produce motor actions as output [3, 4], and emergence, i.e. the importance of viewing behaviour and intelligence as the emergent result of fine-grained interactions between the control system of an agent including its constituent parts, the body structure, and the external environment. An important consequence of this paradigm is that the agent and the environment constitute a single system, i.e. the two aspects are so intimately connected that a description of each of them in isolation does not make much sense [9, 10, 1].

Research in Embodied Cognitive Science often involves simple agents called "reactive agents" [14]. These are agents in which sensors and motors are directly linked and that always react with the same motor action to the same sensory state. In reactive agents internal states (see next section) do not play a role in determining the motor behaviour. The fact that the vast majority of research in this area focuses on simple reactive agents, however, is not due to a rejection of the importance of internal states. Rather, it is due to the difficulty of developing pro-active embodied and situated agents, that is agents able to: (a) extract internal states by integrating sensory-motor information through time and, (b) later use these internal states to modulate their motor behaviour according to the current environmental circumstances. In this paper we will focus on how pro-active agents can be developed and, more specifically, on which are the neural mechanisms that might favour the development of pro-active agents.

Given the difficulty of developing embodied and situated agents through explicit design [13] our attempt to develop pro-active agents will be based on an evolutionary robotics method [16], that is on the attempt to develop these agents through a self-organisation process that allows the evolving robots to develop their skills in interaction with the environment and without human intervention. By comparing the results of five sets of evolutionary experiments in which simulated robots are provided with different types of recurrent neural networks, we will try to understand the relation between the robots' capabilities and the characteristics of their neural controllers. In addition, we will show how special mechanisms for processing information in time facilitate the exploitation of internal states.

The paper is organised as follows. We define the term internal state in §2. In §3 we describe our experimental test bed which consists of a self-localisation problem that cannot be solved through simple reactive strategies. In §4, we review five different neural models described in the literature that are potentially suitable to develop pro-active agents. We describe the results of the experiments and the comparison of the results obtained with the five different neural architectures in §5. Finally, in §6, we discuss the implication of the obtained results and, in particular, the neural mechanisms that seem to constitute a pre-requisite for the emergence of powerful pro-active agents.

2   Internal state

The concept of internal state plays a central role in our investigations. In this section, we define the concept with particular reference to neural network controllers.

An *internal state* is a set of variables of the agent's controller that might be affected by the previous sensory states perceived by the agent and/or the previous actions performed by the agent and that might co-determine, together with the current sensory states, current and future motor actions. By mediating between perception and actions, internal states allow agents to produce behaviour that is decoupled from the immediate circumstances while still remaining sensitive to them.

An internal state can consist of different entities. For example, in the case of a neural controller, they might consist of the activation states of some neurons and/or in the strength of the synaptic weights. It should be noted that there is not a one to one correspondence between the architecture of the controller and the type of strategy adopted by evolving individuals. For instance, although an individual provided with a recurrent neural network controller might potentially develop an ability to integrate information over time, it might also rely on a simple reactive strategy.

As we claimed in the previous section, agents that do not have an internal state are reactive agents, that is agents that always react with the same motor action to the same sensory state. Agents that have an internal state are pro-active instead, that is agents that are able to integrate sensory-motor information through time into internal states that co-determine the agents' motor behaviour.

In the context of neural network controllers, reactive agents are provided with feed-forward neural networks, that is neural networks in which sensory neurons are connected to motor neurons directly or through one or more layers of hidden neurons that do not have a recurrent connection (Fig. 1). In Fig. 1 s1 and s2 represent sensory neurons (also called input units). h1, h2, and h3 represent hidden neurons. o1 represents an output neuron. The bias neuron is a special neuron whose activation state is always 1.0. In these neural networks, neurons are updated in discrete time steps and the activation state of motor neurons and hidden neurons only depends on the activation state of the sensors and on the connection weights that are kept fixed during the lifetime of the agent.

Pro-active agents instead are provided with neural controllers that have an internal state. An internal state can be realised through different neural mechanisms. One possibility, for instance, is to provide a neural controller with recurrent connections. For example, in the neural network shown in Fig. 2, the gray hidden neuron receives connections not only from the sensory neurons but also from the hidden neurons including itself. This implies that the activation state of this hidden neuron is not only a function of the activation of the sensory neurons at time t, but also of the hidden neurons at time t-1. Given that the state of the hidden neuron at time t-1 is also affected by the state at time t-2 and so on, this implies that the activation state of this hidden neuron, that influences the state of the motors at time t, might be influenced by the sensory states previously experienced by the robot.

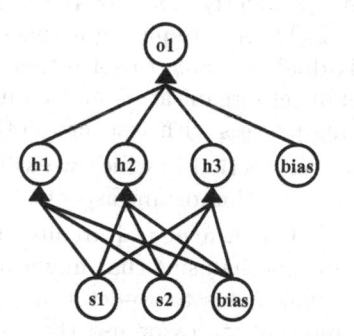

Figure 1: A feedforward neural network

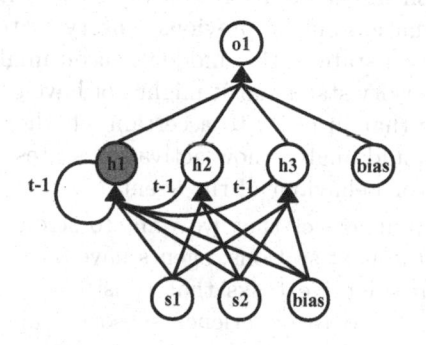

Figure 2: A recurrent neural network.

The recurrent neural connections, however, are only one of the possible neural mechanisms that might realise internal states. Other mechanisms include: (a) dynamical neurons, in which the activation state of a neuron is influenced by its previous activation state; (b) time delayed connections in which the propagation of activation through connections takes time so that the activation state of a neuron might influence the activation of other neuron after some time, (c) networks in which connection weights vary according to learning rules affected by the activation state of the neurons. One of the goals of this paper is indeed to compare the characteristics and the effectiveness of different mechanisms for realising internal states.

Given that internal states might be realised through several different mechanisms and given that these mechanisms might be also combined together we do not pretend to be exhaustive in our analysis. Indeed we will restrict our comparison and analysis only to some possible ways of realising the mechanisms (a) and (b) described above. Moreover, although in some cases different algorithms might be used to set the connection weights, we will restrict our analysis to neural controllers whose connection weights are evolved through a form of evolutionary algorithm [16]. The reason for this choice is twofold: (1) an evolutionary algorithm can be used to evolve the connection weights and other free parameters of the neural controllers independently from the particular neural architecture or neural model used, and (2) by only requiring a general criterion for evaluating how much evolving individuals are able to solve their adaptive task, they allow us to maximise the level of self-organisation and reduce the externally imposed constraints on the learning process with respect to other learning algorithms.

Before moving to the next section in which we will present our experimental setup, we should emphasize two important aspects.

First, an agent provided with a neural architecture with recurrent neural connections or other neural mechanisms that might allow it to extract internal states and use these internal states to co-determine its motor behaviour does not necessarily extract internal states or use them to co-determine its motor behaviour. In the case of the neural controller illustrated in Fig. 2, for example, due to a given configuration of the connection weights, the activation state of the gray hidden neuron might be always off or always on and therefore might not provide any information on the previous sensory states experienced by the robot. Or, the activation state of this hidden neuron might vary and might be affected by previous sensory states but it might not have any affect on the motor neurons. This implies that in order to ascertain whether an agents really is a pro-active agent we should analyze how activation states vary in time and how they influence the motor behaviour of the agent.

The second important aspect that we want to stress here is the fact that, as we will also see in the next sections, agents have often two options available in order to solve their adaptive tasks that consist of: (1) use sensory-motor coordination, that is act so to experience sensory states that allow to solve the problem through a reactive control mechanism [14], and (2) extract internal states and use them to co-determine the way in which the agents react to sensory states. Reactive solutions based on sensory-motor coordination are often simpler and easier to find through artificial evolution and are therefore preferred when available. This means that the emergence of pro-active control strategies only tends to be observed when reactive solutions and sensory-motor coordination are insufficient.

3 Self-localisation task

To investigate the issue described above we evolved the neural controllers of simulated robots that are asked to move and to self-localise in their environment [15] and we compared the results obtained by providing evolving robots with different types of neural controllers. More precisely the agent has to drive around a loopy corridor and to indicate with an output neuron in which room it is currently located. Fig. 3 is a drawing of the environment for this task. The arrows indicate the direction in which the agents are forced to drive. The two rooms are painted in different shades of grey. If the agent is in the top room (light grey), the localisation output neuron has to have a value in the interval $[0, 0.5]$ to be correct. In the bottom room (dark grey) this value has to be in $\langle 0.5, 1]$. Fig. 4 is also a drawing of the environment, displaying the zones in the environment that are used during evolution to stimulate agents to drive around in the environment.

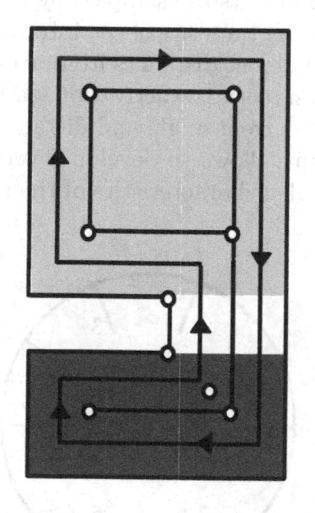

Figure 3: Environment and forced driving direction

The agent we use for the self-localisation task is the Kephera robot [11] (shown in Fig. 5), a miniature mobile robot with a diameter of 55 mm and a weight of 70 g. It is supported by two lateral wheels that can rotate in both directions and two rigid pivots in the front and in the back. By spinning the wheels in opposite directions at the same speed, the robot can rotate without lateral displacement. The sensory system employs eight infrared sensors that

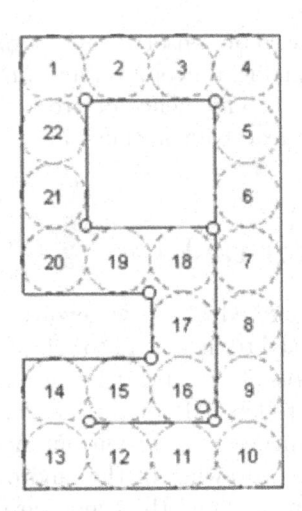

Figure 4: Zones in the environment

are able to detect obstacles up to about 4 cm. Experiments were conducted in simulation by using an extended version of Evorobot [12]. In Evorobot, a sampling procedure is used to compute the activation state of the infrared sensors. Walls and cylindrical objects are sampled by placing one physical robot in front of them and by recording the state of the infrared sensors while the robot is turning 360 degrees at 20 different distances from of each object. These recorded values are used in simulation to set the activation states of the simulated infrared sensors on the basis of the current angle and distance of the robot with respect to obstacles. This procedure allows to develop a very accurate simulation that takes into account the detailed characteristics of the individual robot used in the experiments [16].

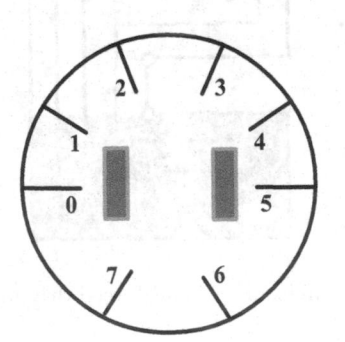

Figure 5: Diagram of a Kephera robot with its sensors

Each evolutionary run begins with an initial population that consists of 100 randomly generated genotypes. A genotype consists of a string of parameters that are encoded in the genotype with 8 bits. During evolution each individual of the population is allowed to 'live' for 4 epochs consisting of 2500 time steps (a time step lasts 100ms). In each epoch, the agent starts at a different position in the environment. The 20 fittest individuals of each generation are allowed to reproduce by generating 5 copies of their genotype with 2% of their bits replaced with a new randomly selected value. The process is repeated for 500 generations.

The fitness function is set up to reward first the ability of the robot to travel in a clockwise direction in the environment and then its ability to indicate in which room in the environment it is located:

$$
F = \begin{cases} \frac{z_c}{z_t} & \text{, if } z_c < z_t \\ 1 + (\text{bottom} * \text{top}) & \text{, if } z_c >= z_t \end{cases} \tag{1}
$$

, where F is the fitness. Furthermore, z_c is the number of zones that the robot has crossed in its lifetime. The zones are illustrated in Fig. 4. z_t is the zone threshold and determines how fast the robot has to drive, before its capacities for localisation are considered to determine its fitness. If the agent crosses more than z_t zones during its lifetime, the extra amount of zones has no effect on its fitness. The self-localisation output of the agent is measured continuously when it is inside one of the two rooms, but is only considered if the agent crosses the zone threshold during the epochs that it is executed. 'bottom' and 'top' are the percentages of good localisations in the bottom and top room, respectively. E.g., 'bottom' is the number of time steps that the self-localisation output is in the interval $\langle 0.5, 1.0]$ and the agent is in the bottom room, divided by the total number of time steps that the agent is in the bottom room. Evidently, the maximal fitness that can be achieved is 2.

The self-localisation task requires the agent to use its internal state. The task is too difficult for a reactive agent, since the two different parts of the environment are largely the same from the viewpoint of the agent. If an agent has the same sensory inputs but it is required to take different actions, it faces a problem of perceptual aliasing [14]. An example of perceptual aliasing is that for the agent's sensors there is no difference between being in the top or bottom horizontal corridor, while the agent has to indicate a different room. A number of factors, such as the forced driving speed and the fact that the corridors are narrow, have as a consequence that reactive agents cannot 'escape' the perceptual aliasing by applying sensory-motor coordination. Experiments performed with reactive agents did not lead to successful individuals [15].

Five classes of experiments in which evolving agents were provided with different type of neural controllers were run. For each neural architecture three experiments with a different driving threshold (22, 23, and 25 rounds, corresponding to $z_t = 440$, $z_t = 506$, and $z_t = 550$, respectively) were run. For each experiment 10 replications were performed. In §4 we describe the five different neural models used. In §5 we describe the obtained results.

4   Five types of recurrent neural controllers

In this section we describe the five types of neural models used to conduct the experiments. All models might allow evolving robots to extract internal states and use these states to co-determine the agents' behaviour. However, different models rely on different neural mechanisms.

4.1   Elman network: EN

The Elman network [6] consists of a neural network with a sensory layer, a layer of hidden neurons, and an output layer. The activation state of the hidden neurons at time t-1 is copied into an additional set of input units at time t.

The architecture used in our experiments consists of 10 sensory neurons, 5 hidden neurons, and 3 output neurons. Two of the output neurons indicate the desired speed of the wheels. We will refer to them as 'motor neurons'. The agent has to indicate with the third output neuron in which room the agent is located. We will refer to this neuron as 'self-localisation output'. The sensory neurons encode the activation state of the 8 infrared sensors, and the activation state of the two motor neurons at time t-1.

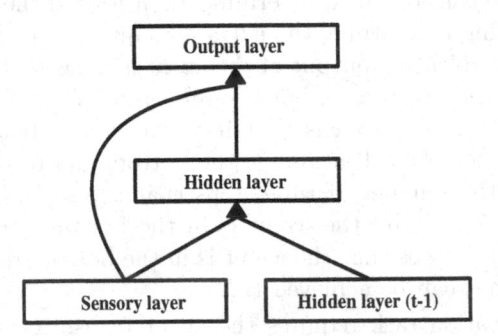

Figure 6: The architecture of the Elman network. Boxes represent collections of neurons. Arrows indicate the connection between collections of neurons (all neurons of the first box are connected to all neurons of the second box).

Hidden and output neurons are activated according to the logistic function. More precisely the activation function of each neuron is:

$$a_i(t) = \sigma(\text{netinput}_i(t) + \text{bias}_i + \text{in}_i(t)) \tag{2}$$

$$\text{netinput}_i(t) = \sum_{j=1}^{N} w_{ji} a_j(t), \tag{3}$$

in which $a_i(t)$ denotes the activation of neuron i at time t and σ is the logistic function, $\sigma(x) = \frac{1}{1+e^{-x}}$. N is the number of neurons connected to neuron i. In the case of the hidden neurons this is the number of sensory neurons plus the number of neurons in the hidden layer. w_{ji} is the weight of the connection from neuron j to neuron i. The external input is represented by $in_i(t)$.

During the evolutionary process the architecture is kept fixed. Only the biases and the synaptic strengths of the connections are encoded in the genotype and allowed to change. All parameters are encoded in the genotype with 8 bits. Connection weights and biases are then normalised in the range [-5.0, 5.0].

4.2 Non-linear autoregressive model with exogeneous inputs: NARX

Nonlinear autoregressive neural networks with exogeneous inputs [7] are an extension of Elman Networks in which the activation state of the sensory neurons at time t, $t-1$, ..., $t-c_{in}$, and the activation state of the output neurons at time $t-1$, $t-2$, ..., $t-c_{out}$ determine the activation of the output and the hidden neurons at time t. The activation of the hidden neurons is also determined by the activation state of the hidden neurons at time t-1.

The architecture used in our experiments consists of 8 sensory neurons, 5 hidden neurons, and 3 output neurons. The sensory neurons encode the activation state of the 8 infrared sensors. The output neurons encode the desired speed of the two wheels and the self-localisation output.

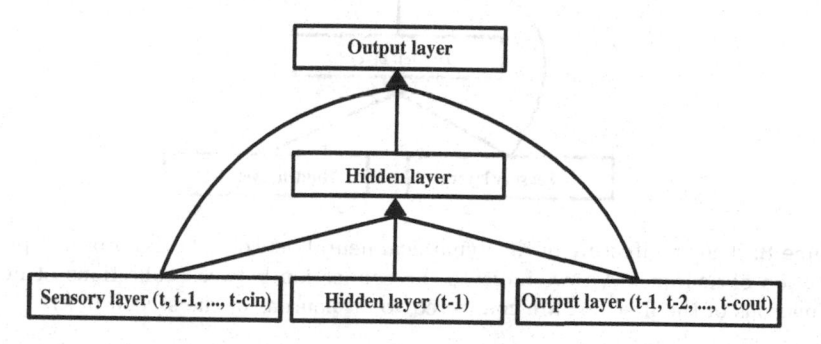

Figure 7: The architecture of the NARX network. Boxes represent collections of neurons. Arrows indicate the connection between collections of neurons (all neurons of the first box are connected to all neurons of the second box).

Hidden and output neurons are activated according to the logistic function. During the evolutionary process the architecture is kept fixed. Only the synaptic strengths of the connections are encoded in the genotype and allowed to change. All parameters are encoded in the genotype with 8 bits. Connection strengths and biases are then normalised in the range [-5.0, 5.0].

4.3 Dynamical Neural Network: DNN

Dynamical neural networks [17, 15] are neural networks constituted by dynamical artificial neurons, that is neurons that tend to vary their activation state at different time rate according to a time constant parameter.

Neurons are updated according to the following activation function:

$$a_i(t) = \mathrm{tc}_i a_i(t-1) + (1 - \mathrm{tc}_i)\sigma(\text{netinput}_i(t) + \text{bias}_i + \text{in}_i(t)), \qquad (4)$$

where tc_i is the parameter adjusted by the evolutionary algorithm that determines the proportion of neural inertia, $\mathrm{tc}_i \in [0,1]$. netinput$_i(t)$ is defined as in equation (3). In the DNNs that we apply to the self-localisation task, only the sensory neurons and the hidden neurons apply activation function (4), the output neurons apply activation function (2). As a consequence, both the activations of the sensory neurons and of the hidden neurons are part of the internal state. The activations of the hidden neurons influence future input-output mappings of the agent in two ways: by serving as neural input (through the recurrent connections) and by imposing a neural inertia.

The architecture used in our experiments consists of 10 sensory neurons, 5 hidden neuron, and 3 output neurons. The sensory neurons encode the activation state of the 8 infrared sensors, and the activation state of the motor neurons at time t-1.

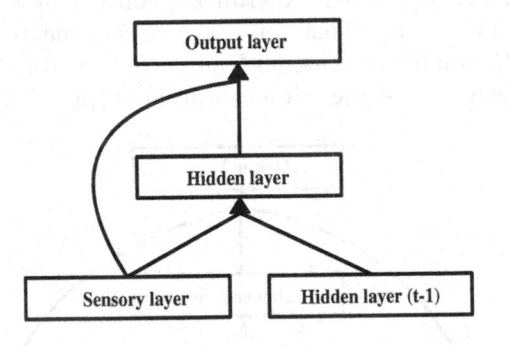

Figure 8: The architecture of the dynamical neural network (DNN). Boxes represents collection of neurons. Arrows indicate the connection between collections of neurons (all neurons of the first box are connected to all neurons of the second box).

During the evolutionary process the architecture is kept fixed. Only the time constants of neurons and the synaptic strengths of the connections are encoded in the genotype and allowed to change. All parameters are encoded in the genotype with 8 bits. Connection weights and biases are then normalised in the range [-5.0, 5.0], time constants are normalised in the range [0.0, 1.0].

4.4   Continuous time recurrent neural network: CTRNN

As dynamical neural networks, continuous time recurrent neural networks [2] are neural networks constituted by dynamical artificial neurons. Also in this case, the activation state of neurons is influenced by their previous activation state. In the case of CTRNN, however, the state of a neuron is characterised by two variables: the activation potential (that corresponds to the depolarisation of the neuron membrane, in the case of real neurons), and the activity of the neuron (that corresponds to the frequency of the spikes produced, in the case of real neurons). The time constant parameter, in this case, determines the rate of change of the activation potential of the neuron.

More precisely, neurons are updated according to the following function:

$$a_i = \sigma(p_i + \text{bias}_i) \tag{5}$$

$$\dot{p}_i = \frac{1}{\text{tc}_i}(-p_i + \text{netinput}_i + g\,\text{in}_i) \tag{6}$$

$$\text{netinput}_i = \sum_{j=1}^{N} w_{ji}a_j, \tag{7}$$

where p_i is the activation potential of a neuron, g the gain of the inputs, and tc_i is the time constant of the neuron. The time constant is adjusted by the evolutionary algorithm, $\frac{1}{\text{tc}_i} \in [0,1]$. Only the sensory neurons and the hidden neurons apply activation function (5), the outputs apply activation function (2). As a result, the activation potentials of the sensory neurons, the activation potentials of the hidden neurons, and the activations of the hidden neurons are part of the internal state. The activation potentials result in a neural inertia, while the activations of the hidden neurons serve as neural input to the hidden layer for the next time step.

We approximate the dynamics of the differential equation by using the standard Euler method (see [8]), with step size 0.1. The architecture and the parameters encoded in the genotype are the same as those described in §4.3, but then with $\frac{1}{\text{tc}_i} \in [0,1]$.

4.5   Time delay recurrent neural network: TDRNN

In a time delay recurrent neural network [5] the propagation of activation through connections takes time and the time delay is controlled by a parameter associated to each connection. For analysis purposes we have used a restricted form of a TDRNN in which each neuron has one common time delay for all its incoming connections.

The architecture used in our experiments consists of 10 sensory neurons, 5 hidden neurons, and 3 output neurons (Fig. 9). The sensory neurons encode the activation state of the 8 infrared sensors, and the activation state of the motor neurons encoding the desired speed at time t-1. The output neurons encode the desired speed of the two wheels and the self-localisation output.

Neurons are updated according to the logistic function. The genotype of evolving individuals encodes the strength of each connection and the time delay associated with each neuron. All parameters are encoded in the genotype with 8 bits. Connection strength and biases are then normalised in the range [-5.0, 5.0]. Time delays are normalised in the range [0, 50] time steps corresponding to a delay in the propagation of the activation ranging from [0.0, 5.0] seconds.

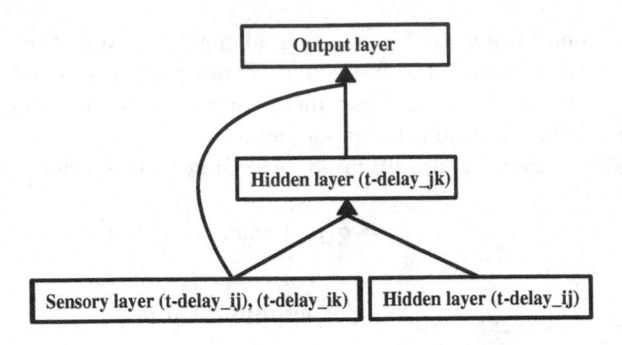

Figure 9: The architecture of the time delay recurrent neural networks (TDRNN). Boxes represents collection of neurons. Arrows indicate the connection between collection of neurons (all neurons of the first box are connected to all neurons of the second box).

5   Results

By running the evolutionary experiments we observed that all evolved individuals were able to travel in the environment at the required speed. However, with respect to the ability to self-localise, performance varied significantly for different neural controllers and for different driving thresholds. The performances of the best evolved agents of the five types of neural controllers applied to the three driving thresholds are shown in table 1. By analysing the performance of the best individuals of each of the five experiments in which evolving agents were provided with five different neural controllers, in fact, we observed that in all cases at least one individual evolved that is able to self-localise more than 75% of the times. At a driving threshold of 23 rounds, instead, in the case of the experiment in which agents were provided with EN networks, no individuals were able to self-localise correctly more than 75% of the times (Table 1). Finally, table 1 shows that only the DNN, CTRNN, and TDRNN neural controllers achieve good performances in the case of a driving threshold of 25 rounds .

Neural controller	20 rounds	23 rounds	25 rounds
EN	91	74	67
NARX	86	89	75
DNN	97	93	92
CTRNN	86	80	94
TDRNN	81	82	79

Table 1: Self-localisation performance in percentages of the best individual of the best replication for experiments with different neural controllers and different driving thresholds. Performances in bold (i.e. performance equal or above 1.56) indicate individuals that are able to correctly self-localise more than 75% of the times. Performances are averaged over 100 runs.

As we will see in the next sub-sections, these results can be explained by considering that evolving agents provided with EN and NARX neural controllers are unable to extract internal states that encode long term regularities and rely on simple quasi-reactive strategies that exploit sensory-motor coordination to solve the self-localisation problem. These simple strategies are based on the fact that, by producing different motor behaviours in different environmental conditions, robots might experience later on different type of sensory states even in identical environmental areas, if these areas are preceded by different environmental structures. These simple strategies however do not allow evolving agents to achieve optimal or close to optimal performance with respect to the self-localisation problem, especially with high driving thresholds that force agents to move quickly in the environment.

On the contrary, TDRNN evolved controllers exploit the time delay on activity propagation, so that past sensory states have long term effects. As a consequence, they still display good performance in the case of high driving thresholds.

Finally, evolved agents that are provided with DNN and CTRNN evolved controllers, are able to extract from sensory states internal states encoding long term regularities that allow these agents to display good and, in some replications, optimal performance.

In the next sections we will describe the control strategies developed by agents provided with different neural networks in detail.

5.1   EN

EN-agents are able to display reasonably good performance only with the lowest driving threshold (20 rounds) and only in two out of ten replications of the experiments. By analysing the behaviour and the activity of internal neurons of the best evolved individuals of the two best replications we realised that they use the same strategy to solve to the problem. Fig. 10 shows the behaviour of the best of these two individuals.

By looking at the activity of hidden neurons we can see that they tend to converge on two rather different equilibrium points corresponding to [0.95, 1.0, 0.02, 0.05, 0.07] and [0.03, 1.0, 1.0, 0.05, 0.84]. The transition between these two equilibrium states occurs very quickly approximately when the agent moves from one room to the other. The former equilibrium state of the internal neurons and the connection weights from these neurons and the self-localisation output neuron assure that the self-localisation output is low, when the robot is in the upper room, and high, when the robot is in the lower room, as requested.

The transition between the two equilibrium states depends on the state of sensors and motors during the few time steps that precede the transition.

More specifically, in the case of the transition between the top and the bottom room, the transition between the former and the latter equilibrium point occurs when the agent negotiates the bottom-right corner of the environment. During the negotiation of this particular corner, in fact, given that the agent reaches the

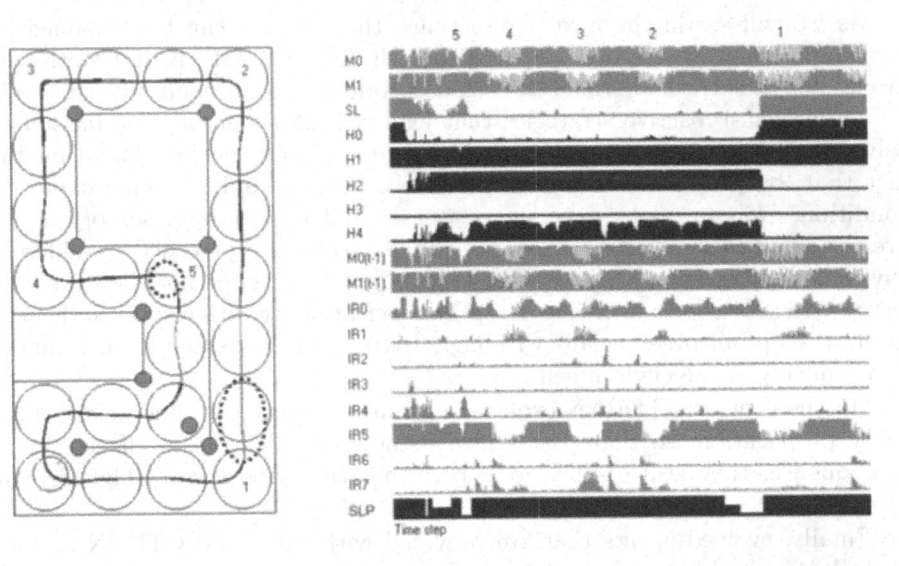

Figure 10: Trajectory and neural activity of the best evolved EN-agent in the case of the experiment with a driving threshold of 20. **Left:** the environment and the robot trajectory during a few laps of the corridor. Dotted circles indicate the areas in which the self-localisation output produced by the agent is wrong. The numbers (from 5 to 1) indicate critical points from the point of view of the ability of this agent to self-localise. **Right:** the activation state of neurons while the robot is performing the last lap of the environment. The activation value is indicated by the height of the graph with respect to the baseline. M0, M1 and SL indicate the activity of the two motor neurons and of the self-localisation output unit. H0-H4 indicate the activity of the 5 internal neurons. M0(t-1), M1(t-1) and IR0-IR7 indicate the activity of the two input units that encode the state of the two corresponding motor neurons at time t-1 and the activity of the 8 infrared sensors. SLP indicates the performance with respect to self-localisation. In this case, the height with respect to the baseline indicates respectively, when the self-localisation is correct (full height), wrong (null height), or when the agents is traveling between the two rooms (half height).

corner by being very close to the wall on its right side, the activation of the two back infrared sensors is almost null. This particular sensory-motor situation, that is the fact that the activity of the right motor neuron decreases in order to turn and negotiate the corner, and the fact that state of the two back infrared sensors is almost null, causes the first transition. This hypothesis has been further verified by freezing the activation state of the two back neurons to 0.05. In this case, in fact, the agent always indicates that it is in the bottom room. This strategy also explains why the agent is unable to self-localise correctly in the bottom room before reaching the bottom-right corner. The fact that the robot reaches the bottom-right room by staying very close to its right-side walls is due to the fact that this evolved agent progressively approaches its right side walls while traveling in a corridor and the fact that this corner is preceded by the longest corridor of the environment.

In the case of the transition between the bottom and the top room, the transition between the latter and the former equilibrium point occurs when the agent negotiates the second left-handed corner which is located just at the beginning of the top room. Also in this case, the agents is able to discriminate between the first and the second left-handed corner on the basis of the state of the sensors and of the motors during the few time steps that precede the transition. More specifically, the transition between the latter and the former equilibrium point occurs when the agent negotiates a left-handed corner (i.e. when the activation state of the left motor neurons is lower than the activation of the right motor neuron) and the activation state of the IR3 and IR4 infrared sensors placed on the frontal-right side of the robot (see Fig. 5) are low. The fact that the activation of IR3 and IR4 tend to be low during the negotiation of the second left-handed corner in turn is the result of the fact that the robot tends to stay close to the wall on its left side after negotiating the first left-handed corner and the presence of an obstacle on the right side of the first left-handed corner. The discrimination between the two left-handed corners, however, is sub-optimal. In fact, due to an increase of the activation of IR4 during the negotiation of the second left-handed corner, wrong self-localisation output are produced (see Fig. 10).

5.2   NARX

Similarly to EN-agents, evolved NARX-agents are able to display reasonably good performance with low driving threshold (20 and 23 turns) but not with high driving threshold (25 turns). Moreover, like in EN-agents, the ability of NARX-agents to self-localise is based on the tendency to converge on two equilibrium states and to move from one to the other equilibrium state on the basis of few sensory and motor states preceding the transition.

Table 2 and 3 show the results obtained by running additional experiments with a driving threshold of 20 and 23 in which we also varied the number of previous sensory and motor states that are copied into additional sensory neurons.

	NARX	NARX	NARX	NARX
c_{in}	0	1	3	4
c_{out}	5	4	2	1
$\overline{F}_{best}$	1.35	1.75	1.61	1.52

Table 2: Performance of the best individual of the best out of ten replications in four experiments with different c_{in} and c_{out} numbers. In all experiments the number of internal neuron is 5 and the driving threshold is 20. $\overline{F}_{best}$ is the performance of the best evolved agent averaged over 100 runs.

By analyzing the behaviour and the internal states of the best evolved individuals we observed that the strategy of the agents evolved with a driving

	NARX	NARX	NARX	NARX
c_{in}	0	1	3	4
c_{out}	5	4	2	1
$\overline{F}_{best}$	1.72	1.80	1.83	1.61

Table 3: Performance of the best individual of the best out of ten replications in four experiments with different c_{in} and c_{out} numbers. In all experiments the number of internal neurons is 5 and the driving threshold is 23. $\overline{F}_{best}$ is the performance of the best evolved agent averaged over 100 runs.

threshold of 20 are similar to the strategy described in the previous section. Below, we only describe the strategy adopted by the best individual obtained in the experiment in which the driving threshold is 23, c_{in} is 1 and c_{out} is 4. The analysis of the other evolved individuals with different values of c_{in} and c_{out}, in any case, revealed that they adopt similar strategies (result not shown).

In this case, as illustrated in Fig. 11, the two equilibrium states are not encoded at the level of the internal neurons but directly in the state of the self-localisation output unit that tends to maintain its activation state close to 0.0 or 1.0. These two states tend to be maintained due to the large positive connection weights of the four connections that link the sensory units encoding the previous activation states of the self-localisation output unit to the unit itself.

The transition between the former and the latter equilibrium points (that correspond to the top and the bottom room, respectively) is triggered by an high activation of the left sensor (IR0), a low activation of the IR5 (i.e. the right sensor), and a null or close to null activation of IR2, IR3 and IR7 (that are activated during the negotiation of a corner). The fact that these conditions are only met when the robot reaches the middle part of the long corridor on the right side of the environment, is due to the particular way of traveling along corridors selected by this agent. As shown in Fig. 11, in fact, this agents produces a curvilinear trajectory in corridors by approaching first the wall placed on the right side of the agent and then, after a certain length, the wall placed on the left side. This curvilinear trajectory assures that the agent approaches the left side walls only in corners or at about the middle part of the long corridor.

The transition between the latter and the former equilibrium points (that correspond to the bottom and the top room, respectively) is triggered by the sensory inputs that are specific to the second turn to the left. The back sensors (IR6 and IR7) are activated in the second turn to the left while IR0 shortly decreases.

As in the case of EN-agents, the fact that NARX-agents are unable to produce reasonably good performance when the driving threshold is 25 can be explained by considering that, by being asked to move at higher speed in the environment, evolving agents cannot select peculiar ways of negotiating corridors and corners (such as moving in corridors by producing curvilinear trajectories) that in turn allow them to identify the location of critical areas of the environment on the basis of a single or few sensory-motor states.

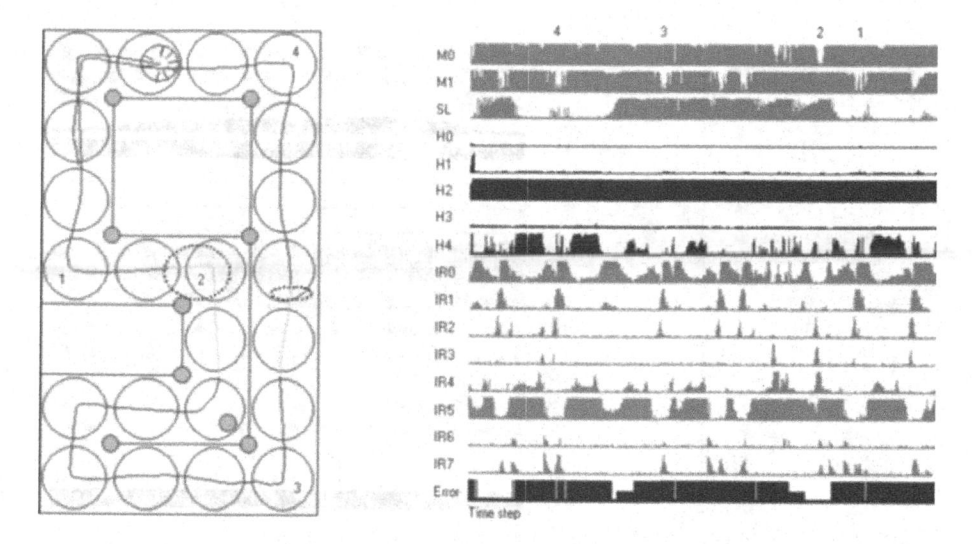

Figure 11: Trajectory and neural activity of one of the best evolved NARX-agents in the case of the experiment with a driving threshold of 23. **Left:** the environment and the robot trajectory during few laps of the corridor. The numbers (from 4 to 1) indicate critical points from the point of view of the ability of this agent to self-localise. Dotted areas indicate the areas in which the self-localisation output produced by the agent is wrong. **Right:** the activation state of neurons while the robot is performing the last lap of the environment. The activation value is indicated by the height of the graph with respect to the baseline. M0, M1 and SL indicate the activity of the two motor neurons and of the self-localisation output unit. H0-H4 indicate the activity of the 5 internal neurons. M0(t-1), M1(t-1) and IR0-IR7 indicate the activity of the two input units that encode the state of the two corresponding motor neurons at time t-1 and the activity of the 8 infrared sensors. SLP indicates performance with respect to self-localisation. In this case, the height with respect to the baseline indicate respectively, when the self-localisation is correct (full height), wrong (null height), or when the agents is traveling between the two rooms (half height).

5.3 DNN

DNN-agents are able to display reasonably good performance, and in some replications close to optimal performance, at all driving thresholds. By analysing the behaviour and the activity of neurons of the best evolved individuals we observed that the self-localisation problem tends to be solved by relying on a few or a single hidden neuron that slowly changes its activation state by always keeping its state below and above a given threshold in the top and in the bottom room respectively, or viceversa. Fig. 12 shows the behaviour and the neural activity of the best evolved individual of the experiment with a driving threshold of 25.

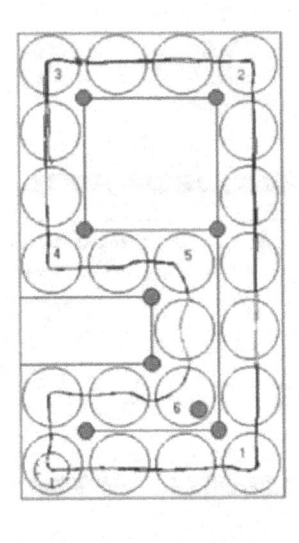
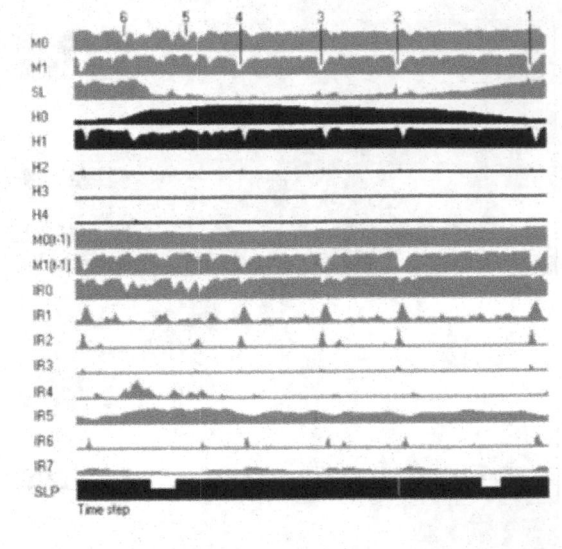

Figure 12: Trajectory and neural activity of the best evolved DNN-agent in the case of the experiment with a driving threshold of 25. **Left:** the environment and the robot trajectory during few laps of the corridor. The numbers (from 6 to 1) indicate critical points from the point of view of the ability of this agent to self-localise. **Right:** the activation state of neurons while the robot is performing the last lap of the environment. The activation value is indicated by the height of the graph with respect to the baseline. M0, M1 and SL indicate the activity of the two motor neurons and of the self-localisation output unit. H0-H4 indicate the activity of the 5 internal neurons. M0(t-1), M1(t-1) and IR0-IR7 indicate the activity of the two input units that encode the state of the two corresponding motor neurons at time t-1 and the activity of the 8 infrared sensors. SLP indicates performance with respect to self-localisation. In this case, the height with respect to the baseline indicates respectively, when the self-localisation is correct (full height), wrong (null height), or when the agents is traveling between the two rooms (half height).

As can be seen in Fig. 12, the activity of the self-localisation output units is mainly affected by the activity of H0 due to a strong inhibitory connection coming from this hidden unit. Given that the time constant parameter of H0 is very high (0.96), the activity of this neuron tends to change slowly in time.

The fact that the activity of H0 tends to decrease slowly while the robot moves along corridors ensures that the activation of this unit progressively decreases while the robot moves along the long corridor located on the right side of the environment. As a consequence, it reaches a value below the critical threshold during the transition from the top to the bottom room.

The fact that the activity of unit H0 tends to increase slightly during the negotiation of right-handed corners assures that the activity of this unit is always above the critical threshold while the robot moves in the top room. Finally, the fact that the activity of H0 tends to increase quickly during the negotiation of left-handed corners assures that the activity of this unit overcomes the critical threshold during the transition from the bottom to the top room.

The ability to integrate information from long sequences of sensory-motor states to detect, for instance, the length of corridors, allows DNN agents to solve their problem without the need to rely on sensory-motor coordination strategies. As we saw above, sensory-motor coordination strategies might allow agents to self-localise correctly on the basis of regularities extracted by few sensory-motor states, but require special ways to negotiate the environment that do not allow to move at high speeds.

5.4 CTRNN

CTRNN-agents are able to display good performances at all driving thresholds. By analysing the behaviour and the activity of neurons of the best evolved individuals we observed that the agents solve the self-localisation problem with strategies very similar to those exhibited by DNN-Agents. Fig. 13 shows the behaviour and the neural activity of the best evolved individual of the experiment with a driving threshold of 25 rounds.

As can be seen in Fig. 13, the activity of the self-localisation output is mainly affected by the activity of H3 due to a strong inhibitory connection coming from this hidden unit. Given that the time constant parameter of H3 is high ($\frac{1}{tc} = 0.02$) , the activity of this neuron tends to change slowly in time. As in the case of DNN, the fact that the activity of H3 tends to decrease slowly while the robot moves along corridors ensures that the activation of this unit progressively decreases while the robot moves along the long corridor located on the right side of the environment by reaching a value below the critical threshold during the transition from the top to the bottom room.

As in the case of DNN, the fact that the activity of unit H3 tends to increase slightly during the negotiation of right-handed corners assures that the activity of this unit is always above the critical threshold while the robot moves in the top room. Finally, the fact that the activity of H3 tends to increase quickly during the negotiation of left-handed corners assures that the activity of this unit overcomes the critical threshold during the transition from the bottom to the top room. In this case however, H3 overcomes the critical threshold already after the first left-handed corner. The combination of this fact and the fact that IR4, that gets activated during the negotiation of left-handed corners, contributes to activate the self-localisation output unit, causes a systematic localisation error during the negotiation of the second left-handed corner (see the dotted area indicated on the right side of Fig. 13).

5.5 TDRNN

TDRNN-agents are able to display reasonably good performance at all driving thresholds (see table 1). The analysis of the best-evolved agents indicates that the time delay on activity propagation plays an important role in the ability of

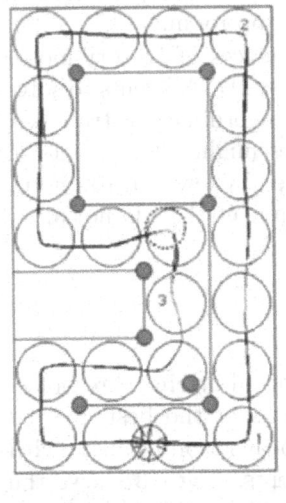 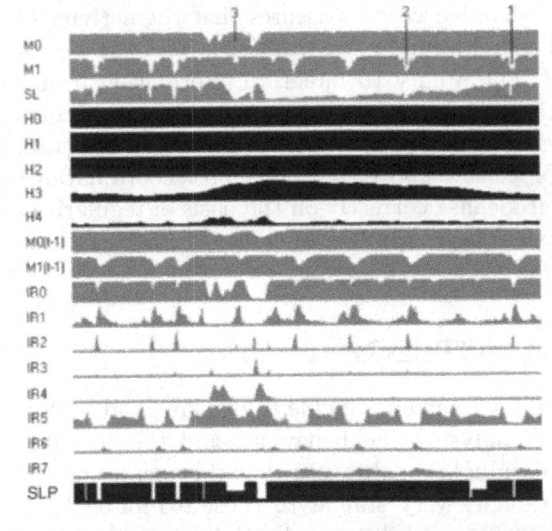

Figure 13: Trajectory and neural activity of the best evolved CTRNN-agent in the case of the experiment with a driving threshold of 25. **Left:** the environment and the robot trajectory during few laps of the corridor. The numbers (from 6 to 1) indicate critical points from the point of view of the ability of this agent to self-localise. Dotted areas indicate the areas in which the self-localisation output produced by the agent is wrong. **Right:** the activation state of neurons while the robot is performing the last lap of the environment. The activation value is indicated by the height of the graph with respect to the baseline. M0, M1 and SL indicate the activity of the two motor neurons and of the self-localisation output unit. H0-H4 indicate the activity of the 5 internal neurons. M0(t-1), M1(t-1) and IR0-IR7 indicate the activity of the two input units that encode the state of the two corresponding motor neurons at time t-1 and the activity of the 8 infrared sensors. SLP indicates performance with respect to self-localisation. In this case, the height with respect to the baseline indicates respectively, when the self-localisation is correct (full height), wrong (null height), or when the agents is traveling between the two rooms (half height).

these agents to self-localise. Fig. 14 shows the trajectory and neural activations of the best TDRNN-agent evolved with a driving threshold of 23 rounds. The values of the time delay parameters are shown in table 4.

The analysis of the evolved connection strengths and time delay parameters indicate that evolved agents use time delay parameters to: (1) detect a sequence of events separated by fixed time intervals, and (2) allow internal states to produce motor effects after a fixed time interval.

Let us consider in particular how the evolved individual shown in Fig. 14 is able to correctly indicate the transition from the bottom to the top room. The activity of the self-localisation output unit of the TDRNN is determined by the delayed activations of the sensory neurons and hidden neurons. In fact, the indication of the transition from the bottom to the top room depends on

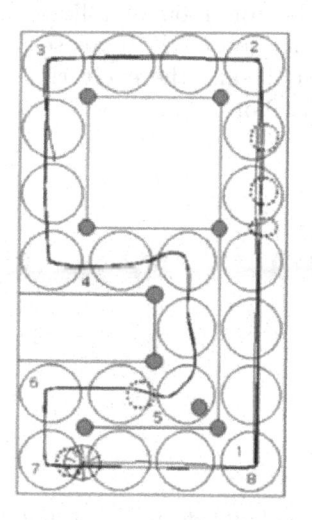 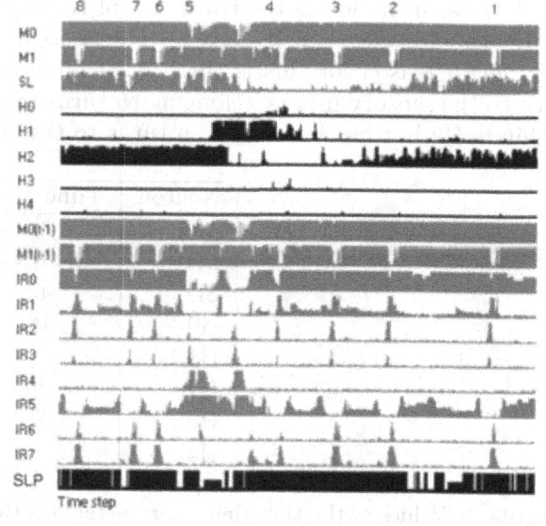

Figure 14: Trajectory and neural activity of one of the best evolved TDRNN-agent in the case of the experiment with a driving threshold of 25. **Left:** the environment and the robot trajectory during few laps of the corridor. The numbers (from 8 to 1) indicate critical points from the point of view of the ability of this agent to self-localise. **Right:** the activation state of neurons while the robot is performing the last lap of the environment. The activation value is indicated by the height of the graph with respect to the baseline. M0, M1 and SL indicate the activity of the two motor neurons and of the self-localisation output unit. H0-H4 indicate the activity of the 5 internal neurons. IR0-IR7 indicate the activity of the 8 infrared sensors. The additional input neurons encoding the state of the sensors and motors in previous time steps are not displayed but can be inferred by previous sensory and motor states. SLP indicate performance with respect to self-localisation. In this case, the height with respect to the baseline indicates respectively, when the self-localisation is correct (full height), wrong (null height), or when the agents is traveling between the two rooms (half height).

the sensory inputs that the agent experiences when it enters the first turn to the left, indicated with the number '5' in Fig. 14. The sensory neurons that are especially involved are I0, I3, I4, and I5. Their particular values have three effects.

1. The self-localisation output decreases after 43 time steps.

2. The activation of H1 increases 22 time steps later.

3. The activation of H2 decreases 36 time steps later.

The increase of H1 and decrease of H2 has as a consequence that SL remains low while the agent is traversing almost the entire top room. The explanation for this is that H1 inhibits H2 and SL, while H2 excites SL.

The agent indicates the transition of the top to the bottom room as follows. After the activation of H1 has been decreased, the activation of H2 increases. The main reason for this is the lack of inhibition from H1 and the excitation due to the sensory inputs belonging to turns to the right. For example, M1(t-1) inhibits H2 but has low values in turns to the right.

Neuron	Time delay
M0	0
M1	0
SL	43
H0	10
H1	22
H2	36
H3	46
H4	6

Table 4: Values of the time delay parameters (in time steps) of all hidden and output neurons. All incoming connections to one neuron have a common delay.

To summarize, the TDRNN-agent extracts an internal state from the sensory signals it experiences between the time steps indicated with '5' and '4'. Besides delaying the signals from the sensory neurons, the agent exploits the recurrency of the hidden layer to let the effects of the sensory signals experienced in the turns to the left fade away. The effect of this strategy is that the agent has a low self-localisation output during most of its traversal of the top room and a high self-localisation output in the bottom room.

6 Discussion

The comparison of the results obtained by providing evolving agents with different types of neural controllers indicate that the use of dynamical neurons and/or time-delayed propagation of activation potentials might constitute a necessary prerequisite for the emergence of the ability to integrate sensory-motor information through time.

In fact, although in principle agents provided with simple recurrent neural networks such as EN or NARX neural networks should be able to develop the same control strategies developed by agents provided with DNN, CTRNN and TDRNN neural networks, in practice they are unable to do so. This failure can be explained by considering that the evolvability (i.e. the probability to produce a better solution through random changes of free parameters) of EN and NARX neural networks is lower than that of DNN, CTRNN and TDRNN.

The fact that agents provided with a DNN or CTRNN neural controller are more evolvable than agents provided with EN or NARX can be explained by considering that the availability of neurons that tend to vary their state at different time rates is a useful prerequisite to solve problems that require to integrate information from sequences of sensory-motor states or to produce motor states lasting several time steps [15]. Although by properly setting the connection weights, any type of recurrent neural network could in principle display neurons that tend to vary their activity at different time rates, neurons that vary their activity at slow time rates (i.e. time rates that are significantly slower from the time rates with which the activity of sensors and neurons are updated) are much more frequent in DNN and CTRNN.

The fact that neurons in DNN tend to vary their activity at a slower time rate than neurons in EN and NARX networks can be demonstrated by considering that the change of activation of a DNN-neuron is always smaller or equal to the change of a neuron in EN and NARX neural networks (i.e. neurons updated according to the standard logistic function). Indeed, the change in activation of a neuron in a DNN is always smaller or equal to the change of activation of a neuron in an EN or NARX, if the neurons have the same bias weight, neural inputs, and past activation, and if $\frac{1}{tc} \in [0, 1]$. First we express the neural activation function of the DNN (equation (9)) in terms of the activation function of the EN (equation (8)), as shown in equation (10).

$$a_{en}(t) = \sigma(\text{netinput}(t) + \text{bias} + \text{in}(t)) \tag{8}$$

$$a_{dnn}(t) = \frac{1}{tc}a(t-1) + (1 - \frac{1}{tc})\sigma(\text{netinput}(t) + \text{bias} + \text{in}(t)) \tag{9}$$

$$a_{dnn}(t) = \frac{1}{tc}a(t-1) + (1 - \frac{1}{tc})a_{en}(t) \tag{10}$$

Since $\frac{1}{tc} \in [0, 1]$ and a_{dnn} is a weighted sum of $a(t-1)$ and $a_{en}(t)$, we can conclude equation (11) and (12) from equation (10).

$$a_{dnn}(t) \in [min\{a_{en}(t), a(t-1)\}, max\{a_{en}(t), a(t-1)\}] \tag{11}$$

$$|a_{dnn}(t) - a(t-1)| \leq |a_{en}(t) - a(t-1)| \tag{12}$$

Equation (12) implies that the change in activation of a neuron in a DNN is always smaller than or equal to that of a neuron in an EN, for the same neural input, bias, past input, and external input.

The fact that agents provided with a TDRNN neural controller are more evolvable than agents provided with EN and NARX can be explained by considering that the availability of neurons that encode the state of sensors and motors at previous time steps within an adaptable time range is a useful prerequisite to integrate information from sequences of sensory-motor states and, as we have seen, to detect sequences of events separated by a given time interval.

Overall the obtained results suggest that a better understanding of the neural mechanisms suitable to process information in time might be an important step towards the development of powerful pro-active agents.

Acknowledgements

This research has been partially supported by the ECAGENTS project founded by the Future and Emerging Technologies program (IST-FET) of the European Community under EU R&D contract IST-2003-1940. The information provided is the sole responsibility of the authors and does not reflect the Community's opinion. The Community is not responsible for any use that may be made of data appearing in this publication.

References

[1] BEER, R.D., "A dynamical systems perspective on agent-environment interaction", *Artificial Intelligence 72* (1995), 173–215.

[2] BEER, R.D., "On the dynamics of small continuous-time recurrent neural networks", *Adaptive Behavior 3, (4)* (1995), 469–509.

[3] BROOKS, R.A., "Intelligence without reason.", *Proceedings of 12th International Joint Conference on Artificial Intelligence* (J. MYLOPOULOS AND R. REITER eds.), San Mateo, CA: Morgan Kaufmann (1991).

[4] CLARK, A., *Being There: Putting Brain, Body and World Together Again.*, Cambridge, MA: MIT Press. (1997).

[5] DURO, R. J., and J. S. REYES, "Ecg beat classification with synaptic delay based artificial neural networks.", *IWANN* (1997), 962–970.

[6] ELMAN, J. L., "Finding structure in time", *Cognitive Science 14* (1990), 179–211.

[7] LIN, T., B. G. HORNE, P. TINO, and C. Lee GILES, "Learning long-term dependencies in narx recurrent neural networks", *IEEE Transactions on Neural Networks* (1996).

[8] MATHAYOMCHAN, B., and R.D. BEER, "Center-crossing recurrent neural networks for the evolution of rhythmic behavior", *Neural Computation* **14** (2002), 2043–2051.

[9] MATURANA, H.R., and F.J. VARELA, *Autopoiesis and cognition: the realization of the living.*, Dordrecht: reidel. (1980).

[10] MATURANA, H.R., and F.J. VARELA, *The tree of knowledge: the biological roots of human understanding*, Shambhala Publications (1988).

[11] MONDADA, R., E. FRANZI, and P. IENNE, "Mobile robot miniaturization: A tool for investigation in control algorithms", *Proceedings of the Third International Symposium on Experimental Robots* (T. YOSHIKAWA AND F. MIYAZAKI eds.), Berlin, Springer-Verlag. (1993).

[12] NOLFI, S., "Evorobot 1.1 user manual", *http://gral.ip.rm.cnr.it/evorobot/ simulator.html.*

[13] NOLFI, S., "Evolving non-trivial behaviors on real robots: A garbage collecting robot", *Robotics and Autonomous System, 22* (1997), 187–198.

[14] NOLFI, S., "Power and the limits of reactive agents", *Neurocomputing, 49* (2002), 119–145.

[15] NOLFI, S., G. BALDASSARE, and D. MAROCCO, "Evolving robots able to self-localize in the environment: The importance of viewing cognition as the result of processes occurring at different time scales", *Proceedings of the Third International Symposium on Homan and Artificial Intelligence Systems. Fukui, Japan* (T. ASAKURA AND K. MURASE eds.), Fukui University Press. (2002).

[16] NOLFI, S., and D. FLOREANO, *Evolutionary Robotics: The Biology, Intelligence, and Technology of Self-Organizing Machines,* Cambridge, MA, MIT Press/Bradford Books (2000).

[17] NOLFI, S., and D. MAROCCO, "Evolving robots able to integrate sensory-motor information over time", *Theory in Biosciences, 120* (2001), 287–310.

[18] PFEIFER, R., and C. SCHEIER, *Understanding Intelligence,* MIT Press, Cambridge, MA (1999).

Chapter 16

Autonomous Discovery and Functional Response to Topology Change in Self-Reconfigurable Robots

Behnam Salemi, Peter Will and Wei-Min Shen
Information Sciences Institue
Department of Computer Science
University of Southern California
{salemi, will, shen}@isi.edu

1. Introduction

A self-reconfigurable system is a special type of complex systems that can autonomously rearrange its software and hardware components and adapt its configuration, such as shape, size, formation, structure, or organization, to accomplish difficult missions in dynamic, uncertain, and unanticipated environments. A self-reconfigurable system is typically made from a network of homogeneous or heterogeneous *reconfigurable modules* (or *agents*) that can autonomously change their physical or logical connections and rearrange their configurations.

Self-reconfigurable robots [3, 7, 8, 10] are examples of such self-reconfigurable systems that consist of many autonomous modules that have sensors, actuators, and computational resources. These modules are physically connected to each other in the form of a configuration network. Since the topology of the configuration network may change from time to time, to accomplish a given task, the controller of the Self-reconfigurable robot must be distributed and decentralized to avoid single-point of failures, and communication bottleneck among modules.

These modules must have some essential capabilities in order to accomplish complex tasks in dynamic and uncertain environments. The capabilities that we addressed in our previous work were: (1) distributed task negotiation [5] – allowing modules to agree on a global task which is to be accomplished, (2) distributed behavior collaboration [6] – allowing modules to "translate" a global task into local behaviors of modules; (3) synchronization – allowing modules to perform local behaviors in a coordinated and timely fashion; In these previous works we assumed the network of modules can have any initial topology but it remains unchanged during the process of accomplishing a selected task.

Here we relax this assumption and allow the topology of the network of modules to change at any time including the duration of accomplishment of the task. Our proposed solution for this problem is a distributed approach inspired by the concept of hormones [1] and is based on 1) giving the ability of detecting local changes in the topology of the network to the modules and 2) Allowing them to select and coordinate new behaviors when the topology of the network changes such that the selected global task can be accomplished.

The related approaches for solving similar problems include Role-based Control [9] and stochastic approaches for self-repair such as [2]. The first approach is based on detecting the changes in local relationships of the immediate neighboring modules and local reaction to these changes. This approach does not require a lot of computational power. However, it is an open-loop approach and might not be suitable for accomplishing complex tasks. The second approach has been applied to lattice-based self-reconfigurable robots, which their configuration space is much smaller than that of the chain-type self-reconfigurable robots such as CONRO.

This chapter is organized as follows: Section 2 defines the problem of autonomous discovery and functional response to topology changes and introduces CONRO self-reconfigurable robots as an illustrative example; Section 3 presents the idea of probing for solving the problem of autonomous discovery and functional response to topology changes; Section 4 presents the FEATURE algorithm and reviews its sub-algorithms; Section 5 gives examples of applying the FEATURE algorithm to real CONRO modules, and Section 6 concludes the chapter with future research directions.

2. Autonomous Discovery and Functional Response to Topology Changes

The problem of autonomous discovery and functional response to topology changes can be defined as follows: Given a global task and a set of self-reconfigurable modules, coordinating global responses to local changes in the topology of the network of modules in order to produce the desired global effects. Local changes include adding or deleting new modules or communication links to/from the network of modules.

This problem is very challenging due to several reasons: relationships among modules may change anytime; changes in configuration is locally detectable but a coordinated global response is required; the number of modules in the robot is not known; modules have no unique global identifiers or addresses; modules do not know the global configuration in advance, and can only communicate with their immediate neighbors.

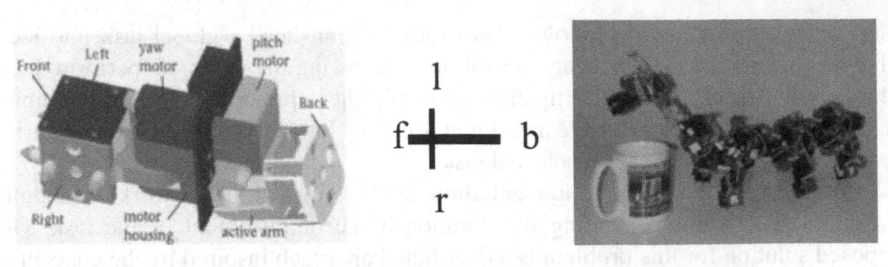

Figure. 1. A CONRO module, the schematic view of one module, and a hexapod configuration with 11 modules.

Generally, accomplishing a given global task is dependent on the topology of the network of modules [6]. Modules can accomplish a global task by selecting correct *behaviors* in coordination with other modules and performing them synchronously. As a result, changes in the topology will directly influence the behaviors that should be selected and the time they should be performed.

Formally, an autonomous discovery and functional response to topology changes problem is a tuple $[G(P, C), Q, T, B]$, where G is the *configuration graph* of the network of modules consisting of P, a list of nodes, p_i and C, a list of labeled physical or logical links, c_j, such that $j \in \{locally\ unique\ labels\}$; Q is the list of the internal states, q_i, associated with each node p_i, such that $i \in \{1,..., N\}$; T is the global task shared by all nodes; and B is a set of behaviors, b_i, available to the nodes. In this situation, an autonomous discovery and functional response to topology changes problem is solved if and only if there is a function, $f(G,Q) \rightarrow B$, that it is a mapping from the current topology of the configuration graph, and the internal state of a node to a sequence of behaviors that can accomplish task T. Here, the nodes and links represent the modules and the communication links between them, respectively. Note that the size of the network is dynamic and unknown to the individual nodes; also the index numbers are only used for defining the problem and not used in the solution.

Under these circumstances, a satisfactory solution to this problem must be distributed. Modules must detect local changes in the configuration graph and inform the rest of the modules in order to let them select the correct behavior.

To illustrate this problem, we use the CONRO self-reconfigurable robot as an example. CONRO is a chain-type self-reconfigurable robot developed at USC/ISI (http://www.isi.edu/robots). Figure 1 shows the schematic views of CONRO module and a six-legged CONRO robot. Each CONRO module is autonomous and contains two batteries, one STAMP II-SX micro-controller, two servo-motors, and four docking connectors for connecting with other modules. Each connector has a pair of infrared transmitters/receivers, called *outgoing-Links* and *incoming-Links*, to support communication as well as docking guidance.

Each module has a set of open I/O ports so that various sensors for tilt, touch, acceleration, and miniature vision, can be installed dynamically. Each module has two Degrees Of Freedom: DOF1 for pitch (about 0-130° up and down) and DOF2 for yaw (about 0-130° left and right). The range of yaw and pitch of a module is divided to 255 steps. The internal state of each module includes the current values of the yaw, pitch of a module, and the number of the sent and received messages. The modules' *actions* consist of moving the two degrees of freedom to one of the 255 positions, attaching to or detaching from other modules, or sending messages to the communication links through the IR senders.

Modules can be connected together by their docking connectors. Connected docking connectors are called *active* connectors. Docking connectors, located at either end of each module. At one end, labeled *back* (*b* for short), there is a female connector, consisting of two holes for accepting another module's docking pins. At the other end, three male connectors of two pins each are located on three sides of the module, labeled *left* (*l*), *right* (*r*) and *front* (*f*).

3. Probing and Communication

The first step for modules in responding to the network topology change consists of detecting local changes. Instances of local changes are: 1) When a new module connects to one of the existing modules in the network, 2) When an existing module disconnects from all other modules in the network, 3) When an existing module establishes a new connection with another module in the network, and 4) When a module disconnects some of its connectors from other modules in the network. In situations 1 and 2 the number of nodes and links and in situations 3 and 4 the number of links in the configuration network changes.

Modules can detect local changes in the topology of the network by periodically monitoring their active docking connectors for disconnections and inactive docking connectors for new connections. This action is called *probing*. In order to detect all the above-mentioned cases of topology change efficiently, we will use two types of probing: 1) Probing when modules are communicating and 2) Probing using *probing signals*.

3.1. Probing by Communication

The communication protocols that use handshaking signals when sending and/or receiving messages between modules can be used for probing the active connection links between modules. A successful communication action over an active connector shows that the connection is still active. Similarly, an unsuccessful communication action shows the disconnection of an already active connector.

Figure 2 shows an asynchronous communication protocol that was implemented in CONRO modules. What follows is a brief description of the handshaking sequence of this protocol: Agent$_1$ is the sender and agent$_2$ is the receiver.

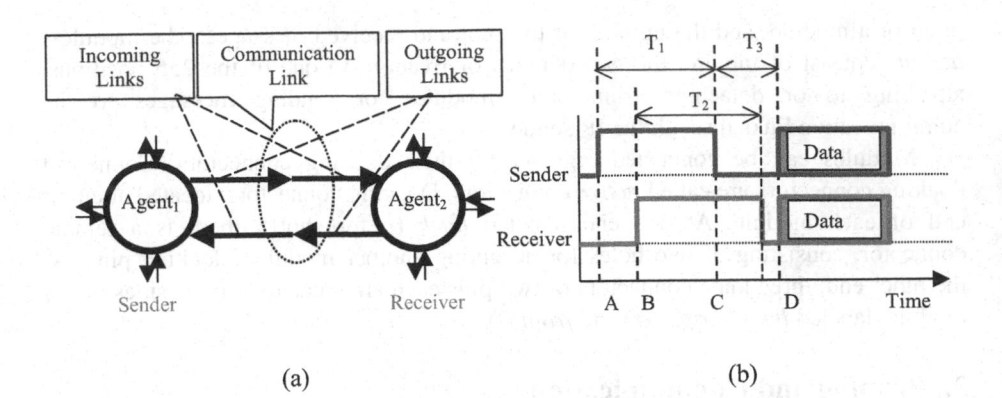

Figure. 2. (a) The communication link between two agents. (b) The asynchronous communication protocol between two agents

1. The sender requests to send a message by making its outgoing link 'High', point A in Figure 2b and then continues checking its incoming link for receiving a 'High' signal.

2. The receiver responds by making its outgoing link 'High', point B, and waits.

3. After receiving the 'High', sender makes its outgoing link 'Low', point C, and starts sending the message (Data) after a short delay, point D. This short delay is called *preparation time* (T_3), which allows the receiver to prepare for receiving Data. Data is communicated using RS232 asynchronous communication protocol. In order to avoid dead-lock, timeouts are added to the sender and receiver to limit their waiting time. T_1 and T_2 are sender's and receiver's timeouts, respectively.

This simple handshaking protocol successfully completes if and only if both modules actively participate. Therefore a successful communication confirms for both modules an active link between the two. In oppositely, an unsuccessful communication confirms that the receiving module is not present and the link is inactive.

This method of probing, however, is not efficient for probing the inactive docking connectors unless one attempts to send a message to an inactive docking connector and waits for the timeout, which could be a quite long time. Also, in situations where two modules do not communicate for longer than a pre-specified duration called 'monitoring period', the communication-based approach will produce a wrong conclusion. In such situations we will use a different type of probing method based on sending *probing signals*.

3.2. Probing Signals

Probing signal are narrow pulses that are periodically sent to inactive connections or active connections if no communication occurs on them for a long time. The width of probing signals is much narrower than the communication protocol signals such that

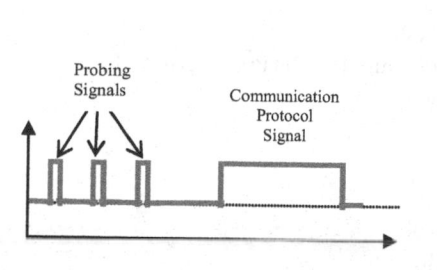

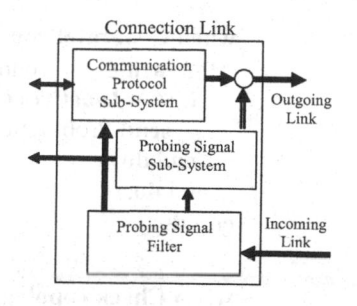

Figure 3. Probing and Communication protocol signals

Figure 4. Merging and separating the probing and communication signals

they can be distinguished and filtered from the communication protocol signals. Figure 3 compares the width of the probing and communication protocol signals.

Figure 4 shows the block diagram of a module's connection link. The 'Probing Signal Filter' on the incoming link separates the probing signals from the communication signals. On the outgoing link, the communication and probing signals are merged to a single output line.

3.3. Probing Algorithm

Figure 5 describes the probing algorithm for detecting local changes in the topology of the network. This algorithm consists of two procedures. The first procedure, **GenerateProbe**, is called for generating probing signals on inactive connectors or the active connectors that have not communicated for longer than the duration of the 'monitoring period'.

The second procedure, **CheckTopology**, is called for detecting changes in local topology of the network based on the received probing signals or the recent communicated messages. This procedure returns a true value if the topology has been changed.

4. Functional Response to Topology Change Using Probing

Our solution for the functional response to topology change problem in self-reconfigurable robots relies on our previous work on distributed control for self-reconfigurable robots. Specifically, the 'distributed task negotiation' and 'distributed behavior collaboration' problems. In this section we will describe these problems and our proposed solutions. Then we will present the FEATURE algorithm that utilizes probing for solving the functional response to the topology change problem.

```
when GenerateProbe () do
  for each C ∈ Connectors do
    if (C = Inactive) or (NoComm (C, Period) = true)  do
      send ProbingSignal to C;
    end do;
  end do;
end do;


when CheckTopology () do
  TempLocalTopology = CurrentLocalTopology;
  TopologyChanged = false;
      for each C ∈ Connectors do        //reset
        CurrentLocalTopology (C) = Inactive; end do;
  for each C ∈ Connectors do
    if (CommOccurred (C) = true) or
       (Probe Signal Received (C) = true) do
       CurrentLocalTopology (C) = active;
    end do; end do;
    if (TempLocalTopology ≠ CurrentLocalTopology) do
       TopologyChanged = true;
    end do;
    return TopologyChanged;
end do;
```

Figure 5. The probing Algorithm

4.1. Distributed Task Negotiation

Distributed Task Negotiation is a process by which modules in a self-reconfigurable robot can negotiate and select a single coherent task among many different and even conflicting choices. This is a very challenging problem due to several reasons: the relationships among modules are not static, but change with configurations; modules have no unique global identifiers or addresses; modules do not know the global configuration in advance, and can only communicate with their immediate neighbors.

In [5] we presented the DISTINCT algorithm as a solution for the distributed task negotiation problem. The main idea is that all modules work together to build global spanning trees and each tree is associated with a task. Initially, all modules that have their own competing tasks start building their own trees, but as they exchange messages for tree building, most modules will give up their selected tasks and "root" status and participate in building trees for other tasks. In this process, modules report their status to their parent module in the tree to which they belong, and the module that does not have parent but received reports from all its children is the root for the entire network of modules. When this happens, this root module can conclude that the negotiation process has succeeded and all modules in the tree have agreed on the same task. An embedded synchronization algorithm detects the termination of the negotiation process.

Formally, a distributed task negotiation problem consists of a tuple (P, L, T, S), where P is a list of nodes, p_i, such that $i \in \{1,..., N\}$; L is a list of communication links, l_{jk}, such that $j,k \in \{1,..., N\}$; T is a list of tasks, t_m, such that $1 \le m \le N$, and S is a set of task selection functions, $S_i: (T') \rightarrow t_i$, such that $i \in \{1,...,N\}$ and $T' \subset T$. Each node has a task selection function that can select a single task from a set of given tasks. A distributed task negotiation problem is solved when all nodes have selected the same task from T, called t^*, and have been notified that the negotiation process is terminated. Note that the index numbers assigned to P are only used for defining the problem and not used in the negotiation process. In addition, the size of the network is unknown to the individual nodes.

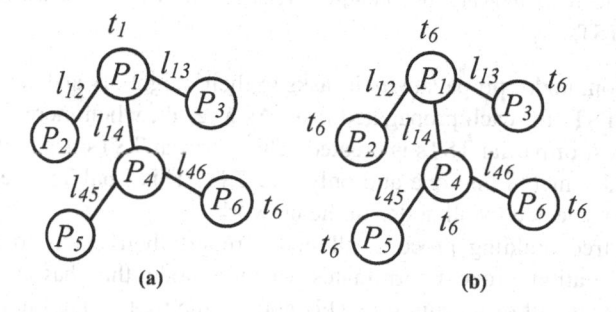

Figure 6. An example of a distributed task negotiation problem. **a)** Initially p_1 and p_6 initiated two tasks (t_1, t_6). **b)** A solution, when all agents have selected $t^* = t_6$.

To illustrate the above definition, consider the example in Figure 6a, where $P = \{p_1, p_2, p_3, p_4, p_5, p_6\}$, $L = \{l_{12}, l_{14}, l_{13}, l_{45}, l_{46}\}$, $T = \{t_1, t_6\}$, and S is a selection function that prefers tasks with greater indexes and shared by all nodes. Initially, node p_1 and p_6 have initiated two tasks, t_1 and t_6, respectively, and the rest of the nodes are waiting to receive tasks. Figure 6b depicts a solution for the given problem where all nodes agreed on task $t^* = t_6$.

Important characteristics of this solution are: 1) modules do not require having unique Ids; 2) ensures that all nodes will select the same task coherently; regardless of the number of competing tasks initiated in the network; and more importantly 3) it is not dependent on the topology of the network of modules.

4.1.1 Negotiation by Creating Spanning Trees

The most obvious solution for the problem is to assign priorities to the competing tasks and force nodes to select tasks that have higher priorities. However, since the importance of tasks cannot be determined statically, it is extremely hard to determine the correct priorities for an arbitrary set of competing tasks.

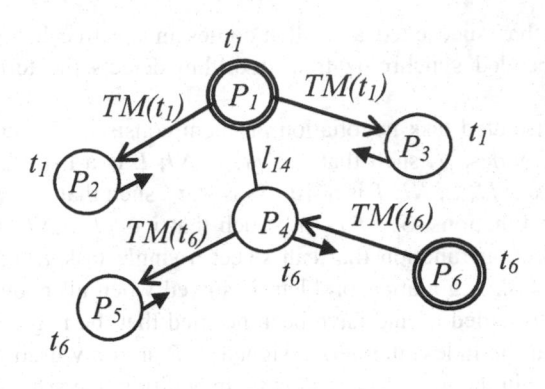

Figure 7. Task message propagation. Arrows on the links indicate messages in transit and arrows parallel to links indicate the "child-of" relationship. Double circles indicate the roots of partial TSTs.

In our solution, nodes propagate their tasks to their neighbors and generate a Task Spanning Tree (TST) for each propagated task. As a result, when more than one task is initiated, a forest of partial TSTs is created. These partial TSTs negotiate with each other and gradually merge into one and only one TST. This final TST represents the task that has been selected by all nodes in the network.

During the tree building process, all nodes report their status to their parent nodes. The negotiation process terminates when a node that has no parent has received reports from all of its children. This node is the root of the final TST, and it then notifies all nodes in the tree with an "end of task negotiation" message and all nodes will select the task associated with the final TST.

4.1.2 Distributed Task Selection

For nodes that have competing tasks to select a single task, the goal is to create a single TST. Each node must decide on two issues: 1) what task to select and propagate, and 2) how to be a part of a TST.

Initially, nodes that have competing tasks propagate their tasks by sending a *task message* (*TM*) to their neighbors and designating themselves as the root of a partial TST. Assuming that the recipient of a *TM* has no tasks for itself and receives only one *TM*, then it will adopt the received task and create a "child-of" relationship toward the sender of the *TM*. The recipient will in turn propagate the received task by sending a new *TM* to the rest of its neighbors. To illustrate this idea, Figure 7 shows an example in which nodes P_1 and P_6 are the initiators of tasks t_1 and t_6 respectively and the rest of the nodes are non-initiator nodes. Node P_2 and P_3 are the recipients of $TM(t_1)$ sent by P_1, and therefore have selected task t_1. Similarly, P_4 and P_5 are the recipients of $TM(t_6)$ sent by P_6, and therefore have selected task t_6. In this situation, parallel arrows show the "child-of" relationships that the nodes have created.

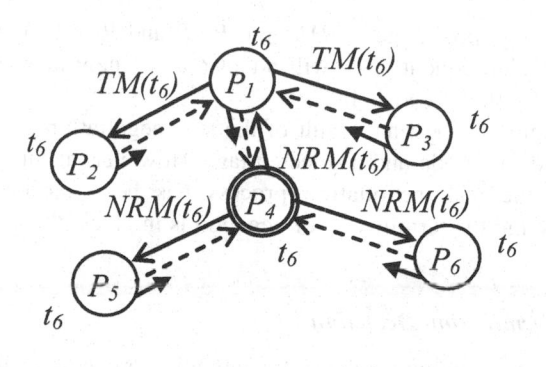

Figure 8. Merging partial TSTs from Figure 2. P_4 is the new root of the merged TST. The dashed arrows indicate the ack messages.

Based on the above assumption, no message has been sent through the link l_{14}. As a result two TSTs have been formed; one rooted at P_1 and the other rooted at P_6. In each TST, all nodes have selected the same task.

At this point, if we relax the above assumption, two cases might occur: **1)** either a root node receives a *TM*, or **2)** a non-root node receives a *TM* from a node that is not its parent. An example of the first case is shown in Figure 8 where P_1, a root node, receives a *TM* from P_4. An example of the second case happens when P_4, a non-root node in the TST rooted at P_6, receives a *TM* from P_1, which belongs to another partial TST.

In the first case, the recipient, which is a root node, drops being a root, adopts the received task, establishes a "child-of" relationship with the sender of the *TM* and propagates new *TM* to the rest of its neighbors, which are its children. In this situation, these nodes adopt the new received task and propagate it to the rest of their neighbors.

In the second case, the received *TM* is a conflicting message since it was received from a non-parent node. To resolve the conflict, the recipient node deletes all of its previous "child-of" relationships, makes a choice between its previous task and the received task (using its task selection function), propagates a *newRoot message* (*NRM*) containing the newly selected task to all of its neighbors, and then promotes itself as a new root for the selected task.

The role of *NRM* is to merge partial TSTs and create a new root for the resulting TST. Therefore, the recipient of a *NRM* adopts the received task, creates a new "child-of" relationship towards the sender of the *NRM*, becomes a non-root node (if previously a root), and propagates a new *NRM* containing the received task to the rest of its children.

Figure 8 shows the result of merging the two partial TSTs in Figure 7 for the situation where P_4 has been the node that has received a conflicting *TM* from P_1. As a result, P_4 chooses a task between t_6 and t_1 (say t_6 is chosen), promotes itself to be the root of the new TST, and propagates $NRM(t_6)$ to P_1, P_5 and P_6, which turns P_1 and P_6 into non-root nodes. Consequently, P_1 will adopt t_6 as its new task and propagate a new TM to P_2 and P_3 for the task switch.

As shown in Figure 8, the final result of the task negotiation process is a single TST with a specified root node and a selected task. However, at this point the nodes do not know that the task negotiation process has been terminated. Unless a mechanism for detecting the termination of negation is in place, the nodes would wait indefinitely.

4.1.3. Distributed Termination Detection

In order to detect the termination of the task negotiation process, we use an approach similar to the "termination detection algorithm for diffusing computation" by Dijkstra and Scholten [1]. For each received TM and NRM, each node must reply with an acknowledge message (AM), after the node receives acknowledges from all its children. For a leaf node, this means that it will acknowledge immediately for every received message. For a non-leaf node, it will send an acknowledge message to its parent after it receives AM from all of its children. If a non-leaf node receives all AM from all its children and it has no parent, then this node is the root for the final TST and it can conclude that the task negotiation process has succeeded.

In Figure 8, dashed arrows indicate the AM messages. The root node, P_4, expects to receive AMs from each of the P_1, P_5, and P_6 nodes. Since P_5 and P_6 do not have any child nodes, they send their AM as soon as they receive $NRM(t_6)$ messages from P_4. However, P_1 sends its AM to P_4 only after it receives AMs from P_2 and P_3. When P_4 receives all of its expected AMs, it detects the termination of the negotiation process and propagates a *taskSelected* message to all of its children. This message will be propagated to all the nodes in the tree and the task negotiation process successfully terminates.

4.2. Distributed Behavior Collaboration

The problem of distributed behavior collaboration can be defined as follows: Given a global task and a group behavior, to select a correct set of local behaviors at each module and coordinate the selected behaviors to produce the desired global effects.

The problem is very challenging due to several reasons: relationships among modules are not static but change with configurations; the number of modules in the robot is not known; modules have no unique global identifiers or addresses; modules do not know the global configuration in advance, and can only communicate with immediate neighbors. Under these circumstances, a satisfactory solution to distributed behavior collaboration must be distributed. Modules must select behaviors through local communication, and the execution of the selected behaviors must be synchronized.

Formally, the problem of distributed behavior collaboration is a tuple (P, Q, C, A, B, t, GB), where P is a list of nodes, p_i; Q is the list of the internal state, q_i, associated with each node p_i, such that $i \in \{1, ..., N\}$; C is a list of labeled physical or logical links, c_j, such that $j \in \{locally\ unique\ labels\}$; A is a set of actions a_s a node can execute, such that $s \in \{1,..., S\}$; B is a set of behaviors in the form of $b_m = (a_x, a_y, a_z,...)$, such that $m \in \{1,..., M\}$; t is the global task given to all nodes, and GB is the desired group behavior in the form of behavior selection rules. These rules are mappings from nodes internal states to behaviors, $Q \to B$. The *configuration graph* of the network of modules is a graph consists of P nodes and C edges. A distributed behavior selection problem is solved if and only if the ordered sequence of the selected behaviors of all nodes over time is equal to the desired group behavior. If β represents the ordered sequence of the selected behaviors or $\beta = \beta + b_{pi}$ where, $i \in \{1,..., N\}$, $\beta = GB$, should hold. Note that the size of the network is dynamic and unknown to the individual nodes; also the index numbers are only used for defining the problem and not used in the solution.

4.2.1. Extended Neighborhood Topology

When modules in a self-reconfigurable robot have negotiated and decided on a global task, they must then generate a group behavior to accomplish the task. A group behavior is the result of the coordinated performance of local behaviors of individual modules, while the local behaviors are selected based on the location of the modules relative to other modules. In [4], we represented the module's location in a configuration as the *type* of the module. Table 1 lists 32 local types of CONRO module, which reflects the ways that a module can connect to its immediate neighbors. Figure 9 shows some example types in various CONRO configurations.

The type information in Table 1 could provide modules with the necessary information to uniquely determine their location in most cases and select the appropriate local behaviors for the global task accordingly (see details in [4]). However, these types are not enough to guarantee determining modules' location in any complex configuration. For example, consider the T-shape and the snake configurations in Figure 9. Modules A, and A' are both of type T2, yet they must

Table 1. 32 local topological types of CONRO module

Connected to other modules	This Module					This Module				
	b	*f*	*r*	*l*	Type	*b*	*f*	*r*	*l*	Type
					T0	*f*	*b*			T16
	f				T1	*f*		*b*		T17
		b			T2	*f*			*b*	T18
			b		T3		*b*	*b*	*b*	T19
				b	T4	*f*	*b*	*b*		T20
	l				T5	*f*		*b*	*b*	T21
	r				T6	*f*	*b*		*b*	T22
		b	*b*		T7	*l*	*b*	*b*		T23
			b	*b*	T8	*l*		*b*	*b*	T24
		b		*b*	T9	*l*	*b*		*b*	T25
	l	*b*			T10	*r*	*b*	*b*		T26
	l		*b*		T11	*r*		*b*	*b*	T27
	l			*b*	T12	*r*	*b*		*b*	T28
	r	*b*			T13	*f*	*b*	*b*	*b*	T29
	r		*b*		T14	*l*	*b*	*b*	*b*	T30
	r			*b*	T15	*r*	*b*	*b*	*b*	T31

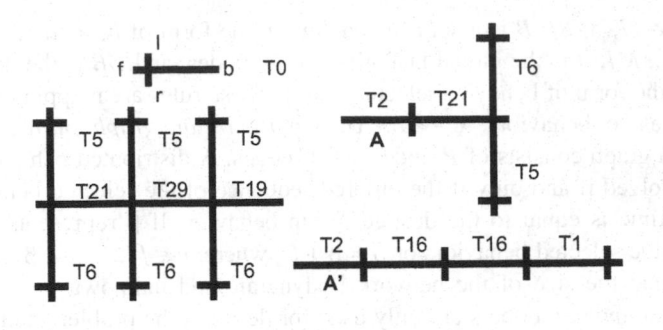

Figure 9. Immediate neighborhood topological types of CONRO modules in different configurations: a single module, a hexapod, a T-shape, and a snake.

behave differently in the two different configuration. The module A' must perform a sinusoidal behavior in the snake configuration, while the A module must keep still in the T-shape "butter-fly" locomotion (i.e., the leg modules move in a cycle of up, left, down, and right, while the body modules keep still).

To solve this problem, we extend module's type from the immediate neighborhood to neighbors that are n modules away. We call this extended type, type(n), and define it as how the active connection links of a given module are connected to the connection links of the modules of distance n. For example, in Figure 10, type(0) for module A is [(bf)] because module B is the only one of distance zero from A, and the b connector of B is connected to the f connector of A. However, type(1) of module A is [(br,bf),(bl,bf)] because this is the way A is connected to module C and D, which are one module away ($n = 1$). Similarly, the type(0) of module A' is [(bf)] and its type(1) is [(bf,bf)]. Note that module B has only immediate neighbors (distance = 0) and therefore it only has type(0) information.

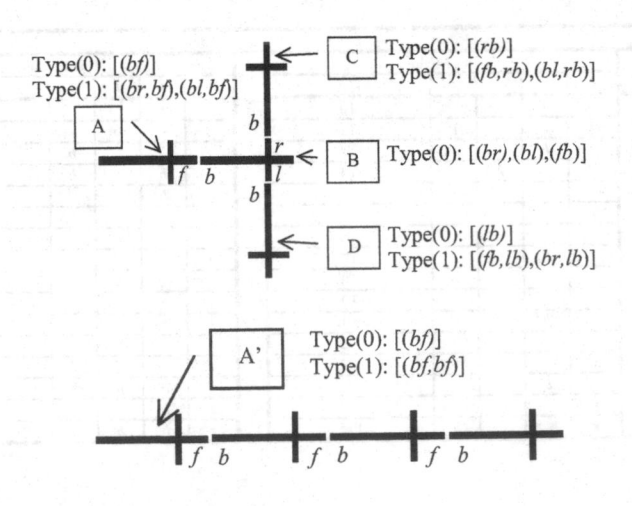

Figure 10. Examples of the extendable type(n).

As we can see, although type(0) values of module A in the T-configuration and module A' in the snake configuration are the same, but they have different type(1) value. It can be proven that using the extended types, modules can always uniquely identify themselves in a configuration as long as the labels of connection links of modules are locally unique. The proof is based on having unique path between any two nodes in a tree. Later we will show that modules can use the extended type information to select the appropriate behavior based on the given task.

It can be shown that the *type* definitions in [4, 9] are special cases of the extended type and equivalent to type(0). In addition, global representation of the entire network for each agent is equivalent to [type(0),type(1), . . ., type(d-1)], where d is the diameter of the network.

It is possible for the modules to autonomously dynamically discover their extended types and. The solution is based on the characteristics of hormone-inspired messages described in [7]. Each hormone message contains a path field that records a list of connector-pairs (ex. *bf*) through which the message has been propagated. When a module receives a hormone message with $|path|=m$, it will insert the path into its extended type(m-1) values. As more and more messages are received, the extended type information will be built up. Since messages are propagated through the network, each module will eventually build up the correct type values for itself. See [6] for more details.

4.2.2. Selecting Local Behaviors Via Type(n) Values

A straightforward approach for behavior collaboration in a modular system is the centralized 'gait control table' [11], in which a designated module, called the central controller, is given the information about behaviors of other modules in the form of a table. Each column of this table contains the sequence of actions that a module, identified by its identifier, has to perform over time based on its location in the configuration (equivalent to the behavior of the module). The central controller job is to send each row of the table specifying the actions that all modules should perform at a time.

The 'gait control table' approach, however, is not an ideal approach for controlling the self-reconfigurable system for the following reasons: first, requiring the central controller to send actions to the rest of the modules in the configuration creates a communication bottleneck. In addition, if the central controller becomes faulty the entire system will be disabled. More importantly, when the network of modules restructures themselves, the pre-specified behaviors of the modules in the table might not be valid anymore. The source of this difficulty is that modules do not know how their behaviors are chosen so they cannot select new behaviors as they re-locate in the configuration.

Our approach for solving this problem is based on using the extended types to uniquely identify the location of modules in a configuration, and use them to select the correct local behaviors by the modules for the given global task. For example, as shown in Figure 11, for a quadruped to accomplish the 'Move forward' task, the 'front left leg' module and 'back right leg' will select 'Swing Backward' behavior, while the 'front right leg' module and the 'back left leg' module will select the 'Lift and Swing Forward' behavior and modules of types 'front spine' and 'back spine' will select 'bend left' and 'bend right', respectively. Figure 11b shows the robot after the modules have performed their selected behaviors. In general, the group behavior to accomplish "Move forward" consists of the following behaviors: the 'front left leg' and the 'back right leg' perform the 'Lift and Swing Backward' behavior, while the 'front right leg' and the 'back left leg' is performing the 'Swing Forward' behavior, and then the two groups switch their behaviors. In the next section we present a flexible algorithm called Distributed BEhavior SelecTion (D-BEST) for behavior selection, which is based on the extended neighboring types.

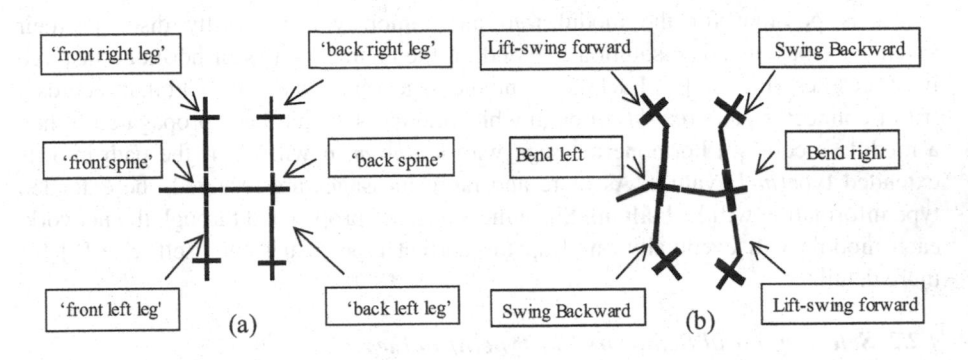

Figure 11. (a) The module types in a four-legged self-reconfigurable robot; (b) The selected local behaviors of each module for "move forward".

4.2.3. The D-Best Algorithm

Using the extended types as the condition for selecting behaviors will provide the modules with the information they require to autonomously collaborate to select new behaviors. Figure 12 illustrates the basic idea of the behavior collaboration based on the extended types. Initially, modules communicate their currently selected behaviors to their neighbors by sending hormone messages and wait for receiving new hormone message. This initial behavior could be a **Null** behavior. The communicated hormone message content consists of the type of the message, in this case of type **<Behavior-selection>**, and an initially empty path field, represented by **<(path)>**.

When a new hormone message is received, the module updates the path field of the message, updates its extended type based on the received path and propagates the message to its neighbors. Then it uses the current extended type to select a behavior from a lookup table representing the desired group behavior. This process will continue until all modules receive and propagate the initiated hormone messages.

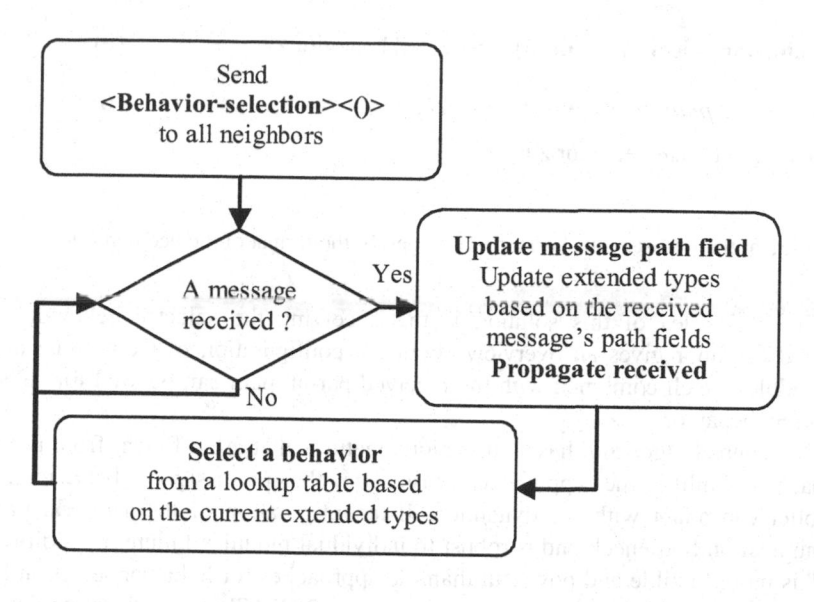

Figure 12. Basic idea of behavior selection based on the extended types

Although this approach can dynamically adapt to the changes in the topology of the network, it has two problems. First, an initiated message from a module will be propagated to all other modules in the configuration. This means that the total number of communicated messages will be $O(N^2)$, where N is the number of modules. The second problem is that when the diameter of the configuration is large, the size of the path field, and therefore the size of the message, will be large. This will considerably slow down the communication when the bandwidth is narrow.

To solve these two problems, we limit the maximum length of the path field in the messages. For example, if the maximum length of path is set to k, a message will stop being propagated after k hops. In this situation, if there are N agents in the network, and each agent has the average number of a active connectors, the number of communicated messages will be $O(N)$ (since at most $a*N$ messages will be initiated and each message will be communicated k times therefore $k*a*N$ messages). The tradeoff of this solution is that the created extended type will be partial, and therefore might not be enough for some modules to select the correct local behaviors.

This problem can be solved by including the modules' selected behaviors in the communicated hormone messages and representing the group behavior as a set of *decision rules* based on the partial paths and received behaviors. In this situation, the content of the hormone messages and decision rules are shown in Figures 13a and 13b, respectively.

<**Behavior-selection**><(**path**)><Selected behavior>   **(a)**

if (received *path* == X) **and** (received *behavior* == Y)   **(b)**

then (select local behavior Z)

Figure 13. a) the format of the hormone messages. b) the format of the decision rules

The basic idea of this solution is that receiving the selected behavior of an extended neighbor gives an overview about the configuration of the module around that module, which combined with the received partial path can be used for selecting the correct behavior.

This control algorithm has some unique features that are different from previous approaches. Unlike the approaches based on the pre-assigned behaviors, this controller can adapt with the dynamic self-reconfiguration of the network, prevent communication bottleneck and is robust to individual modules failure. In addition, D-BEST is more flexible and powerful than the approaches for behavior selection based on immediate neighboring connection patterns as D-BEST uses both immediate and extended neighboring modules connection pattern for behavior selection.

It can be seen that the shorter the maximum size of the path results the smaller number of communicated message. This feature can be utilized at the design time of the decision rules for a desired group behavior in the following way. Starting from the smallest maximum length ($k = 0$) the designer of the group behavior will write the rules that can uniquely select the correct behaviors for the modules. If there is ambiguity in selecting behaviors for the possible configurations, the k will be increased to provide the modules with more information such that the ambiguity is resolved. This characteristic of the D-BEST algorithm allows the number of the communicated message to be a function of the complexity of the desired group behavior and/or possible configurations.

Figure 14 shows an example of the behavior collaboration for generating a Butterfly locomotion in a T-shape CONRO robot. In this example Butterfly_Spine, Move_East, Move_West and CAT0 are different behaviors and four decision rules are used. The maximum path length is chosen to be zero, $k = 0$, specifying that the messages that immediate neighboring modules communicate will not be propagated to other modules.

According to the rule 1, if a module receives a message from one of its left or right connectors, it will be a spine module otherwise it can be a spine or a module in a snake configuration. Based on this rule, module B can select the correct behavior, Butterfly_Spine. However, module A does not know if it is a spine module or in the snake configuration. Rule 4 can resolves this issue by determining the module cannot be part of a snake configuration if the neighboring module B has selected

Butterfly_Spine. Rules 2 and 3 will be used by the side legs to select the correct direction for their movements. If module A applies rule 1, it will consider the possibility of being part of a snake by selecting the CAT0 (sinusoidal motion starting from angle zero for the caterpillar move). This selection will be corrected if at some point rule 4 is applicable.

1) **If** path = ((*bl*) **or** path = (*br*))
then select Butterfly_Spine
else select CAT_0

2) **If** path = (*rb*)
and behavior = Butterfly_Spine
then select Move_West

3) **If** path = (*lb*)
and behavior = Butterfly_Spine
then select Move_East

4) **If** (path = (*fb*) **or** path = (*bf*))
and behavior = Butterfly_Spine
then select Butterfly_Spine

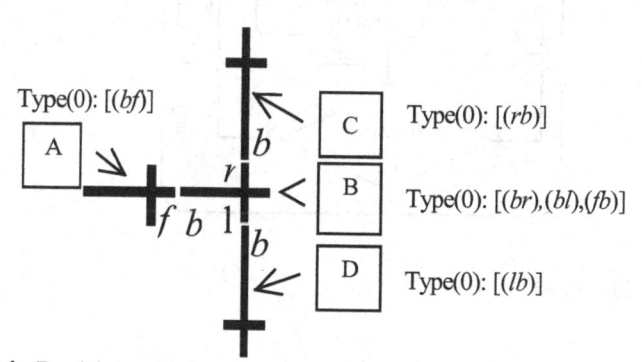

Figure 14. Decision rule for Butterfly locomotion.

4.3. The FEATURE Algorithm

In this section, we will describe the FEATURE algorithm that brings all the above-mentioned pieces together and solves the problem of functional response to topology change in self-reconfigurable robots. This will be the algorithm that will run on all modules of the self-reconfigurable robot to ensure the homogeneity of all modules. Figure 15 depicts this algorithm.

Initially all modules will wait to receive a new task. The new task can be initiated by an outside controller or one of the sensors on a module. Receiving a new task initiates a distributed negotiation process among all modules in the robot. This is necessary to ensure that 1) all modules know what task they accomplishing and 2) in cases where multiple modules have initiated more than one task, all modules will agree on accomplishing the same task that has the highest priority. The above-

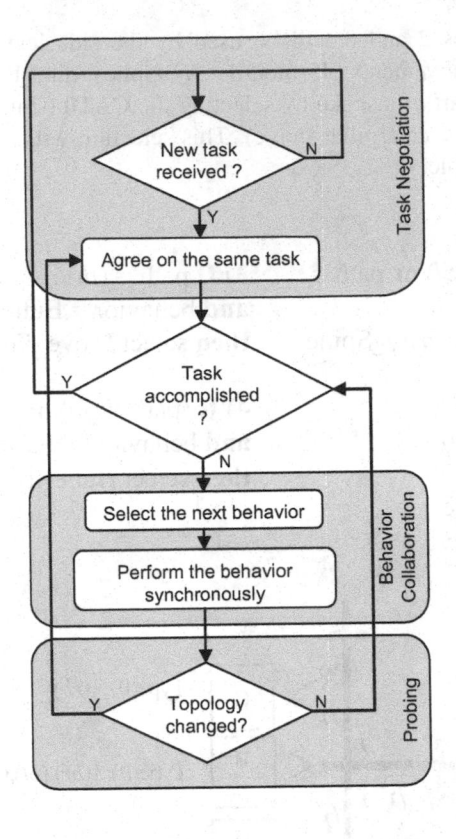

Figure 15. The FEATURE algorithm

mentioned process is controlled by the DISTINCT, task negotiation algorithm shown on top part of the Figure 15.

If the selected task is not already accomplished, modules will generate a set of relevant behaviors. The relevant behaviors are represented by a set of decision rules that have been downloaded in all modules. The execution of the selected behaviors will be coordinated by an embedded distributed synchronization mechanism. The above-mentioned process is controlled by the D-BEST, behavior collaboration algorithm shown in the middle section of the Figure 15.

If the topology of the network of module changes while modules are performing their behaviors, the modules that detected the local changes, will initiate the DISTINCT algorithm using the current selected task in order to dynamically create a new spanning tree for synchronizing and initiating new sets of behaviors based on the current topology of the network. The topology detection process will be controlled by the probing algorithm shown on bottom of the Figure 15.

5. Examples

We have implemented and tested the FEATURE algorithm and all of it sub–algorithms on the CONRO self-reconfigurable robots. All modules are running as autonomous systems without any off-line computational resources and are loaded with the same control program and decision rules. For economic reasons, the power of the modules is supplied independently through cables from an off-board power supply.

In our experiment we gave a 'Move' task to a quadruped CONRO robot. The robot initiated a 'Four-Legged Walking' gait. While performing the gait, we detached the two spine modules. The resulting configuration was two separate T-shape robots. In this situation, each T-shape robot continued the locomotion by executing the 'Butterfly Stroke' gait. Later, two T-shape robots were re-connected and the resulting four-legged robot re-initiated the 'Four-Legged Walking' gait.

For the snake configuration, we have experimented with caterpillar movement with different lengths ranging from 1 module to 10 modules. With no modification of programs, all these configurations can move and snakes with more than 3 modules can move properly as caterpillar. The average speed of the caterpillar movements is approximately 30cm/minute. In this experiment, we have dynamically "cut" a 10-module running snake into three segments with lengths of 4, 4, and 2. All these segments adapt to the new configuration and continue to move as independent caterpillars. We also dynamically connected two or three independent running caterpillars with various lengths into a single and longer caterpillar. The new caterpillar adapted to the new configuration and continued to move in the caterpillar gait. These experiments show that the described approach is robust to changes in the length of the snake configuration.

To test this approach for self-reconfiguration from a Snake to T-shape, a self-reconfiguration task was manually given to one the middle modules of a snake-shape robot consisting of seven modules. After completion of the self-reconfiguration task the new topology of the robot was detected and a butterfly gait for the T-shape robot was generated. The videos of these experiments are available at http://www.isi.edu/robots.

6. Conclusion and Future work

This chapter presented the FEATURE algorithm that combines a set of distributed algorithms for accomplishing global tasks in a chain-type self-reconfigurable robots consisting of multiple modules with dynamic topology. These combined algorithms were DISTINCT for distributed task negotiation, and D-BEST for distributed behavior collaboration. The FEATURE algorithm used probing for detecting the local topology changes and this information was used for global coordinated selection of the new behaviors in the modules. The FEATURE algorithm was implemented on the real CONRO self-reconfigurable robot modules and the experimental results were presented.

As the future work, we will study the conditions for performing successful self-reconfiguration and locomotion tasks based on the received messages and develop a complete set of decision rules for performing all possible self-reconfiguration and locomotion tasks.

References

[1]   Dijkstra, E.W., C.S. Scholten, *Termination Detection for Diffusing Computations.* Information Processing Letters, 1980. **11**.

[2]   Murata, S., H. Kurokawa, E. Toshida, K. Tomita, and S. Kokaji, , *A 3-D self-reconfigurable structure* in ICRA,1998.

[3]   Rus, D., Z. Butler, K. Kotay, M. Vona, *Self-Reconfiguring Robots.* ACM Communication, 2002

[4]   Salemi, Behnam, WM. Shen and P. Will, *Hormone Controlled Metamorphic Robots.* in *ICRA* 2001.

[5]   Salemi Behnam, Peter Will, and Wei-Min Shen. *"Distributed Task Negotiation in Modular Robots".* Robotics Society of Japan, Special Issue on "Modular Robots", 2003.

[6]   Salemi Behnam, Wei-Min Shen *Distributed Behavior Collaboration for Self-Reconfigurable Robots.* International Conference on Robotics and Automation, New Orleans, LA, USA, 2004.

[7]   Shen, W.-M., B. Salemi, and P. Will., *Hormone-Inspired Adaptive Communication and Distributed Control for CONRO Self-Reconfigurable Robots.* IEEE Transaction on Robotics and Automation,. 18(5): p. 700-712, 2002a.

[8]   Shen, W.-M. and M. Yim (editors), *Special Issue on Self-Reconfigurable Modular Robots*, IEEE Transactions on Mechatronics, 7(4), 2002b.

[9]   Stoy, K., Shen, WM., Will, P., *Using Role-Based Control to Produce Locomotion in Chain-Type Self-Reconfigurable Robots.* IEEE/ASME Transactions on Mechatronics,. 7(4): p. 410.M. Young, The Technical Writer's Handbook. Mill Valley, CA: University Science, 2002.

[10]   Yim, M., Y. Zhang, D. Duff, *Modular Robots.* IEEE Spectrum, 2002.

[11]   Yim, M., *Locomotion with a unit-modular reconfigurable robot* (Ph.D. Thesis), in Department of Mechanical Engineering. 1994, Stanford University.

Understanding Complex Systems

Jirsa, V.K.; Kelso, J.A.S. (Eds.)
Coordination Dynamics: Issues and Trends
XIV, 272 p. 2004 [3-540-20323-0]

Kerner, B.S.
The Physics of Traffic:
Empirical Freeway Pattern Features,
Engineering Applications, and Theory
XXIII, 682 p. 2004 [3-540-20716-3]

Kleidon, A.; Lorenz, R.D. (Eds.)
Non-equilibrium Thermodynamics
and the Production of Entropy
XIX, 260 p. 2005 [3-540-22495-5]

Kocarev, L.; Vattay, G. (Eds.)
Complex Dynamics in Communication Networks
X, 361 p. 2005 [3-540-24305-4]

McDaniel, R.R.Jr.; Driebe, D.J. (Eds.)
Uncertainty and Surprise in Complex Systems:
Questions on Working with the Unexpected
X, 200 p. 2005 [3-540-23773-9]

Ausloos, M.; Dirickx, M. (Eds.)
The Logistic Map and the Route to Chaos:
From The Beginnings to Modern Applications
XVI, 411 p. 2006 [3-540-28366-8]

Kaneko, K.
Life: An Introduction to Complex Systems Biology
VIII, 312 p. 2006 [3-540-32666-9]

Braha, D.; Minai, A.A.; Bar-Yam Y. (Eds.)
Complex Engineered Systems: Science Meets Technology
VIII, 394 p. 2006 [3-540-32831-9]